DATE DUE

MINNESOTA STUDIES IN THE PHILOSOPHY OF SCIENCE

Minnesota Studies in the
PHILOSOPHY OF SCIENCE

RONALD N. GIERE, GENERAL EDITOR

HERBERT FEIGL, FOUNDING EDITOR

VOLUME XV
Cognitive Models of Science

EDITED BY
RONALD N. GIERE

UNIVERSITY OF MINNESOTA PRESS
MINNEAPOLIS

Published by the University of Minnesota Press
2037 University Avenue Southeast, Minneapolis, MN 55414
Printed in the United States of America on acid-free paper

Library of Congress Cataloging-in-Publication Data

Cognitive models of science / edited by Ronald N. Giere.
    p. cm. — (Minnesota studies in the philosophy of science ; v. 15)
    "Volume grew out of a workshop ... held in October 1989 under the sponsorship of the Minnesota Center for Philosophy of Science"—P.
    Includes bibliographical references and index.
    ISBN 0-8166-1979-4 (hc)
    1. Science—Philosophy. 2. Cognition. I. Giere, Ronald N.
II. Minnesota Center for Philosophy of Science.   III. Series.
Q175.M64 vol. 15
[Q175.3]
501 s—dc20
[501]   91-34528
                                CIP

Contents

PART I
MODELS FROM COGNITIVE PSYCHOLOGY

PART IV
BETWEEN LOGIC AND SOCIOLOGY

PART V
CRITIQUE AND REPLIES

CRITIQUE

REPLIES TO GLYMOUR

Acknowledgments

Major support for the Center Workshop on Implications of the Cognitive Sciences for the Philosophy of Science was provided by the National Science Foundation through its program in History and Philosophy of Science. Additional funding was provided by the Scholarly Conference Fund of the College of Liberal Arts and by the Graduate School of the University of Minnesota. The editor owes particular thanks to the secretary of the Center for Philosophy of Science, Steve Lelchuk, who oversaw the whole process from sending out initial invitations to contributors, through local arrangements for the workshop, to compiling the final manuscript for the University of Minnesota Press. He also prepared the index.

Introduction:
Cognitive Models of Science

This volume grew out of a workshop on implications of the cognitive sciences for the philosophy of science held in October 1989 under the sponsorship of the Minnesota Center for Philosophy of Science. The idea behind the workshop was that the cognitive sciences have reached a sufficient state of maturity that they can now provide a valuable resource for philosophers of science who are developing general theories of science as a human activity. The hope is that the cognitive sciences might come to play the sort of role that formal logic played for logical empiricism or that history of science played for the historical school within the philosophy of science. This development might permit the philosophy of science as a whole finally to move beyond the division between "logical" and "historical" approaches that has characterized the field since the 1960s.

There are, of course, philosophers of science for whom the very idea of a "cognitive approach" to the philosophy of science represents a regression to ways of thinking that were supposed to have been decisively rejected by the early decades of the twentieth century. From the time of the classical Greek philosophers through the nineteenth century, logic and psychology were closely related subjects. Nineteenth-century writers spoke easily of the principles of logic as "the laws of thought." Under the influence of Frege and Russell, that all changed. Logic became an autonomous, normative discipline; psychology an empirical science. The idea that how people *actually* think might have any relevance to the question of how they *should* think was labeled "psychologism," and cataloged as an official "fallacy." This point of view was incorporated into logical empiricism, the dominant Anglo-American philosophy of science until the 1960s. But it has persisted in various forms even among the strongest critics of logical empiricism, Imre Lakatos (1971) being a primary example.

Part of Kuhn's (1962) contribution to the philosophy of science was to challenge the separation of psychology from the philosophy of science.

His account of science invoked notions from gestalt psychology and the early "new look" psychologists associated with Jerome Bruner (Bruner, Goodnow, and Austin 1956). N. R. Hanson (1958), inspired mainly by Wittgenstein, had reached similar conclusions somewhat earlier. By the end of the 1960s, Quine (1969) had made psychology the basis for a "naturalized epistemology." Similar ideas were championed in related fields by such pioneers as Donald Campbell (1959) in social psychology and Herbert Simon (1977) in economics, psychology, and computer science. Although very influential, these works did not quite succeed in making psychology a fundamental resource for the philosophy of science.

One of the main reasons the constructive psychologism of Kuhn and Quine did not have more impact was simply that neither of the psychological theories to which they appealed — gestalt psychology and behaviorism, respectively — was adequate to the task. That has changed. Since the 1960s the cognitive sciences have emerged as an identifiable cluster of disciplines. The sources of this development are complex (Gardner 1985) but include the development of computers and transformational approaches in linguistics.

The emergence of cognitive science has by no means escaped the notice of philosophers of science. Within the philosophy of science one can detect an emerging specialty, the philosophy of cognitive science, which would be parallel to such specialties as the philosophy of physics or the philosophy of biology. But the reverse is also happening. That is, the cognitive sciences are beginning to have a considerable impact on the content and methods of philosophy, particularly the philosophy of language and the philosophy of mind (Dennett 1983; Fodor 1987; P. M. Churchland 1989; P. S. Churchland 1986), but also on epistemology (Goldman 1986). The cognitive sciences are also now beginning to have an impact on the philosophy of science. Inspired by work in the cognitive sciences, and sometimes in collaboration with cognitive scientists, a number of philosophers of science have begun to use the cognitive sciences as a *resource* for the philosophical study of science as a cognitive activity.

The unifying label "cognitive science" in fact covers a diversity of disciplines and activities. For the purposes of this volume, I distinguish three disciplinary clusters: (1) artificial intelligence (itself a branch of computer science), (2) cognitive psychology, and (3) cognitive neuroscience. These clusters tend to be thought of as providing three different levels of analysis, with the functional units becoming more abstract as one moves "up" from neuroscience to artificial intelligence. Each of these disciplinary clusters provides a group of models that might be used in approaching problems that are central to the philosophy of science. I begin with cognitive psychology because it seems to me that the models being developed within cognitive psychology are, at least for the moment, the most useful for a cognitive approach to the philosophy of science.

1. Models from Cognitive Psychology

Nancy Nersessian provides a prototype of someone drawing on research in the cognitive sciences to solve problems in the philosophy of science. The focus of her research is a problem originating in the historical critique of logical empiricism. Logical empiricism made science cumulative at the observational level while allowing the possibility of change at the theoretical level. But any noncumulative changes at the theoretical level could only be discontinuous. The historical critics argued that science has not been cumulative even at the empirical level. But some of these critics, such as Kuhn and Feyerabend, also ended up with a view of theoretical change as being discontinuous, though for different reasons. Thus was born the problem of "incommensurability." Nersessian's project is to dissolve the problem of incommensurability by showing how the theoretical development of science can be continuous without science as a whole being cumulative.

The historical focus of her study is the development of electrodynamics from Faraday to Einstein. She argues that neither philosophers nor historians have yet done justice to this development. This includes those like Hanson and Kuhn who explicitly appealed to gestalt psychology. Nersessian argues that more recent work in cognitive psychology provides tools that are more adequate to the task, particularly for understanding the use of analogy, mechanical models, thought experiments, and limiting case analysis. Here I will note only two features of her study that have general implications for a cognitive approach to the philosophy of science.

Most historically minded critics of logical empiricism took over the assumption that scientific theories are primarily *linguistic* entities. The main exception is Kuhn, who gave priority to concrete exemplars over linguistically formulated generalizations. Nersessian adopts a theory of "mental models" as elaborated, for example, by Johnson-Laird (1983). On this approach, language, in the form of propositions, may be used not to describe the world directly but to construct a "mental model," which is a "structural analog" of a real-world or imagined situation. Once constructed, the mental model may yield "images," which are mental models viewed from a particular perspective. This interplay of propositions, models, and images provides a richer account of the rep-

xviii   *Ronald N. Giere*

resentational resources of scientists than that employed by either logical empiricists or most of their critics. It may be thought of as an extension of the model-theoretic approach to the nature of scientific theories as elaborated, for example, by Suppe (1989), van Fraassen (1980, 1989), and myself (Giere 1988). In any case, the cognitive theory of mental models provides the main resource for Nersessian's account of the dynamics of conceptual change in science. Some such account of representation seems sure to become standard within a cognitive approach to the philosophy of science.

Another assumption shared by logical empiricists and most of their historically based critics is that the basic entities in an account of science are abstractions like "theories," "methods," or "research traditions" (which for both Lakatos [1970] and Laudan [1977] are explicitly characterized in terms of laws, theories, and methodological rules). Nersessian, by contrast, insists on including the *individual scientist* as an essential part of her account. Her question is not simply how the theory of electrodynamics developed from the time of Faraday to that of Einstein, but how Faraday, Maxwell, and Einstein, as individual scientists, developed electrodynamics. Theories do not simply develop; they are *developed* through the cognitive activities of particular scientists. It is the focus on scientists, as real people, that makes possible the application of notions from cognitive psychology to questions in the philosophy of science.

Nersessian's insistence on the role of human agency in science is strongly reinforced by David Gooding's analysis of the path from actual experimentation, to the creation of demonstration experiments, to the development of theory. Insisting that all accounts of scientific activity, even those recorded in laboratory notebooks, involve reconstruction, Gooding distinguishes six types, or levels, of reconstruction. Standard philosophical reconstructions, which Gooding labels "normative," are last in the sequence. The first are "cognitive" reconstructions, with "rhetorical" and "didactic" reconstructions being among the intermediate types. Gooding is particularly insistent on the importance of "procedural knowledge," such as laboratory skills, in the cognitive development of science.

Gooding focuses on the sequence of experiments that led to Faraday's invention and development of the world's first electric motor. He develops a notation for representing the combined development of experimental procedures and theory that led to Faraday's discovery. This notation is an elaboration of one developed earlier by Ryan Tweney. Utilizing this notation, Gooding exhibits a multiplicity of possible experimental pathways leading to the electric motor, noting that Faraday himself presented different reconstructions of the pathway in different contexts.

Gooding concludes his essay by arguing that the power of thought experiments derives in part from the fact that they embody tacit knowledge of experimental procedures. This argument complements Nersessian's analysis of how an "experiment" carried out in thought can have such an apparently powerful empirical force. She argues that conducting a thought experiment is to be understood as using a mental model of the experimental situation to run a simulation of a real experiment. The empirical content is built into the mental model, which includes procedural knowledge.

Ryan Tweney was among the first of recent theorists to advocate a cognitive approach to the study of science. And he has pursued this approach in both experimental and historical contexts. Here he explores some implications of the recent vogue for parallel processing for the study of science as a cognitive process. Tweney acknowledges the importance of having models that could plausibly be implemented in a human brain, but is less impressed by neuroscientific plausibility than by the promise of realistic *psychological* models of perception, imagery, and memory — all of which he regards as central to the process of science.

Rather than joining the debate between advocates of serial models and of parallel models, Tweney takes a third route that focuses attention on cognitive activities in natural contexts — leaving the question of which sort of model best fits such contexts to be decided empirically on a case-by-case basis. But it is clear that Tweney is impressed with the promise of parallel models, even though, as he points out, they have yet to be applied successfully to higher-level cognitive processes. Here he considers two applications: (1) an account of the memory aids used by Michael Faraday to index his notebooks, and (2) Paul Thagard's analysis of scientific revolutions using a parallel network implementation (ECHO) of a theory of explanatory coherence. He finds the concept of parallel processing useful in the first case but superfluous in the second.

For nearly two decades, sociologists of science have been gathering under a banner labeled "The Social Construction of Scientific Knowledge." The aforementioned essays suggest that we can equally well speak of "The Cognitive Construction of Scientific Knowledge." There are, however, two important differences in the ways these programs are conceived. First, unlike social constructionists, cognitive constructionists make no claims of exclusivity. We do not insist that cognitive construction is all there is. Second, social constructionists typically deny, or claim indifference to, any genuine representational connection between the claims of scientists and an independently existing world. By contrast, connections with the world are built into the cognitive construction of scientific knowledge. This is particularly clear in Gooding's paper, which emphasizes the role of procedural knowledge in science.

The historical movement in the philosophy of science made conceptual change a focus of research in the history and philosophy of science. It has subsequently become a research area within cognitive psychology as well, although Piaget had already made it a focus of psychological research in Europe several decades earlier. Indeed, one major strand of current research may be seen as an extension of Piaget's program, which used conceptual development in children as a model for conceptual development in science (Gruber and Vonèche 1977). This line of research is represented here by Susan Carey.

Carey works within the "nativist" tradition, which holds that at least some concepts are innate, presumably hard-wired as the result of our evolutionary heritage. The question is what happens to the conceptual structure possessed by a normal human in the natural course of maturation, apart from explicit schooling. An extreme view is that conceptual development consists only of "enrichment," that is, coming to believe new propositions expressed solely in terms of the original set of innate concepts. Another possible view is that humans also form new concepts by differentiation and combination. Objects become differentiated to include animals, then dogs. Colors become differentiated into red, green, blue, and so forth. Combination produces the concept of a red dog (an Irish setter). Carey argues that normal development also produces conceptual systems that are, in Kuhn's (1983) terms, "locally incommensurable" with earlier systems.

As an example of local incommensurability, Carey cites the differentiation of the concepts weight and density within a system originally possessing only a single undifferentiated concept. The earlier (child's) conceptual system and the later (adult) system are incommensurable because there remains no counterpart of the undifferentiated weight/density concept in the adult conceptual system. Carey presents evidence showing that children do indeed begin with an undifferentiated weight/density concept and gradually develop a system with the differentiated concept. She also describes earlier historical research (Wiser and Carey 1983) showing that seventeenth-century scientists at the Academy of Florence possessed a similarly undifferentiated concept of temperature/heat. She thus links research in the history of science with research in cognitive development.

Carey takes pains to argue that local incommensurability between children's and adults' concepts does not mean that adults and children cannot understand one another, that children do not learn language by interacting with adults, or that psychologists cannot explain the child's conceptual system to others. So the concept of incommensurability employed here has none of the disastrous implications often associated with philosophical uses of this notion. It seems, therefore, that philosophers and psychologists may at last have succeeded in taming the concept of incommensurability, turning it into something that can do useful work.

The shift from novice to expert provides another model recently exploited by cognitive psychologists to study conceptual change in science. Michelene Chi has been a leader in this research. Here, however, she treats conceptual change in more general terms. She argues that even Carey's notion of change between incommensurable conceptual systems is not strong enough to capture the radical nature of the seventeenth-century revolution in physics. That revolution, she argues, involved a more radical conceptual shift because there was a shift in *ontological* categories. In particular, the conceptual system prior to the scientific revolution mainly employed concepts within the ontological category of "material substance" whereas the new physical concepts were mainly relational, covering what she calls "constraint-based events." According to Chi's analysis, therefore, the difficulty people have moving beyond an undifferentiated weight/density concept is due to difficulty in conceiving of weight as relational rather than substantial. Density, being an intrinsic property of objects (mass per unit volume), is developmentally the more primitive concept.

Chi criticizes other studies of conceptual change, such as Carey's, for giving only a kinematics of change and not providing any account of the dynamical mechanisms of change. She locates the engine of change at the early stages of a scientific revolution when anomalies first begin to be recognized as such. This leads not so much to a change in concepts as to the construction of a rival conceptual system that for a while coexists with the original system. Completing a scientific revolution, or a process of conceptual change in an individual, is a matter of increasing reliance on the new system and disuse of the old.

The final two essays in Part I employ a cognitive approach to problems that were prominent among logical empiricists. Questions about the nature of observation and, more technically, measurement were high on the agenda of logical empiricism. That was in large part because of the foundational role of observation in empiricist epistemology. But even if one abandons foundationalist epistemology, there are still interesting questions to be asked about observation and measurement. Richard Grandy explores several such issues from the general perspective of cognitive agents as information processors.

One question is whether there is any place for the notion of an "observation sentence" in a cognitive philosophy of science. Developing suggestions first explored by Quine, Grandy argues that there is, although whether a sentence can serve as an observation sentence cannot be solely a matter of either syntax or semantics. It depends, among other things, on the typical reactions of a relevant community in a variety of contexts. That is something that can itself only be determined empirically.

Another topic Grandy explores is the relative information provided by the use of various types of measurement scales. Grandy demonstrates that the potential information carried by a measurement typically increases as one moves from nominal, to ordinal, to ratio scales. More surprising, he is able to show that what he would regard as observation sentences typically convey more information than ordinal-scale measurements, though not as much as ratio-scale measurements. This is but one step in a projected general program to analyze the contributions of new theories, instruments, and methods of data analysis in terms of their efficiency as information generators or processors. Such an analysis would provide a "cognitive" measure of scientific progress.

In the final essay of Part I, Wade Savage explores the possibility of using recent cognitive theories of perception to develop a naturalized foundationalist empiricism. He begins by distinguishing strong from weak foundationalism. Strong foundationalism is the view that some data provided by sensation or perception are both independent (not based on further data) and infallible (incapable of error). Weak foundationalism holds that only some data of sensation or perception are more independent and more reliable than other data. Savage's view is that weak foundationalism provides a framework for a naturalistic theory of *conscious* human knowledge and strong foundationalism provides a framework for a naturalistic theory of *unconscious* human knowledge. The mistake of the classical foundationalists, he claims, is to have assumed that strong foundationalism could be a theory of conscious knowledge.

Recent work in cognitive science, Savage claims, is particularly relevant to the theory of unconscious human knowledge. Drawing inspiration from works by people such as Marr (1982) and MacArthur (1982), Savage argues for a *presensational foundationalism* in which the basic data are unconscious "ur-sensations." His main arguments are directed toward showing that ur-sensations are both independent and infallible in an appropriate sense. Savage concludes by sketching a theory reminiscent of Minsky (1987) that explains the relationship between unconscious ur-sensations and the consciousness of higher cognitive agents.

2. Models from Artificial Intelligence

Among the many crosscurrents within the fields of computer science and artificial intelligence (AI) is a tension between those who wish to use the computer as a means to study the functioning of *human* intelligence and those who see the computer primarily as a tool for performing a variety of tasks quite apart from how humans might in fact perform those same tasks. This tension is evident in the original work on "discovery programs" inspired by Herbert Simon (1977) and implemented by Pat Langley, Simon, and others (Langley et al. 1987). This work has demonstrated the possibility of developing programs that can uncover significant regularities in various types of data using quite general heuristics. Among the prototypes of such programs are BACON, GLAUBER, and KEKADA (Kulkarni and Simon 1988). BACON, for example, easily generates Kepler's laws beginning only with simple data on planetary orbits.

One way of viewing such programs is as providing "normative models" in the straightforwardly instrumental sense that these models provide good means for accomplishing well-defined goals. This use of AI is exhibited in this volume by Gary Bradshaw and Lindley Darden. For a variety of other examples, see Shrager and Langley (1990).

Bradshaw, who began his career working with Simon and Langley, applies Simon's general approach to problem solving to invention in technology. He focuses on the much-discussed historical question of why the Wright brothers were more successful at solving the problem of manned flight than their many competitors. Dismissing a variety of previous historical explanations, Bradshaw locates the crucial difference in the differing heuristics of the Wright brothers and their competitors. The Wright brothers, he argues, isolated a small number of functional problems that they proceeded to solve one at a time. They were thus exploring a relatively small "function space" while their competitors were exploring a much larger "design space."

Darden proposes applying AI techniques developed originally for diagnosing breakdowns in technological systems to the problem of "localizing" and "fixing" mistaken assumptions in a theory that is faced with contrary data. Here she outlines the program and sketches an application to the resolution of an empirical anomaly in the history of Mendelian genetics. Darden is quite clear on the goal of her work: "The goal is not the simulation of human scientists, but the making of discoveries about the natural world, using methods that extend human cognitive capacities."

Programs like those of Darden and others are potentially of great scientific utility. That potential is already clear enough to inspire many people to develop them further. How useful such programs will actually prove to be is not something that can be decided a priori. We will have to wait and see. The implications of these sorts of programs for a cognitive philosophy of science are mainly *indirect*. The fact that they perform as well as they do can tell us something about the structure of the domains in which they are applied and about possible strategies for theorizing in those domains.

Others see AI as providing a basis for much more far-reaching philosophical conclusions. The essay by Greg Nowak and Paul Thagard and that by Eric Freedman apply Thagard's (1989) theory of explanatory coherence (TEC) to the Copernican revolution and to a controversy in psychology, respectively. Nowak and Thagard hold both that the objective superiority of the Copernican theory over the Ptolemaic theory is shown by its greater overall explanatory coherence, and that the triumph of the Copernican theory was due, at least in part, to the intuitive perception of its greater explanatory coherence by participants at the time.

Thagard, who advocates a "computational philosophy of science" (Thagard 1988), implements his theory of explanatory coherence in a connectionist program, ECHO, which utilizes localized representations of propositions. It has been questioned (for example, by Tweney and Glymour in this volume) whether ECHO is doing anything more than functioning as a fancy calculator, with all the real work being done by TEC. If so, it is a very fancy calculator, performing nonlinear optimization with several constraints, and containing various adjustable parameters that can materially affect the outcome of the calculation.

Freedman's study provides a further illustration of the operations of TEC and ECHO. He analyzes the famous controversy between Tolman and Hull over the significance of Tolman's latent-learning experiments. Applying TEC and ECHO, Freedman finds that Tolman's cognitive theory is favored over Hull's behaviorist theory. Yet Hull's approach prevailed for many years. By varying available parameters in ECHO, Freedman shows several ways in which ECHO can be made to deliver a verdict in favor of Hull. For example, significantly decreasing the importance of the latent-learning data can tip the balance in favor of Hull's theory. To Freedman, this provides at least a suggestion for how the actual historical situation might be explained. So ECHO does some work. But this study also makes it obvious that to decide among the possibilities suggested by varying different parameters in ECHO, one would have to do traditional historical research. ECHO cannot decide the issue.

Thagard's work shows that a deep division between cognitive psychologists and AI researchers carries over into the ranks of those advocating a cognitive approach to the philosophy of science. Most cognitive psychologists would insist that cognitive psychology is fundamentally the study of the cognitive capacities of human agents. For a cognitive philosophy of science this means studying how scientists actually represent the world and judge which representations are best. In Thagard's computational philosophy of science, all representations are conceptual and propositional structures, and judgments of which representations are best are reduced to computations based primarily on relationships among propositions. There is as yet little evidence that the required propositional structures and computations are psychologically real.

3. Models from Neuroscience

In Part III the relevance of models from the neurosciences to the philosophy of science is argued by the primary advocate of the philosophical relevance of such models, Paul Churchland. It is Churchland's (1989) contention that we already know enough about the gross functioning of the brain to make significant claims about the nature of scientific knowledge and scientific reasoning. In his essay he argues that a "neurocomputational" perspective vindicates (more precisely, "reduces") a number of claims long advocated by Paul Feyerabend. For example: "Competing theories can be, and occasionally are, *incommensurable*," and "the long-term best interests of intellectual progress require that we proliferate not only theories, but research *methodologies* as well."

It is interesting to note that Churchland's analysis supports Chi's conclusion that the dynamical process in theory change is not so much transformation as it is the construction of a new representation that then replaces the old representation. That, Churchland claims, is just how neural networks adapt to new situations.

Whatever one's opinion of Churchland's particular claims, I think we must all agree that the neurosciences provide a powerful and indisputable constraint on any cognitive philosophy of science. Whatever cognitive model of scientific theorizing and reasoning one proposes, it must be a model that can be implemented by humans using human brains.

4. Between Logic and Sociology

Except during momentary lapses of enthusiasm, no one thinks that a cognitive theory of science could be a *complete* theory of science. The cognitive activities of scientists are embedded in a social fabric whose contribution to the course of scientific development may be as great as that of the cognitive interactions between scientists and the natural world. Thus cognitive models of science need to be supplemented with social models. The only requirement is that the two families of models fit together in a coherent fashion.

There are those among contemporary sociologists of science who are not so accommodating. Latour and Woolgar, for example, are now famous for suggesting a ten-year moratorium on cognitive studies of science, by which time they expect to have constructed a complete theory of science that requires no appeal to cognitive categories. Such voices are not directly represented in this volume, but they do have supporters nonetheless.

Arthur C. Houts and C. Keith Haddock agree with the sociological critics of cognitivism in rejecting the use of cognitive categories like representation or judgment in a theory of science. But they insist there is need for a genuine psychology of science. From a cognitivist point of view, these are incompatible positions. For Houts and Haddock these positions are not incompatible because their psychology of science is based on the behaviorist principles of B. F. Skinner. In Skinnerian theory, the determinants of behavior are to be found in the environment, both natural and social, which provides the contingencies of reinforcement. There is no need for any appeal to "mental" categories such as representation or judgment. Several commentators, for example, Slezak (1989) and myself (Giere 1988), have criticized behaviorist tendencies in the writings of sociologists of science. For Houts and Haddock, these tendencies are not a basis for criticism but a positive virtue. They make possible a unified approach to both the psychology and the sociology of science.

Within cognitive psychology there is a tradition, already several decades old, in which scientific reasoning tasks are simulated in a laboratory setting. Michael E. Gorman reviews this tradition and compares it with the more recent tradition of computational simulation pioneered by Simon and represented in this volume by Thagard. He relies heavily on the distinction between externally valid and ecologically valid claims. A claim is *externally valid* if it generalizes well to other well-controlled, idealized conditions. A claim is *ecologically valid* if it generalizes well to natural settings, for example, to the reasoning of scientists in their laboratories. Gorman argues that while both laboratory and computer simulations may be externally valid, laboratory studies are more ecologically valid. Granting this conclusion, however, does little to remove doubts about the ecological validity of laboratory studies themselves.

Gorman proposes bridging the gap between cognitive and social studies of science by designing experimental simulations that include social interactions among the participants. Here experimental paradigms from social psychology are merged with those that have been used in the experimental study of scientific reasoning. Gorman's hope is that one might eventually develop experimental tests of claims made by sociologists as well as by more theoretical "social epistemologists" such as Steve Fuller.

Fuller himself questions a central presupposition of most cognitive approaches to the philosophy of science, namely, that the individual scientist is the right unit of analysis for any theory of science. Not that he advocates returning to abstract entities like theories. Rather he thinks that the appropriate unit will turn out to be something more like a biological species than an individual scientist. Bruno Latour's (1987) "actor network" may be a good example of the kind of thing Fuller expects might emerge as the proper unit of study. Fuller's argument is both historical and critical. He sketches an account of how the individual scientist came to be regarded as the basic entity for epistemology generally, and why this assumption has led to difficulties in several areas, particularly in analytic epistemology, but also in Churchland's neurocomputational approach.

5. Critique and Replies

Clark Glymour was among the first philosophers of science to grasp the possibility of deploying methods and results from the cognitive sciences, particularly artificial intelligence, to the philosophy of science itself. (Herbert Simon, who I definitely would wish to claim as a philosopher of science, must surely have been the first.) But as his contribution to this volume makes crystal clear, Glymour is quite disappointed with what some other philosophers of science have been doing with this strategy. In his essay in Part V he expresses his disappointment with work by three of the participants in the Minnesota workshop, Churchland, Thagard, and myself. By mutual agreement, Glymour's comments appear as he wrote them. They are followed by replies from each of the three named subjects of his remarks. Since my own reply is included, I will say no more here in my role as editor.

References

Bruner, J. S., J. Goodnow, and G. Austin. 1956. *A Study of Thinking*. New York: Wiley.
Campbell, D. T. 1959. Methodological Suggestions from a Comparative Psychology of Knowledge Processes. *Inquiry* 2:152–82.
Churchland, P. M. 1989. *A Neurocomputational Perspective*. Cambridge, Mass.: MIT Press.
Churchland, P. S. 1986. *Neurophilosophy*. Cambridge, Mass.: MIT Press.
Dennett, D. 1983. *Brainstorms*. Cambridge, Mass.: MIT Press.

Fodor, J. 1987. *Psychosemantics.*Cambridge, Mass.: MIT Press.

Gardner, H. 1985. *The Mind's New Science.* New York: Basic Books.

Giere, R. N. 1988. *Explaining Science: A Cognitive Approach.* Chicago: University of Chicago Press.

Goldman, A. 1986. *Epistemology and Cognition.* Cambridge, Mass.: Harvard University Press.

Gruber, H., and J. J. Vonèche, eds. 1977. *The Essential Piaget.* New York: Basic Books.

Hanson, N. R. 1958. *Patterns of Discovery.* Cambridge: Cambridge University Press.

Johnson-Laird, P. N. 1983. *Mental Models.* Cambridge, Mass.: Harvard University Press.

Kuhn, T. S. 1962. *The Structure of Scientific Revolutions.* Chicago: University of Chicago Press.

———. 1983. Commensurability, Comparability, Communicability. In *PSA 1982*, ed. P. Asquith and T. Nickles, 669–87. East Lansing, Mich.: Philosophy of Science Association.

Kulkarni, D., and H. Simon. 1988. The Processes of Scientific Discovery: The Strategy of Experimentation. *Cognitive Science* 12:139–75.

Lakatos, I. 1970. Falsification and the Methodology of Scientific Research Programmes. In *Criticism and the Growth of Knowledge*, ed. I. Lakatos and A. Musgrave, 91–195. Cambridge: Cambridge University Press.

———. 1971. *History of Science and Its Rational Reconstructions.* Boston Studies in the Philosophy of Science, ed. R. C. Buck and R. S. Cohen, vol. 8. Dordrecht: Reidel.

Langley, P., H. A. Simon, G. L. Bradshaw, and J. M. Zytkow. 1987. *Scientific Discovery.* Cambridge, Mass.: MIT Press.

Latour, B. 1987. *Science in Action.* Cambridge, Mass.: Harvard University Press.

Laudan, L. 1977. *Progress and Its Problems.* Berkeley: University of California Press.

MacArthur, D. J. 1982. Computer Vision and Perceptual Psychology. *Psychological Bulletin* 92:283–309.

Marr, D. 1982. *Vision.* San Francisco: Freeman.

Minsky, M. 1987. *The Society of Mind.* New York: Simon & Schuster.

Quine, W. V. O. 1969. Epistemology Naturalized. In *Ontological Relativity and Other Essays*, 69–90. New York: Columbia University Press.

Shrager, J., and P. Langley, eds. 1990. *Computational Models of Scientific Discovery and Theory Formation.* Hillsdale, N.J.: Erlbaum.

Simon, H. A. 1977. *Models of Discovery.* Dordrecht: Reidel.

Slezak, P. 1989. Scientific Discovery by Computer as Empirical Refutation of the Strong Programme. *Social Studies of Science* 19:563–600.

Suppe, F. 1989. *The Semantic Conception of Theories and Scientific Realism.* Urbana: University of Illinois Press.

Thagard, P. 1988. *Computational Philosophy of Science.* Cambridge, Mass.: MIT Press.

———. 1989. Explanatory Coherence. *Behavioral and Brain Sciences* 12:435–67.

van Fraassen, B. C. 1980. *The Scientific Image.* Oxford: Oxford University Press.

———. 1989. *Laws and Symmetry.* Oxford: Oxford University Press.

Wiser, M., and S. Carey. 1983. When Heat and Temperature Were One. In *Mental Models*, ed. D. Gentner and A. Stevens. Hillsdale, N.J.: Erlbaum.

PART I

MODELS FROM COGNITIVE PSYCHOLOGY

How Do Scientists Think?
Capturing the Dynamics
of Conceptual Change in Science

The Scene

August 19, 1861, a cottage in Galloway, Scotland.

The young Clerk Maxwell is sitting in a garden deep in thought. On the table before him there is a sheet of paper on which he sketches various pictures of lines and circles and writes equations.

The Question

What is he thinking? Is he trying to cook up a model to go with the equations he has derived already by induction from the experimental data and electrical considerations alone? Is he concerned that his mathematical results are not quite right and so is thinking how to fudge his analysis to make it look right in terms of the model? Is he searching for a way to make the notion of continuous transmission of actions in an electromagnetic "field" meaningful? And if so, what resources is he drawing upon? What is he doing?

The Problem

Do we have the means to understand what Maxwell is doing? What scientists like him are doing when they are creating new conceptions? Based on the record they leave behind, can we hope to fathom the creative processes through which scientists articulate something quite new? Or are these processes so mysterious that we are wasting our time by trying to understand them? And if we could, what possible profit could such understanding yield for the philosopher of science? the historian of science? others?

The Path to Solution

I hope to persuade the reader that we can formulate a more rigorous analysis of the creative processes of scientific discovery and give more satisfactory answers to long-standing, unresolved puzzles about the nature of conceptual change in science than we have now by combining two things that are usually kept apart. One is fine-structure examinations of the theoretical and experimental practices of scientists who have created major changes in scientific theory. The other is what we have been learning about the cognitive abilities and limitations of human beings generally. Creative processes are extended and dynamical, and as such we can never hope to capture them fully. But by expanding the scope of the data and techniques allowed into the analysis we can understand more than traditional approaches have permitted so far.

Recent developments in psychology have opened the possibility of understanding what philosophers and historians have been calling "conceptual change" in a different and deeper way. Through a combination of new experimental techniques and computer modeling, new theories about human cognitive functioning have emerged in the areas of representation, problem solving, and judgment. An interdisciplinary field of cognitive science has recently formed — a loose confederation of cognitive psychology, artificial intelligence, cognitive neurology, linguistics, and philosophy. It offers analyses and techniques that, if used with proper respect for their scope and their limitations, can help us develop and test models of how conceptual change takes place in science.

In this essay I set myself the following aims: (1) to propose a fresh method of analysis; (2) to recast the requirements of a theory of conceptual change in science; (3) to draw on new material from a heuristically fertile case study of major conceptual change in science to analyze some processes of conceptual change — analogical and imagistic reasoning and thought experiments and limiting case analyses; (4) to examine these in light of some work in cognitive science; and (5) to argue, more generally, for what philosophers and historians of science and cognitive scientists might gain from further application of the proposed method of analysis.

1. What Is "Cognitive-Historical" Analysis?

"Cognitive-historical" analysis in the sense employed here is not quite the same as what historians of science do in their fine-structure historical examinations of the representational and problem-solving practices scientists have employed to create new scientific representations of phenomena. Rather, it attempts to enrich these further by means of investigations of ordinary human representational and problem-solving practices carried out by the sciences of cognition. *The underlying presupposition is that the problem-solving strategies scientists have invented and the representational practices they have developed over the course of the history of science are very sophisticated and refined outgrowths of ordinary reasoning and representational processes.* Thus, the method combines case studies of actual scientific practices with the analytical tools and theories of the cognitive sciences to create a new, comprehensive theory of how conceptual structures are constructed and changed in science. The historical dimension of the method has its origins in the belief that to understand scientific change the philosophy of science must come to grips with the historical processes of knowledge development and change. This is the main lesson we should have learned from the "historicist" critics of positivism. Equally as important as problems concerning the rationality of acceptance — which occupy most philosophers concerned with scientific change — are problems about the construction and the communication of new representational structures. The challenging methodological problem is to find a way to use the history of scientific knowledge practices as the *basis* from which to develop a theory of scientific change.

The cognitive dimension of the method reflects the view that our understanding of scientific knowledge practices needs to be psychologically realistic. Putting it baldly, creative scientists are not only exceptionally gifted human beings — they are also human beings with a biological and social makeup like all of us. In a fundamental sense, science is one product of the interaction of the human mind with the world and with other humans. We need to find out how human cognitive abilities and limitations constrain scientific theorizing *and this cannot be determined a priori.* This point is not completely foreign to philosophers. It fits into a tradition of psychological epistemology beginning with Locke and Hume and making its most recent appearance with the call of Quine for a "naturalized epistemology." Why did earlier "psychologizing" endeavors fail? The main reason was their reliance on inadequate empiricist/behaviorist psychological theories. The development of cognitive psychology has paved the way for a much more fruitful synthesis of psychology and epistemology. Suggestions for how to frame such a synthesis are to be found, for example, in the work of Alvin Goldman (1986). Insights from cognitive psychology are beginning to make their way into investigations of scientific reasoning (see, e.g., Giere 1988; Gooding 1990; Gorman and Carlson 1989; Langley et al. 1987; Thagard 1988; and Tweney 1985). What is needed now is to integrate these with the historical findings about the representational and problem-solving practices that have actually brought about major scientific changes.[1]

Philosophers this century have mostly been working under the assumption that analysis of science takes place within two contexts: justification and discovery. The former is traditionally within the province of philosophers; the latter, of historians and psychologists. Cognitive-historical analysis takes place in a new context — that of development — which is the province of all three. The context of development[2] is the domain for inquiry into the processes through which a vague speculation gets *articulated* into a new scientific theory, gets *communicated* to other scientists, and comes to *replace* existing representations of a domain. These processes take place over long periods of time, are dynamic in nature, and are embedded in social contexts.

This new context of development, in actuality, was opened up by the work of Hanson, Kuhn, and Feyerabend nearly thirty years ago, but they lacked the analytical tools to pursue it in depth. True, they attempted to integrate insights from psychology into their analyses. However, cognitive psychology, in the form of the "new look" psychology of Bruner and others, was in its infancy, whereas Gestalt psychology offered no understanding of the processes underlying the "gestalt switch." Since their vision predated the kind of psychological theory that would have helped them better express it, it would be unfair to fault them for not completing what they had begun. It is important, however, to see the continuing repercussions of the inadequate insights they did use.

First, drawing from these psychological theories led to an unfortunate identification of knowledge change with hypothesized aspects of visual perception. Second, and more important, the metaphor of the "gestalt switch" led them astray in a way that has had deeper and more lasting consequences. The metaphor does not support the extended nature of the conceptual changes that have actually taken place in science.[3] Thus, while calling for a historicized epistemology, Kuhn's and Feyerabend's own historical analyses offered in support of the "incommensurability" hypothesis were decidedly unhistorical in the following sense. By emphasizing the endpoints of a conceptual change (e.g., Newtonian mechanics and relativistic mechanics), the change of gestalt was made to appear artificially abrupt and discontinuous. Historically, however, we did not get to relativity without at least passing through electromagnetism and the theory of electrons; and this developmental process is central to understanding such questions as the nature of the relationship — or reason for lack of relationship — between, e.g., the different concepts called by the name "mass" in each theory. My earlier study of the construction of the concept of electromagnetic field, *Faraday to Einstein: Constructing Meaning in Scientific Theories* (1984), was an attempt to show how incorporating the dimension of development into the analysis gives a quite different picture of the nature of meaning change in science.

Significantly, although Kuhn does talk about discovery as an "extended process" (Kuhn 1965, pp. 45ff.) and, in his role as historian of science, has provided detailed examinations of such processes, in his role as philosopher of science he identifies conceptual change with "the last act," when "the pieces fall together" (Kuhn 1987).[4] Thus portrayed, conceptual change appears to be something that happens to scientists rather than the outcome of an extended period of construction by scientists. A "change of gestalt" may be an apt way of characterizing this last point in the process, but focusing exclusively on that point has — contrary to Kuhn's aim — provided a misleading portrayal of conceptual change; has reinforced the widespread view that the processes of change are mysterious and unanalyzable; and has blocked the very possibility of investigating how precisely the new gestalt is related to its predecessors. *In short, the metaphor has blocked development of the historicized epistemology being advocated.*

The ultimate goal of the cognitive-historical method is to be able to reconstruct scientific thinking by means of cognitive theories. When, and if, we reach that point, we may decide to call the method "cognitive analysis of science." At present, however, cognitive theories are largely uninformed by scientific representational and problem-solving practices, making the fit between cognitive theories and scientific practices something that still needs to be determined.[5] Cognitive-historical analysis is reflexive. It uses cognitive theories to the extent that they help interpret the historical cases — at the same time it tests to what extent current theories of cognitive processes can be applied to scientific thinking and indicates along what lines these theories need extension, refinement, and revision. In other words, the method is a type of bootstrapping procedure commonly used in science.

2. What Would a Cognitive Theory of Conceptual Change in Science Look Like?

2.1. Background

Much philosophical energy this century has been spent on the problem of conceptual change in science. The major changes in physical theory early in the century thrust the problem of how to understand the seemingly radical reconceptualizations they offered into the spotlight for scientists, historians, and philosophers alike. As we know, the comforting picture of conceptual change as continuous and cumulative offered by logical positivism itself suffered a revolutionary upheaval in the mid-1960s. The critics of positivism argued that major changes in science are best characterized as "revolutions": they involve overthrow and replacement of the reigning conceptual system with one that is "incommensurable" with it. The infamous "problem of incommensurability of meaning" dominated the literature for over a decade. Philosophers have by and large abandoned this topic. Those who work on scientific change tend now to focus on the problem of rational choice between competing theories. This shift in focus did not, however, come from a sense of having a satisfactory solution to the infamous problem, but more from a sense of frustration that the discussion and arguments had become increasingly sterile.

The crux of the original problem, however, is still with us. How, if in any manner at all, are successive scientific conceptualizations of a domain related to one another? The instinctive response of critics of incommensurability has always been that even though they are not simply extensions, the new conceptual structures must somehow grow out of the old. The view of knowledge change as a series of unconnected gestalt switches has a high intuitive implausibility. *In recasting the problem of conceptual change, the cognitive-historical method furnishes the means through which to turn these intuitions into solid analyses.*

In cognitive-historical analysis the problem of conceptual change appears as follows. It is the problem of understanding how scientists combine their human cognitive abilities with the conceptual resources available to them as members of scientific communities and wider social contexts to create and communicate new scientific representations of a domain. For example, the problem posed in the opening scene becomes that of understanding how Maxwell joins his human cognitive endowment with the conceptual resources of a Cambridge mathematical physicist living in Victorian England to construct and communicate a field representation of electromagnetic forces. Admittedly, this is a quite complex problem and we are only beginning to have the means to attack it. Nevertheless, I shall show in some detail that the cognitive-historical approach offers more possibility of achieving a solution than any we have yet considered.

Where the traditional philosophical approach views conceptual change as static and ahistorical, cognitive-historical analysis is able to handle the dynamic and historical process that it is. Customarily, conceptual change is taken to consist of the replacement of one linguistic system by another, and understanding conceptual change requires analyzing the logical relationships between propositions in the two systems.

In a cognitive theory conceptual change is to be understood in terms of the people who create and change their representations of nature and the practices they employ to do so. This opens the possibility of understanding *how it is that scientists build on existing structures while creating genuine novelty*. That is to say, a route is opened toward explaining the continuous and noncumulative character of conceptual change that is amply supported by results of individual studies of scientific creativity undertaken by historians of science.

2.2. Outline of a Cognitive Theory of Conceptual Change

Further on in this essay I will turn to an examination of some of the processes of conceptual change. We need first, however, to have some sense of how a full theory would look.

A scientific theory is a kind of representational system. Several forms of representation have been proposed by cognitive scientists. Although it is a point of some controversy as to whether there is any form of representation other than strings of symbols, I will be following an authoritative account by Johnson-Laird (1983) in assuming the existence of at least three: (1) "propositional" representation (strings of symbols such as "the cat is on the mat"), (2) "mental models" (structural analogs of real-world or imagined situations, such as a cat being on a mat or a unicorn being in the forest), and (3) "images" (a mental model from a specific perspective, such as looking down on the cat on the mat from above). I will also be assuming with him that even if at the level of encoding all representations are propositional, in reasoning and understanding people construct mental models of real and imaginary phenomena, events, situations, processes, etc. One value of having a mental-models form of representation is that it can do considerable inferential work without the person having to actually compute inferences and can also narrow the scope of possible inferences. For example, moving an object immediately changes all of its spatial relationships and makes only specific ones possible. The hypothesis that we do such inferencing via mental models gains plausibility when we consider that, as biological organisms, we have had to adapt to a changing environment. In fact, artificial intelligence researchers have run into considerable problems handling the widespread effects of even small changes in knowledge-representational systems that are represented propositionally.

To continue, in a cognitive theory of conceptual change, a scientific theory will, itself, be construed as a structure that picks out classes of models, which accords better with the semantic view of theories (van Fraassen 1980) than with the Carnapian view of a theory as linguistic framework. Thinking about and in terms of a theory necessitates the construction of mental models. While scientific concepts may be encoded propositionally, understanding them involves interpretation, i.e., the construction of a mental model of the entities or processes they represent. Thus, what philosophers have been calling "meaning" and "reference" (i.e., the interplay between words, minds, and the world), is, on this view, mediated by the construction of mental models that relate to the world in specified ways.

Like science itself, a theory of conceptual change in science needs to provide both descriptive and explanatory accounts of the phenomena. These dimensions of the theory will here be called, respectively, the "kinematics" and "dynamics" of conceptual change. Kinematics is concerned with problems of how to represent change, and dynamics with the processes through which change is created.

2.2.1. The Kinematics of Change

Any analysis of how to represent conceptual change in science must be solidly informed by the actual representational practices scientists use in developing and changing conceptual systems. Examinations of the conceptual changes that have been part of "scientific revolutions" yield the following insights. New concepts are created, such as 'spin' in quantum mechanics, and existing ones disappear, such as 'phlogiston' from chemistry. Some concepts in the new system, e.g., 'mass' and 'field' in relativity, are what can only be called "conceptual descendants" of existing ones. And, finally, some, such as 'ether', while appearing to be eliminated, have significant aspects absorbed by other concepts, in this case 'field' and 'space-time'.

If the situations of creation and disappearance were all we had, handling the problems would be far less complex. In that case, "conceptual change" could be characterized as the replacement of one concept or structure by another. *Given the reality of descendants and absorption*, though, in addition to representing change of conceptual systems, we need to be able to represent conceptual change at the level of individual concepts. Some philosophers have trouble countenancing what it could possibly mean for a concept to *change* its meaning. As I have argued in my book (Nersessian 1984), the failure of existing theories of 'meaning' and 'meaning change' even to allow for this possibility has led to many of the various conundrums associated with the so-called problem of incommensurability of meaning. If, as has been traditionally held, concepts are represented by neatly bundled and individuated units (i.e., sets of necessary and sufficient conditions), only replacement, not change, is possible. Therefore, *a different form of representation is needed to accommodate the data of change.*

Psychological research on categorization supports the view that in many instances people do not represent concepts by means of sets of necessary and sufficient conditions (see, e.g., Rosch and Mervis 1975; Smith and Medin 1981; and Murphy and Cohen 1984). Examination of cases from the history of science also substantiates the view that for numerous scientific concepts — or even for a concept within a single theory — it is not possible to specify a set of necessary and sufficient conditions that will take in all their historical usages. For example, I have shown in my book how there is no set of necessary and sufficient conditions defining 'electromagnetic field'. Yet there is a traceable line of descent between the Faradayan and the Einsteinian concepts and a significant overlap in salient features between successive field concepts.

The question now is: Can we assume that how scientists structure *mental* representations is reflected in their *external* representations? That is, Does what they write give a clue to how they think? Psychological studies all start from the assumption that how people represent mentally is reflected in their use of language, and there is good reason to make the same assumption here. Since I have dealt with the kinematics of conceptual change extensively, though far from exhaustively, in my book and in a number of articles (Nersessian 1985, 1986, and in press[a]), I want the focus of this essay to be its dynamics. We need to keep in mind, though, that the two problems are connected. In order to be complete a cognitive theory will have to determine how the historical data on the individual units of change — "concepts" — mesh with those from psychology and also with attempts at constructing psychologically realistic representational systems in artificial intelligence.

2.2.2. The Dynamics of Change

By what processes are new scientific representations constructed? The prevailing view among philosophers is that the discovery processes are too mysterious to be understood. This view receives support from numerous stories of discovery through flashes of insight of geniuses, such as Kekulean dreams and Archimedean eureka-experiences. What is omitted from such renderings are the periods of intense and often arduous thinking and, in some cases, experimental activity that precede such "instantaneous" discoveries. There again, the rendering of conceptual changes as "gestalt switches" reinforces the prevailing prejudice. Even Kuhn substitutes the phrase "exploitation by genius" for analysis of actual constructive processes when discussing how Galileo formed his concept of 'pendulum' (Kuhn 1965, p. 119). There is, however, no *in-*

herent conflict between the view that discovery processes are creative and the view that they are reasoned. We need to give up the notion that "creativity" is an *act* and try to fathom it as a *process*.

Historical evidence supports the conviction that conceptual change in science is at heart a problem-solving process. While pragmatist philosophers, such as Dewey, Mead, and Popper, have strongly defended this view of science, conceptual change has not been included in their analyses. Laudan (1977) does introduce "conceptual problems" into the realm of the scientist's concerns, but offers no account of the specific processes of conceptual change. I want to extend the conception of science as a problem-solving enterprise to include what has traditionally been called "conceptual change." New representations do not emerge fully grown from the heads of scientists but are constructed in response to specific problems by systematic use of heuristic procedures. Problem solving in science does differ from much of "ordinary" problem solving in that scientific problems are more complex and often less well-defined, and the solution is not known in advance to anyone. A cognitive theory of conceptual change assumes the position long advocated by Herbert Simon that "the component processes, which when assembled make the mosaic of scientific discovery, are not qualitatively distinct from the processes that have been observed in simpler problem-solving situations" (Simon, Langley, and Bradshaw 1981, p. 2). The plausibility of this assumption is not diminished by the fact that the computer "discovery" programs implemented by Simon and his co-workers to model these processes thus far have tackled only the simplest of problem-solving heuristics. While the ability to model the problem-solving techniques that have brought about major conceptual changes seems a long way off, the type of cognitive analysis advocated here is within our grasp and is also a necessary preliminary step to more realistic computer modeling.

The next section of this essay will be devoted to examination of how a selection of problem-solving heuristics create new representations in science. As in my earlier work, I will draw largely, though not exclusively, on historical data from the construction of the field representation of forces from Faraday to Einstein. These are very rich data and as yet have been far from exhausted in their fertility for our purposes. Extending my earlier work, I subject mostly novel data to fresh layers of analysis.

Throughout the history of scientific change we find recurrent use of (1) analogical reasoning, (2) imagistic reasoning, (3) thought experiment, and (4) limiting case analysis. These are all modeling activities, and although they constitute a substantial portion of scientific method, none except analogy has received more than scant attention in the philosophical literature. The main problems philosophers have had in countenancing these as methods are that they are nonalgorithmic and, even if used correctly, may lead to the wrong solution or to no solution. This very feature, however, makes them much more realistic from a historical point of view.

Limiting scientific method to the construction of inductive or deductive arguments has needlessly blocked our ability to make sense of many of the actual constructive practices of scientists. I call the particular subset of practices I will be discussing "abstraction techniques." As we will see, they are strongly implicated in the explanation of how existing conceptual structures play a role in constructing new, and sometimes radically different, structures.

3. Abstraction Techniques and Conceptual Change

3.1. Analogical and Imagistic Reasoning

3.1.1. Background

There are numerous cases where analogy has played a central role in the construction of a new scientific concept: Newton's analogy between projectiles and the moon ('universal gravitation'), Darwin's analogy between selective breeding and reproduction in nature ('natural selection'), and the Rutherford-Bohr analogy between the structure of the solar system and the configuration of subatomic particles ('atom') are among the more widely known. Also, although less well known, there are numerous cases that establish the prominence of reasoning from pictorial representations in the constructive practices of scientists who were struggling to articulate new conceptualizations. Such imagistic representations have often been used in conjunction with analogical reasoning in science.

The major problem, as was noted above, is that while amply documented, these constructive practices have received scant attention from analysts of scientific method. Analogy has received the most attention, but the thrust of those analyses has been to conceive of it as a weak form of inductive argument. Following Campbell's (1920) lead, Hesse (1966) broke some ground in stressing the importance of analogy in giving meaning to new theoretical terms and in trying to formulate how it could be an inductive method without being a *logic*, i.e., algorithmic. Sellars (1965; see also H. Brown 1986) argued that, in general, analogical reasoning creates a bridge from existing to new conceptual frameworks through the mapping of relational structures from the old to the new.

I intend to show here that the insights of those who have recognized the importance and power of analogical reasoning in concept formation and change can be furthered by cognitive-historical analysis. To do this I will go beyond my earlier analysis of the construction of the field representation of forces to discuss how it is that imagistic and analogical reasoning were used by Faraday and Maxwell in their constructive efforts. I will also draw from current cognitive theories to show how it is possible that such reasoning *generates* new representations from existing ones.

3.1.2. Case Study: Faraday, Maxwell, and the Field

This analysis will focus on Maxwell as we find him in our opening scene. Naturally, like all scientists, Maxwell was working in a context. His analysis depends heavily, among other things, on a method he claims to have "poached" from William Thomson, on specific representations he takes from Faraday, and on a certain mathematical approach to analyzing continuum phenomena being developed at the time by Cambridge mathematical physicists.

The field representation of electromagnetic forces had its origin in vague speculations that there might be physical processes in the regions surrounding bodies and charges that could account for the apparent action of one body on another at some distance from it. Faraday was the first to attempt to construct a unified representation for the continuous transmission and interconversion of electric and magnetic actions. His formulation is primarily in qualitative form, and reasoning from a specific imagistic representation figures predominantly in its construction. He constructed his field concept by reasoning from representations of the "lines of force" such as those that form when iron filings are sprinkled around a magnetic source (see Figure 1). Many linelike features are incorporated into his representation. He characterized the lines as "expanding," "collapsing," "bending," "vibrating," "being cut," and "turning corners," and attempted to devise experiments to capture the diverse motions of the lines. Thus, he transformed the static visual representation of the lines into a qualitative dynamical model for the transmission and interconversion of electric and magnetic forces, and, ultimately, for all the forces of nature and matter. As Maxwell ([1855] 1991) remarked, although this model is qualitative, it embodies within it a great deal of mathematical understanding, which Maxwell himself was able to extract from it.

In the most complete formulation of Faraday's field representation nothing exists but a "sea" of lines of force: all the forces of nature are unified and interconvertible through various motions of the lines, with matter itself being nothing but point centers of converging lines of force. The centrality of the image in his reasoning can also be seen in the only quantitative relationship he formulated explicitly: that between the number of lines cut and the intensity of the induced force. This re-

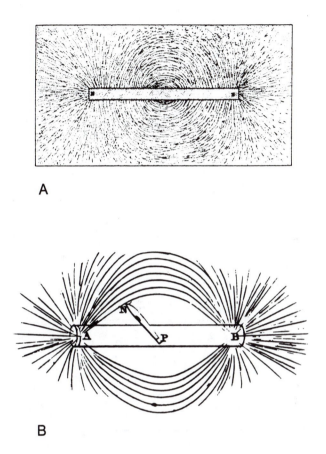

A

B

Figure 1. **A** *Actual pattern of lines of force surrounding a bar magnet (from Faraday 1839–55, vol. 3);* **B** *Schematic representation of lines of force surrounding a bar magnet (vol. 1).*

lationship is incorrect because "number of" is an integer, while "field intensity" is a continuous function. With our hindsight we can trace the "mistake" directly to the fact that lines are discrete entities and the image represents the filings as such, whereas, except in rare cases, the actual lines of force spiral indefinitely in a closed volume.

Near the end of his research Faraday introduced another image that was to play a key role in Maxwell's construction of a quantitative field representation. That image was of interlocking curves that represent the dynamical balance between electricity and magnetism (see Figure 2A) —

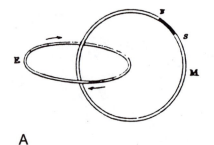

A

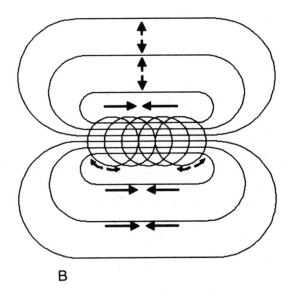

B

Figure 2. **A** *Faraday's representation of the interconnectedness of electric currents and magnetic force (from Faraday 1839–55, vol. 3);* **B** *My schematic representation of the reciprocal relationship between magnetic lines of force and electric current lines.*

what Faraday called "the oneness of condition of that which is apparently two powers or forms of power" (Faraday 1839–55, vol. 3, para. 3268). The image represents the structural relations between the electric and magnetic lines of force and is, thus, itself abstracted from the lines of force image. We can see this from Figure 2B. Here the outer lines represent magnetism and the inner lines, current electricity. The lateral repulsion of the magnetic lines has the same effect as a longitudinal expansion of the electric current lines, and the lateral attraction of the current lines has the same effect as a longitudinal contraction of the magnetic lines. This dynamical balance is captured in the image of interlocking curves.

Maxwell called the reciprocal dynamical relations embodied in the image "*mutually embracing* curves" (Maxwell 1890, vol. 1, p. 194n). While he does not include a drawing of this visual representation in the paper, Maxwell does describe it while referring the reader to the appropriate section of Faraday's *Experimental Researches.* Wise (1979) offers a convincing account of exactly how the image plays a role in the mathematical representation Maxwell constructed in his first paper on electromagnetism (Maxwell 1890, vol. 1, pp. 155–229) and throughout his work in his complicated use of two fields each for electric and magnetic forces: one for "intensity," a longitudinal measure of power, and one for "quantity," a lateral measure. Wise further provides a plausible argument that the image could even have provided a model for propagation of electromagnetic actions through the ether. If we expand Figure 2A into a "chain," then summations of the quantities and intensities associated with the electric and magnetic fields would be propagated link-by-link through interlocking curves.

While Maxwell constructed his full quantitative representation over the course of three papers, his central analysis is in the second, "On Physical Lines of Force" (Maxwell 1890, vol. 1, pp. 451–513). It is in this paper that he first derived the field equations, i.e., gave a unified mathematical representation of the propagation of electric and magnetic forces with a time delay, and calculated the velocity of the propagation of these actions. He achieved this by using a method he called "physical analogy" to exploit the powerful representational capabilities of continuum mechanics in his analysis. According to Maxwell, a physical analogy provides both a set of mathematical relationships and an imagistic representation of the structure of those relationships drawn from a "source" domain to be applied in analyzing a "target" domain about which there is only partial knowledge. In this case the method worked as follows.

Maxwell began by transforming the problem of analyzing the production and transmission of electric and magnetic forces into that of analyzing the potential stresses and strains in a mechanical electromagnetic medium ("target" domain) and then constructed an analogy between these and well-formulated relationships between known continuum mechanical phenomena ("source" domain). The process of application of the method of physical analogy comprised identifying the electromagnetic quantities with properties of the continuum mechanical medium; equating the forces in the electromagnetic ether with mechanical stresses and strains; abstracting what seemed to be appropriate relationships from the source domain and fitting them to the constraints of the target domain. In all this Maxwell explicitly provided imagistic representations to accompany the mathematical analysis.

Maxwell first constructed a simple representation consistent with a set of four constraints: the physical observations that (1) electric and magnetic actions are at right angles to one another and (2) the plane of polarized light is rotated by magnetic action, plus Faraday's speculative notions that (3) there is a tension along the lines of force and (4) there is a lateral repulsion between them. A mechanical analogy consistent with these constraints is a fluid medium, composed of vortices and under stress (see Figure 3B). With this form of the analogy Maxwell was able to provide a mathematical representation for various magnetic phenomena.

The problem of how to construe the relationship between electric current and magnetism led to an elaboration of the analogy. As we can see from Figure 3A, the vortices are all rotating in the same direction, which means that if they touch, they will stop. Maxwell argued that mechanical consistency requires the introduction of "idle wheels" to keep the mechanism going. He thus enhanced the source by surrounding the vortices with small spherical particles revolving in the direction opposite to them. There is a tangential pressure between the particles and the vortices, and for purposes of calculation Maxwell now had to consider the fluid vortices to be rigid pseudospheres. Maxwell's own imagistic representation of this enhanced source is seen in Figure 3B. He represented the dynamical relationships between current and magnetism mathematically by expressing them in terms of those between the particles and the vortices.

At this point Maxwell submitted the paper for publication. It took him several months to figure out the last — and most critical — piece of the representation: electrostatic actions. This is the point at which we joined him in the garden in Galloway. He found that if he made the vortices elastic and identified electrostatic polarization with elastic displacement he could calculate a wave of distortion produced by the polarization, i.e., what he called the "displacement current." He now had a unified, quantified representation of the continuous transmission of electromagnetic actions with a time delay. A testable consequence followed: electromagnetic actions are propagated at approximately the speed of light.

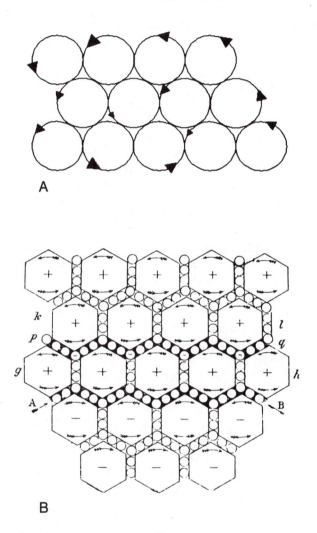

Figure 3. **A** *My schematic representation of initial crude source retrieved by Maxwell;* **B** *Maxwell's representation of his fully elaborated "physical analogy"* *(from Maxwell 1890, vol. 1).*

3.1.3. Cognitive Analysis: Analogical and Imagistic Reasoning

Analogical problem solving has been the subject of much recent work in cognitive psychology and artificial intelligence. It is widely recognized that analogy is a primary means through which we transfer knowledge from one domain to another. Knowing how this process works is essential for understanding learning and for developing expert systems. While still undergoing formulation, most cognitive theories of analogy agree that the creative heart of analogical reasoning is a modeling process in which relational structures from existing modes of representation and problem solutions are abstracted from a source domain and are fitted to the constraints of the new problem domain. A complete theory must give an account of the processes of retrieval, elaboration, mapping, transfer, and learning and of the syntactic, semantic, and pragmatic constraints that operate on these processes. Further, these processes need to be hooked up with other cognitive functions, such as memory. Most computational proposals are variants of three major theories of analogical problem solving: "structure mapping" (Gentner 1989), "constraint satisfaction" (Holyoak and Thagard 1989), and "derivational analogy" (Carbonell 1986).

Since my purpose is to show how we might conceive of what scientists like Faraday and Maxwell were doing with their images and analogies, I will not give detailed descriptions and evaluations of these theories. Rather, I will note some pertinent results from the empirical studies that inform them. Many psychological studies have been undertaken to understand how analogy functions in problem solving, especially in learning science. The results most germane to the issues of this essay are as follows. First, productive problem solving has the following features: (1) "structural focus": preserves relational systems; (2) "structural consistency": isomorphic mapping of objects and relationships; and (3) "systematicity": maps systems of interconnected relationships, especially causal and mathematical relationships. Second, the analogical reasoning process often creates an abstraction or "schema" common to both domains that can be used in further problem solving. Finally, in investigations of analogies used as mental models of a domain, it has been demonstrated that inferences made in problem solving depend significantly upon the specific analogy in terms of which the domain has been represented. For example, in one study where subjects constructed a mental model of electricity in terms of either an analogy with flowing water or with swarming objects, specific inferences — sometimes erroneous — could be traced directly to the analogy (Gentner and Gentner 1983). This result gives support to the view that analogies are not "merely" guides to thinking, with logical inferencing actually solving the problem, but *analogies themselves do the inferential work and generate the problem solution.*

Do these findings lend support to the interpretation I gave the case study, i.e., that analogical and imagistic reasoning are generating the respective field representations of Faraday and Maxwell? Can we model

Maxwell's use of physical analogy in cognitive terms? And what about the imagistic representations used by both him and Faraday?

While no current cognitive theory is comprehensive enough even to pretend to be able to handle all of the complexity of this case study, using what we believe we understand, cognitively, thus far, does enhance our understanding of it. It enables us, e.g., to fathom better what Maxwell was doing that summer day in Galloway and why he presented the physical analogy to his peers. Furthermore, this case points to areas of needed investigation in the cognitive sciences as well. I will first outline a cognitive analysis of Maxwell's generation of the field equations via the method of physical analogy and will then discuss the role of the imagistic dimension of that analogy along with the function of Faraday's imagistic representations.

Figure 4 provides a chart of Maxwell's modeling activities. His overall goal was to produce a unified mathematical representation of the production and continuous transmission of electromagnetic forces in a mechanical ether. The obvious source domain lay within continuum mechanics, a domain Maxwell was expert in. Continuous-action phenomena, such as fluid flow, heat, and elasticity, had all recently been given a dynamical analysis consistent with Newtonian mechanics and it was quite plausible to assume that the stresses and strains in an electromagnetic ether could be expressed in terms of continuum mechanical relationships. Using this source domain — if the analogy worked — he could presume to get: (1) assurance that the underlying forces are Newtonian; (2) continuity of transmission with the time delay necessary for a field theory; and (3) unification through finding the mathematical expression for the dynamical relations through which one action gives rise to another. He got all three, but the first was a false assurance. As we will discuss, the electromagnetic field equations represent a non-Newtonian dynamical system.

Maxwell retrieved a crude source from this domain by applying the four constraints we discussed above. He broke the overall goal down into subproblems, namely, representing magnetic induction, electricity, electromagnetic induction, and electrostatic induction. He then produced mappings between the electromagnetic quantities and mechanical properties of the fluid vortex medium and between the presumed stresses and strains and those known to take place in fluid medium under stress. The mappings are isomorphic and maintain causal interconnectedness. He reiterated the process twice, altering and enhancing the source to fit the constraints of the target domain. In the process he made "mistakes," most of which can be explained in terms of the model. For example, he takes the "displacement current" to be in the direction opposite from

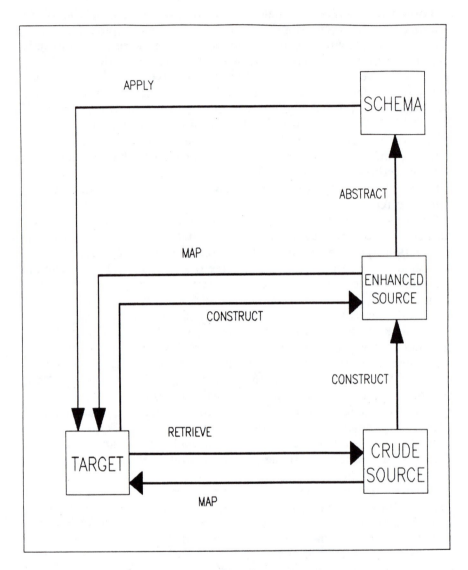

Figure 4. Maxwell's use of "the method of physical analogy."

the field intensity, and not — as is now customary and as he later held — in the same direction as the field intensity. In terms of the analogy, the equation is correct as it stands; i.e., in an elastic medium the restoring force and the displacement would have opposite orientation.

Through this process he abstracted a schema (i.e., a set of general dynamical [Lagrangian] relations) that could then be applied back to the target domain directly, without the need for any specific analogy. That is, at the end of the process he knew how to represent velocity, momentum, potential energy, and kinetic energy. With this knowledge he could re-derive the field equations using solely electromagnetic considerations, which is what he went on to do in his next paper on electrodynamics, "A Dynamical Theory of the Electromagnetic Field" (Maxwell 1890, vol. 1, pp. 526–97). But he only obtained this knowledge — the necessary general dynamical relations — *by abstracting it from the analogy he constructed with continuum mechanics.* This interpretation of what he actually did is consistent with what we have been learning about analogical reasoning in general. The physical analogy was generative and the process consistent with what are thought to be good analogical problemsolving techniques. But it also points to an area in need of investigation by cognitive theorists. None of the current theories addresses the kind of reiteration process Maxwell carried out: the modification and enhancement of the analogy to fit the constraints of the target domain. Yet one would expect such re-representation to be common in ordinary problem solving as well.

I want to underscore the point of my message by returning briefly to the historical story. In Maxwell's eyes his problem solution was never complete. To be so he would have to give an account of the actual underlying forces in the ether. The real power of analogical reasoning in creating new representations is driven home quite forcefully when we realize that he could *never* have done this: Maxwell's laws are those of a non-Newtonian system. How could it have turned out this way? That is, How could he have created a representation for a non-Newtonian system using a Newtonian system as the source of his analogy?

Maxwell did not take an existing physical system and plug the electromagnetic parameters into the equations for that system to solve the problem, as Thomson had done in finding a potential field representation for electrostatics in terms of Fourier's analysis of heat. Rather, he used various aspects of continuum mechanics and constraints from electromagnetism *to put together* a system, which *in its entirety* even he maintained did not exist — *possibly could not exist* — in nature. And *by drawing inferences from this abstracted model* he was able to extract a mathematical structure of greater generality than Newtonian mechanics. Contrary to what was believed at the time, Newtonian mechanics and general dynamics are not coextensive. This belief made Maxwell think that by giving a general dynamical formulation of electromagnetism he had shown the underlying forces to be Newtonian in nature. We now know that many different kinds of dynamical systems can be formulated in general dynamical terms, e.g., relativity and quantum mechanics.

Interpreted in the way I have been proposing, concept formation by analogical reasoning can, thus, be characterized as a process of abstraction from existing representations with increasing constraint satisfaction. This interpretation leads to a novel interpretation of continuity and commensurability between representations in cases like the one we just analyzed: these are to be found in the abstracted relationships common to the source and the target structures.

What about the imagistic representation of Maxwell's analogy and Faraday's earlier ones? While there are differences between the kinds of mappings made by Maxwell and those made by Faraday, both treated the imagistic representations as embodiments of an analogical source. I suggest that we construe this kind of imagistic reasoning as a species of analogical reasoning. In the early, qualitative phase, i.e., Faraday's work, the imagistic representation was taken as more nearly identical with what it represents than in Maxwell's quantitative analysis. That is, the schematic lines were identified with the physical lines they were to represent, and too many aspects of the specific image were incorporated into the field representation. A possible interpretation for this is that when there are insufficient constraints from the target domain to guide the process, analogies can be too generative. In the quantitative phase, i.e., in Maxwell's work, the function of the image is more abstract. It serves primarily to make certain structural relationships visualizable, and, thus, it is possible that any imagistic representation that embodied these relationships would serve the same purposes.

What further cognitive purposes might these physical embodiments serve? There is very little in the cognitive science literature on the possible role of imagistic representations like these in problem solving. I will offer some speculations by putting together recent work by Larkin and Simon (1987) on diagrams, and myself and James Greeno (1990) and Roschelle and Greeno (1987) on abstracted models in the hope of opening some avenues for research into possible roles these representations on paper play in constructing and reasoning with mental models, especially in the process of constructing a mathematical representation.

The main idea centers on the fact that perceptual inferences are easy for humans to make. By clustering connected information and making visual a chain of interconnected inferences the imagistic representations support a large number of immediate perceptual inferences. The representations on paper are presented in a form that already focuses on and abstracts specific aspects of phenomena. A great deal of mathematical information is implicit in such representations. By embodying structural relations thought to underlie phenomena they could facilitate access to the quantifiable aspects of the phenomena. As such, they provide an intermediate level of abstraction between phenomena and mathematical forms of representation (formulae). Additionally, they stabilize the image for the reasoner and make various versions accessible for direct comparison, in a way not available for internal images, and may thus take some of the load off memory in problem solving. Finally, they potentially play an important role in communicating new representations by providing a stable embodiment that is public. The imagistic representation could make it easier for others to grasp parts of the new representation than text and formulae alone. For example, Maxwell did not comprehend all the subtleties of the field concept Faraday articulated, but he did grasp the mathematical structures inherent in the lines of force representation and in the dynamical balance embodied in the interlocking curves. And Maxwell, in trying to communicate his new field representation to his colleagues, felt it necessary to provide them with the *physical* (i.e., embodied) analogy and extensive commentary on how to understand the method, in addition to leading them through the reasoning and rather than just presenting the mathematical arguments.

3.2. Thought Experiments and Limiting Case Analysis

3.2.1. Background

Another heuristic that occurs frequently in cases of major conceptual change is thought experimentation. The notion that an experiment can take place in thought seems paradoxical. Earlier scholarship presented two poles of interpretation of the role of thought experiments in creating conceptual change. Duhem dismissed them as bogus precisely because they are "not only not realized but incapable of being realized" (Duhem 1914, p. 202); i.e., they are not "experimental" in the customary sense. Koyré (1939, 1968), in contrast, argued that their logical force is so compelling that they supplant real experimentation in the construction of new representations for phenomena. The "thought" part of the experiment predominates and shows the synthetic a priori nature of scientific knowledge.

Contemporary historians and philosophers of science by and large reject both the extreme empiricism and rationalism of their forerunners. They acknowledge that thought experiment, while not eliminating the need for real experiment, is an important heuristic for creating conceptual change in science. Despite the consensus, based on ample historical documentation, there has been little theoretical analysis of just how such experiments function. As with the heuristics of analogy and imagery, I will attempt a cognitive-historical analysis of thought experiment, subsuming limiting case analysis as a species of this form of reasoning.

There have been a few recent attempts to analyze the function of thought experiments. While these are sketchy and limited, some do yield useful insights. Kuhn's analysis provides the starting point for most of these discussions. Kuhn claims that "thought experiment is one of the essential analytical tools which are deployed during crises and which then help to promote basic conceptual reform" (1964, p. 263). The importance of his analysis is that it is the first to try to come to grips with both the experimental and the thought dimensions of thought experiments. He argues that thought experiments show that there is no consistent way, *in actual practice*, of using accepted existing conceptions. That is, the thought experiment reveals that it is not possible to apply our conceptions consistently to real-world phenomena, and this practical impossibility translates into a logical requirement for conceptual reform. Gooding (1990), in his analysis of Faraday's experimental practices, picks up on Kuhn's analysis, rendering the empirical force of thought experiments in terms of their demonstration of what he calls the "impracticability of doing." Gooding is concerned to show how the real-world experimenter's knowledge of practical skill is utilized in the construction and manipulation of thought experiments.

My point will be that cognitive-historical analysis, by placing thought experiments within the framework of mental-models theory, offers the possibility of explaining *how* an experiment made in thought can have both the logical and empirical force Kuhn and Gooding argue for. In an unpublished work, Manukian (1987) has begun a cognitive-historical analysis of Galileo's thought experiments focusing on thought experiment as a species of information processing that employs mental models. He is the only one thus far to attempt to understand thought experiments as a form of "world modeling," as he calls constructing idealizations. Since his attention, however, is directed toward concerns in the field of artificial intelligence, he restricts his analysis to what I see as only a special case of thought experiment, the limiting case analysis.

Two other analyses need mentioning for the contrast in approach they provide. First, Brown (J. Brown 1986) claims that thought experiments are a species of a priori reasoning and that they get at something that cannot be derived from logical argumentation. His positive suggestion is that they provide a special window through which the mind grasps universals. The most cogent rendering I can give this claim is that Brown is trying to capture the idealizing function of thought experiments. Unlike Manukian, however, his approach through linguistic analysis does not afford the possibility of understanding their experimental nature.

Second, Norton's (1986) view that thought experiments can, in essence, be reconstructed as and replaced by arguments is the most sympathetic for philosophers who wish to restrict reasoning to logical argumentation — whether deductive or inductive. Certainly thought experiments contain an argument. As Norton himself acknowledges, however, the argument can be constructed only *after the fact*; i.e., it is not evident until after the thought experiment has been executed. By concentrating exclusively on this aspect he misses the importance of their experimental dimension. Additionally, while his claim that the presentation contains particulars irrelevant to the generality of the conclusion is correct, this emphasis reveals that he has also failed to see the *constructive function of the narrative form* in which thought experiments are customarily presented.

Both of these analyses take the traditional philosophical route of construing thought experiments in terms of propositional representations. My contention is that propositional representations cannot do the trick. *Mental simulation is required for a thought experiment to be both thought and experimental.* The original thought experiment is the construction of a mental model by the scientist who imagines a sequence of events. She or he then uses a narrative form to describe the sequence in order to communicate the experiment to others. Considerations of space will not permit here an analysis of comparable depth with that on analogical and imagistic reasoning. I do want, however, to put an analysis of a different heuristic before the reader to underscore the power of cognitive-historical analysis and to show that its utility is not restricted to what it offers for understanding the constructive practices of nineteenth-century British field theorists.

3.2.2. Case Studies: Galileo and Einstein

Although thought experiment occurs repeatedly in conceptual change, the thought experimenters who have attracted the most attention are Galileo and Einstein. I will give a brief presentation of a few of their thought experiments to convey some sense of the variety such experiments display and to elicit some common features for further analysis.

Galileo's importance as a pivotal figure in the transition from the qualitative categories of Aristotelian and medieval theories of motion to the quantitative representation of motion provided by Newton's mechanics is widely recognized. As shown in analyses by Koyré (1939) and Clavelin (1968), among others, Galileo drastically transformed the problem of how to go about constructing a mathematical representation of the phenomena of motion. That process, which Koyré called "mathematization," required constructing an idealized representation, quantifying this representation, and mapping the quantified representation back onto the real world. While it is now clear that Galileo must have performed many more real-world experiments than Koyré would have liked (see, e.g., Drake 1973; Naylor 1976; and Settle 1961), no one would deny the importance of thought experiment and limiting case analysis in the mathematization process. Take, as example, his analysis of falling bodies (Galilei 1638, pp. 62–86).

According to the Aristotelian theory, heavier bodies fall faster than lighter ones. This belief rests on a purely qualitative analysis of the concepts of 'heaviness' and 'lightness'. Galileo argued against this belief and constructed a new, quantifiable representation through a sustained analysis using several thought experiments and limiting case analyses. The outline of his use of these procedures is as follows. He calls on us to imagine we drop a heavy body and a light one, made of the same material, at the same time. We would customarily say that the heavy body falls faster and the light body more slowly. Now suppose we tie the two bodies together with a very thin — almost immaterial — string. The combined body should both fall faster and more slowly. It should fall faster because a combined body should be heavier than two separate bodies and should fall more slowly because the slower body should retard the motion of the faster one. Clearly something has gone amiss in our understanding of 'heavier' and 'lighter'. Having pinpointed the problem area, Galileo then goes on to show that it is a mistake to extrapolate from what is true at rest to what happens when bodies are in motion. That is, he has us consider that when the two bodies are at rest, the lighter will press on the heavier, and therefore the combined body is heavier. But when we imagine the two bodies are in motion, we can see the lighter does not press on the heavier and thus does not increase its weight. What Galileo has done up to this point is use the thought experiment to reveal the inconsistencies in the medieval belief, the ambiguities in the concepts, and the need to separate the heaviness of a body from its effect on speed in order to analyze free fall. He then goes on, using the methods of thought experiment and limiting case analysis in tandem, to show that the apparent difference in the speed of falling bodies is due to the effect of the medium and not to the difference in heaviness between bodies.

As Clavelin has pointed out, it is crucial for quantifying the motion of falling bodies that 'heaviness' not be the cause of the difference in speed because then we could not be sure that motion would be the same for all bodies. Galileo again used a thought experiment to demonstrate that the observed differences in speed should be understood as being caused by the unequal way media lift bodies. He asks us to suppose, e.g., that the density of air is 1; that of water, 800; of wood, 600; and of lead, 10,000. In water the wood would be deprived of 800/600ths of its weight, while lead would be deprived of 800/10,000ths. Thus, the wood would actually not fall (i.e., would float) and the lead would fall more slowly than it would in a less dense medium, such as air. If we extrapolate to a less dense medium, such as air, we see that the differential lifting effect is much less significant (e.g., 1/600 to 1/10,000 in air). The next move is to consider what would happen in the case of no medium, i.e., in extrapolating to the limiting case. With this move, Galileo says, "I came to the opinion that if one were to remove entirely the resistance of the medium, all materials would descend with equal speed" (Galileo 1638, p. 75). Having performed the extrapolation in this way we can quantify this idealized representation of the motion of a falling body and know that it is relevant to actual physical situations; we need only add back in the effects of a medium.

Galileo repeatedly used thought experiments and limiting case analyses in tandem as shown by this example both in constructing a quantifiable representation of bodies in motion and in attempting to convey this new representation to others. Later I will propose the cognitive functions of thought experiments and of limiting case analysis are much the same. Before getting to the cognitive analysis, though, I want to lay out somewhat different thought experiments used by Einstein in the development of the special and general theories of relativity.

Einstein began his paper "On the Electrodynamics of Moving Bodies" (1905) with the following thought experiment. Consider the case of a magnet and a conductor in relative motion. There are two possibilities. In the first case, the magnet is at absolute rest and the conductor moving. According to electromagnetic theory, the motion of the conductor through the magnetic field produces an electromotive force that creates a current in the conductor. In the second case, the conductor is at rest and the magnet moving. In this case, again according to electromagnetic theory, the motion of the magnet creates a changing magnetic field that induces an electric field that in turn induces a current in the conductor. With respect to the relative motions, however, it makes no difference whether it is the magnet or the conductor that is considered to be in motion. But according to the Maxwell-Lorentz electromagnetic theory, the absolute motions create a difference in how we would explain the production of a current in the conductor. Since the explanatory asymmetry could not, in principle or in practice, be accounted for by the observable phenomena — the measurable current in the conductor — Einstein argued that this supported his conclusion that "the phenomena of electrodynamics as well as of mechanics possess no properties corresponding to the idea of absolute rest" (p. 37).

Although I will not discuss them at any length, two more thought experiments figure crucially in this analysis. The second thought experiment in the paper is the famous one in which Einstein constructed an "operational" definition for the concept of simultaneity. According to Newtonian theory it is possible for distant events to occur simultaneously. Indeed this is necessary for there to be action at a distance. The thought experiment shows that we can define 'simultaneity' and thus what is meant by 'time' only for a particular reference system, and not in general. This experiment feeds into the next one, in which Einstein established the relativity of length; that is, he established the incorrectness of the Newtonian assumption that a body in motion over a specific time interval may be represented by the same body at rest in a definite location.

In a similar manner many thought experiments figured in Einstein's constructing and communicating the general theory of relativity, i.e., the field representation of gravitational action. We will just consider one he presented in various formats but claims to have first conceived in 1907. Einstein (1917, pp. 66–70) asks us to imagine that a large opaque chest, the interior of which resembles a room, is located in space far removed from all matter. Inside there is an observer with some apparatus. In this state, the observer would not experience the force of gravity and would have to tie himself with strings to keep from floating to the ceiling. Now imagine that a rope is connected to the outer lid of the chest and a "being" pulls upward with a constant force, producing uniform acceleration. The observer and any bodies inside the chest would now experience the very same effects, such as a pull toward the floor, as in a gravitational field. The experiment demonstrates that the behavior of a body in a homogeneous gravitational field and one in a uniformly accelerated frame of reference would be identical. Once we see that there is no way of distinguishing these two cases we can understand the importance of the Newtonian law that the gravitational mass of a body equals its inertial mass: these are just two manifestations of the same property of bodies. That is, we have a different interpretation for something we already knew.

3.2.3. Cognitive Analysis:
Thought Experiments as Mental Modeling

Will rendering thought experiments as a species of mental modeling support the interpretation that when they are employed in conceptual change, they are "essential analytical tools" in the process? We can only speculate about what goes on in the mind of the scientist in the original thought experiment. Scientists have rarely been asked to discuss the details of how they went about setting up and running such experiments. As stated previously, it is quite possible that the original experiment involves direct construction, without recourse to language. However, reports of thought experiments are always presented in the form of narratives that call upon the reader/listener to simulate a situation in his or her imagination. Thus, drawing on what we think we know about both the process through which we imagine in general and the process through which we comprehend any narrative may help us to answer that most perplexing question about thought experiments: *How can an "experiment" carried out in thought have such powerful empirical force?* As was the case in the analysis of analogy and imagistic reasoning above, much research and development need to be done in this area, but I hope the sketch I present will persuade the reader that following this direction does offer good prospects of accounting for both the "thought" and the "experiment" aspects of thought experiments.

The most pertinent aspect of mental-models theory for this analysis is the hypothesis that understanding language involves the construction of a mental model. In the case of thought experiments we need to understand how: (1) a narrative facilitates the construction of an experimental situation in thought, and (2) thinking through the experimental situation has real-world consequences. Framed in mental-models theory, the "thought" dimension would include constructing a mental model and "running" a mental simulation of the situation depicted by the model, while the "experimental" dimension comprises the latter and the connection between the simulation and the world.

Briefly, the mental-models thesis about text comprehension is that understanding involves relating linguistic expressions to models; i.e., the relationship between words and the world is mediated by the construction of a structural analog to the situations, processes, objects, events, etc. depicted by a text (Franklin and Tversky 1990; Mani and Johnson-Laird 1982; Johnson-Laird 1983; McNamara and Sternberg 1983; Morrow et al. 1989; and Tversky 1990). What it means for a mental model to be a "structural analog" is that it embodies a representation of the spatial and temporal relationships between, and the causal structure connecting, the events and entities of the narrative. In constructing and updating a representation, readers would call upon a combination of conceptual and real-world knowledge and would employ the tacit and recursive inferencing mechanisms of their cognitive apparatus.

That the situation is represented by a mental model rather than by an argument in terms of propositions is thought to facilitate inferencing.

We can actually generate conclusions without having to carry out the extensive computations needed to process the same amount of background information propositionally. The conclusions drawn are limited to those that are directly relevant to the situation depicted. The ease with which one can make inferences in such simulative reasoning has suggested to some that mechanisms either used in — or similar to those used in — perception may be involved. As we saw in the discussion of the function of imagistic representations in problem solving, if we do employ perceptionlike mechanisms here many inferences would be immediate.

To date, most empirical investigations of the hypothesis have focused on the representation of spatial information by means of mental models. The main disagreement has been over whether the representation involves a perceptionlike image or is "spatial," i.e., allows different perspectives and differential access to locations. Although there is some research investigating the use of knowledge of causal structure in updating, on the whole there is little research investigating how knowledge and inferencing mechanisms are employed in running and revising the simulation.

Before beginning to sketch a way of understanding thought experiments and their role in conceptual change in terms of mental-models theory, we need first to glean some common features of thought experiments from the narratives presented above. While there is great variety among thought experiments, in general they do share important salient features. The Galileo and Einstein examples help us to see some of these. First, as noted, by the time a thought experiment is public it is in the form of a narrative. The narrative has the character of a simulation. It calls upon the reader/listener to imagine a dynamic scene — one that unfolds in time. The invitation is to follow through a sequence of events or processes *as one would in the real world*. That is, even if the situation may seem bizarre or fantastic, such as being in a chest in outer space, there is nothing bizarre in the unfolding: objects float as they would in the real world in the absence of gravity. The assumption is that if the experiment could be performed, the chain of events would unfold according to the way things usually take place in the real world.

Second, a thought experiment embodies specific assumptions — either explicit or tacit — of the representation under investigation. It usually exposes inconsistencies or exhibits paradoxes that arise when we try to apply certain parts of that representation to a specific situation, such as 'heavy' and 'light' to falling rocks. The paradox can take the form of a contradiction in the representation (e.g., it requires that an object be both heavy and light) or of something being not physically possible (e.g., observing the asymmetry electromagnetic theory requires).

Third, by the time a thought experiment is presented it always works and is often more compelling than most real-world experiments. We rarely, if ever, get a glimpse of failed thought experiments or avenues explored in the construction of the one presented to us.[6] Some experiments, such as Galileo's second, could potentially be carried out — at least until the analysis extrapolates to the limit. Others, such as Einstein's first, underscore that doing a real-world experiment could not provide the data the theory requires. Still others, such as Einstein's "chest" in space, are impossible to carry out in practice, either in principle or because we do not yet have the requisite level of technological achievement. Once understood, however, a thought experiment is usually so compelling in itself that even where it would be possible to carry it out, the reader feels no need to do so. The constructed situation is apprehended as pertinent to the real world either by revealing something in our experience that we did not see the import of before (e.g., the measurable current in the stationary and in the moving conductor is the same, so on what basis can we support the difference in theoretical explanation?) or by generating new data (e.g., in the case of no medium lead and wood would fall at the same speed) or by making us see the empirical consequences of something in our existing representation (e.g., the attributes called 'gravitational mass' and 'inertial mass' are the same property of bodies).

Finally, the narrative presentation has already made some abstraction from the real-world phenomena. For example, certain features of objects that would be present in a real experiment are eliminated, such as the color of the rocks and the physical characteristics of the observers. That is, there has been a prior selection of the pertinent dimensions on which to focus, which evidently derives from our experience in the world. We know, e.g., that the color of a rock does not effect its rate of fall. This feature strengthens our understanding of the depiction as that of a prototypical situation of which there could be many specific instances. In more colorful narratives there may be more irrelevant features in the exposition, but these most often serve to reinforce crucial aspects of the experiment. For example, Einstein's characterization, in one version of the chest — or "elevator" — experiment, of a physicist as being drugged and then waking up in a box served to reinforce the point that the observer could not know beforehand if he were falling in outer space or sitting in a gravitational field. In the version discussed above, the opacity of the chest is to prevent the observer from seeing if there are gravitational sources around.

We can outline the function of thought experiments in terms of mental-models theory as follows. The first performance of a thought experiment is a direct mental simulation. This hypothesis gains plausibility when we realize the likelihood that direct mental simulation precedes even real-world experiments; i.e., the scientist envisions and unfolds a sequence of steps to be carried out in the experiment.[7] The cognitive function of the narrative form of presentation of a thought experiment to others is to guide the construction of a structural analog of the prototypical situation depicted in it. Over the course of the narrative, we are led to run a simulation that unfolds the events and processes by constructing, isolating, and manipulating the pertinent dimensions of the analog phenomena. The process of constructing and running the model gives the thought experiment its applicability to the world. *The constructed situation inherits empirical force by being abstracted from both our experiences and activities in, and our knowledge, conceptualizations, and assumptions of, the world.* In running the experiment, we make use of various inferencing mechanisms, existing representations, and general world knowledge to make realistic transformations from one possible physical state to the next. In this way, the data that derive from a thought experiment, while constructed in the mind, are *empirical* consequences that at the same time pinpoint the locus of the needed representational change.

Limiting case analysis can be construed as a species of thought experiment. In this species the simulation consists of abstracting specific physical dimensions to create an idealized representation, such as of a body falling in a vacuum. The isolation of the physical system in thought allows us to manipulate variables beyond what is physically possible. Just what dimensions produce the variation and how to extrapolate from these may be something we determine initially in real-world experimentation, but the last step can only be made in the imagination. In physics, it is the idealized representation that is quantifiable. The idealized representation, however, is rooted in and relevant to the real world because it has been created by controlled extrapolation from it. We get from imagination to the real world by adding in some of the dimensions we have abstracted, again in a controlled process.

3.3. Summary: Abstraction Techniques and Conceptual Change

What we have seen in our discussion of the dynamics of conceptual change in science is the potential for acquiring a deeper understanding of the processes through which new scientific representations are constructed and communicated by joining historical analysis with our developing insights into how human beings represent, reason, and solve problems generally. By linking the conceptual and the experiential dimensions of human cognitive processing, mental-models theory offers the possibility of capturing and synthesizing theoretical, experimental, and social dimensions of scientific change. Our investigation demonstrated in some detail how cognitive-historical analysis helps us to fathom how the heuristics of analogy, imagistic reasoning, thought experiment, and limiting case analysis, of which we see recurrent use in what has been called "radical" and "revolutionary" conceptual change, could function to create genuinely novel representations by increasing abstraction from existing representations in a problem-solving process.

4. Wider Implications

To conclude I would like to underscore the potential fertility of the cognitive-historical method outlined and illustrated above by considering its wider implications for the disciplines it comprises. To do this we need to return to our title query: How do scientists think?

4.1. Implications for Philosophy of Science

1. While philosophers would be comfortable with the generality of the question, the detour through language taken by many philosophers of science has prevented its asking. Those who would ask it would prefer it transformed into: How ought scientists to think? And, quite generally, the "creative processes" are deemed by philosophers to be too mysterious to be understood. The main point of the investigations above is to show how cognitive-historical analysis opens the possibility of construing the representational and constructive practices of scientists as part of scientific method, and as such within the province of philosophical investigation. The analysis supports and elaborates upon the intuitions of those who have argued that *reasoning comprises more than algorithms* and for the generative role of such heuristics as analogy. Further, developing criteria for evaluating good and bad uses of heuristics will enable us to show why it is rational to believe inferences resulting from good heuristics are worth testing, holding conditionally if testing is not feasible, etc.

2. A major problem for historicist philosophers has been *how to go from a case study to a more general conclusion*. Those who want to use scientific knowledge practices as a basis from which to develop a historicized epistemology recognize the dangers of generalizing from one case, no matter how salient. Placing discovery processes within the framework of human representational and problem-solving abilities enables us to extend from case studies without committing the serious injustices of past reconstructive approaches.

3. As discussed in the body of this essay, cognitive-historical analysis

offers a way of recasting many of the puzzles associated with "incommensurability of meaning." By focusing on the people and practices that create and change representations of a domain, rather than on static linguistic representations that change dramatically from time to time, we open the possibility of understanding how scientists build new and sometimes radically different representations out of existing ones, and thus for explaining the continuous and noncumulative character of scientific change. And, lastly, we even have a way of making sense of that most paradoxical of all Kuhnian claims: Postrevolutionary scientists quite literally understand and experience the world in a manner incommensurable with their prerevolutionary counterparts (or selves in some cases). If we do negotiate the world by constructing mental models, prerevolutionary and postrevolutionary scientists would construct different mental models and would, thus, truly have different experiences of the world.

4.2. Implications for History of Science

1. Historians do not pose the question this way. Those who still do address such issues ask: How did my individual scientist, such as Maxwell, think? However, every historian — no matter how scrupulously he or she has tried to reconstruct the mosaic of a discovery process in a manner faithful to the historical record — must have experienced the nagging doubt: But did they *really* think this way? In the end we all face the recorded data and know that every piece is in itself a reconstruction by its author. The diaries and notebooks of a Faraday may be the closest we will ever get to "on-line" thinking, and even these cannot be taken as involving no reconstruction and as capturing all the shifts and strategies employed. If we can show, however, that what the particular scientist claims and/or seems to be doing is in line with what we know about human cognitive functioning generally, we can build a stronger case for our interpretation and fill in missing steps in a plausible manner, as I have done in the Maxwell example.

2. In claiming a generative role for heuristics such as those discussed above one often has the sense of "preaching to the converted" when talking with historians of science. *But historians do not always come down on the side of taking apparent uses of problem-solving heuristics at face value.* Witness the Maxwell case. It is still controversial as to whether or not he was reasoning through the analogical model he presented.[8] Historians who see such models as off to the side, while some other process is actually generating the representation, have at least tacitly bought the philosopher's assumption that reasoning is only by means of inductive or deductive algorithms.

Cognitive-historical analysis provides support for the idea that such heuristics are not "merely suggestive" (Heimann 1970) or an "unproductive digression" (Chalmers 1986) *but are fundamental, i.e., they constitute the essence of the reasoning process.* When they are taken in this way we get a better fit with the data and have less of a need to throw inconvenient pieces, such as Maxwell's "errors," away. Further, insights into how cognitive abilities and limitations contribute to and constrain scientific theorizing, experimentation, assessment, and choice can enrich the analyses of those who do take such heuristics seriously, irrespective of whether the scientists in question go down dead ends, contribute to "winning" science, employ different strategies to get to the same point, etc.

3. Controversies have often arisen within the history of science over such questions as: Do we find the concept of inertia in Galileo's physics? or Did Faraday have a field concept that guided his research from early on, or did he formulate it only in his last year, or never? The metatheoretical question that lies at the heart of such seemingly irresolvable disputes — *indeed at the core of historical method* — has scarcely been noticed by historians. While as historians of science we must not attribute present-day views to past scientists, even though the concepts may look quite familiar, *there is no explicit guidance on a theoretical level for how to do this.* Intuitive strategies for avoiding the problem are learned with the craft. What is missing is an explicit metatheoretical notion of what constitutes the meaning of a scientific concept.

Now why would an explicit notion yield better results than mere intuitions acquired in the craft? The answer is that underlying many of these controversies is the tacit assumption that a concept is represented by a set of necessary and sufficient conditions. As we have seen in the discussion of kinematics above, this form of representation cannot accommodate a substantial body of historical data. With it we have no criteria other than the modern for determining whether or when Faraday's concept is a field concept; no means for justifying the intuitive sense of "family resemblance" between what Galileo is discussing and what Newton called 'inertia' or for making sense of the fact that we seem able to trace out a distinct line of descent and a pattern of progress over time in a conceptual domain. *Seeing these problems as part of the wider representational problem opens a new avenue for their resolution.* In an article on Faraday's field concept, I have shown in detail how such an analysis can help to resolve standing controversies among historians (Nersessian 1985).

4. Many contemporary historians are concerned with issues about the form and rhetoric of presentation of novel ideas. These are usually

framed in terms of how scientists adopt certain modes and conventions of writing in order to persuade others of their ideas. What tends to be left out of the analysis is that in order to *persuade* one has to get one's colleagues to *comprehend* the new ideas and, again, in order to *negotiate,* one has to *comprehend* what is being proposed. Cognitive-historical analysis allows us to take the public communications of scientists presenting new representations as attempts at trying to get others to understand them. That is, we can view such communications as presented in ways that the creators find meaningful for their own construction and understanding. Success at communication does not, of course, entail success at persuasion.

Looked at in terms of the rhetoric of persuasion, Maxwell's analogies might seem utter failures. His presentation was quite out of line with the modes and conventions of his contemporaries who were publishing in *Philosophical Transactions* at that time. And even the person from whom he claims to have "poached" the technique, William Thomson, did not accept the method of analysis as transformed by Maxwell and therefore Maxwell's results. We can make better sense of what Maxwell was doing in presenting the work in that format if we assume he was trying to get his colleagues to understand his new field representation of forces by leading them through the modeling processes he used to construct the electromagnetic field concept, as well as trying to convince them of its potential.

5. Finally, we repeatedly find claims in the historical literature about the influence of wider cultural factors on a person's science; notable examples from physics are the influence of Faraday's Sandemanian religion on his belief in the unity of forces and acausality in quantum mechanics deriving from Weimar culture. Cognitive-historical analysis can capture the *locality* that is essential to historical understanding. It offers the potential for determining how it is that representational resources that are part of the scientist's local culture — whether these be construed as within a community of scientists, such as Cambridge mathematical physicists, or as within a wider Weltanschauung, such as obtained in Victorian England — are incorporated into scientific representations. *The problem becomes that of how it is that scientists, working individually or collectively, combine the cognitive abilities they have in virtue of their biology with the conceptual resources they acquire from the various facets of their lives in a wider community.*

4.3. Implications for Psychology

1. Cognitive psychologists do ask the question, but by themselves lack the means to answer it fully. "Revolutionary" science is rare and so is the

possibility of catching it "on-line." Cognitive-historical analysis greatly increases the database for psychological research. A cognitive theory of problem solving, in order to be adequate, needs to take into account the practices of scientists who have created major innovations in its formulation. This has scarcely been done to date. Combining the resources of cognitive psychologists, artificial intelligence researchers who work on modeling human cognitive processes, and historians and philosophers of science will lead to a more realistic portrayal of the complexities of scientific reasoning than current cognitive models and "discovery programs" provide.

2. Are the conceptual changes that take place in development and/or learning like those in scientific revolutions? This question is acquiring an important place in the contemporary psychological literature. A growing contingent of cognitive psychologists has been arguing that the processes of cognitive development and conceptual change (or "restructuring") in learning are indeed like those of major scientific revolutions (see, e.g., Carey 1985; Keil 1989). The main support for the psychological hypothesis comes from research that describes the initial states of children and students and compares those states with the desired final state. The kinds of changes necessary to get from one state to the other seem to resemble those that have taken place in scientific revolutions *as they have been construed by Kuhn*, whose "gestalt switch" metaphor of a scientific revolution many psychologists have uncritically adopted. And, as in history and philosophy of science, the nature of the *processes* through which conceptual change is brought about has not been explored in any depth in psychology. As demonstrated in this essay, cognitive-historical analysis points the way to a quite different understanding of the kinds of conceptual changes that take place in scientific revolutions and opens an avenue for examining the processes that bring them about.

4.4. Implications for Science Education

There is growing interest in how all three disciplines might work together on the problem of how to help students learn science. Cognitive-historical analysis opens a new area of exploration. In fact, it offers a way of fundamentally recasting the old position — proposed by Dewey, Bruner, and Conant, among others — that developing an appreciation for the historical roots of scientific ideas will facilitate learning because students will have a context in which to place them.

A cognitive theory of conceptual change views scientific "discovery" as a process in which scientists actively construct representations by employing problem-solving procedures that are on a continuum with those we employ in ordinary problem solving. With such a "constructionist" conception of discovery the cognitive activity of the scientist becomes directly relevant to learning. *The historical processes provide a model for the learning activity itself,* and, thus, have the potential to assist students in constructing representations of extant scientific theories. The history of science becomes, in this domain, a repository not of case studies, but rather of strategic knowledge of how to go about constructing, changing, and communicating scientific representations. As I have proposed elsewhere (Nersessian 1989, in press[b]), we should "mine" historical data — publications, diaries, notebooks, correspondence, etc. — for these strategies and then devise ways of integrating and transforming these more realistic exemplars of scientific problem solving into instructional procedures.

5. Return To Galloway

Coming back once more to our opening scene, what would cognitive-historical analysis have Maxwell doing that summer day? It would have him searching for a way to make electrostatic phenomena meaningful in terms of mechanical phenomena he believes he understands.

Constructing an analogical model allowed Maxwell to gain access to a systematic body of knowledge: a structure of causal and mathematical relationships with its own constraints. Maxwell generated a continuous-action representation for electromagnetism through a process of fitting the model to the constraints of the new domain. The physical embodiment facilitated his exploration of the kind of mechanical forces that could be capable of producing electromagnetic phenomena. This was done at an intermediate level of abstraction: concrete enough to give substance to the relationships he was examining and indicate possible paths to solution, and yet abstract enough to generate a novel representation — one that is dynamical but not mechanical. Once he had understood the dynamical relations through this process he was able to rederive the mathematical representation without this — or any — specific model, but just with the assumption of a general mapping between electromagnetic and dynamical variables.

The Scene

May 18, 1863, 8 Palace Garden Terrace, Kensington West.

Maxwell is sitting in a parlor sipping tea. "Aha!" he thinks, "now I know how to do it without the model."

The Problem

If we were to take this as our starting point, we could never hope to fathom the nature of conceptual change in science.

Notes

The research undertaken for this paper and its preparation were supported by NSF Scholars Award DIR 8821422. The author wishes to thank David Gooding, James Greeno, Mary Hesse, Simon Schaffer, and Robert Woolfolk for valuable discussions of the material in this paper and Floris Cohen, Richard Grandy, Larry Holmes, Philip Johnson-Laird, Paul Thagard, and Norton Wise for critical and editorial comments on the penultimate version.

1. Giere (1988) argues against a role for history of science in a cognitive theory of science. His main point is that historians of science have — by and large — not done the kind of analysis of the scientific record that a cognitive theory requires. While there are notable exceptions, such as Holmes (1985), he is correct on this point. What I am arguing here is that the historical record, itself, does contain material of great importance to a cognitive theory of science. It contains much information about the cognitive activity of scientists over the history of science.

2. I owe the name "context of development" to Richard Grandy.

3. Holmes (1985, pp. 119–20) argues in a similar vein.

4. Cohen (1990) contains an interesting discussion of the "two Kuhns" — the philosopher of science and the historian of science — and of the repercussions of the split for his analysis of the scientific revolution of the seventeenth century.

5. Notable exceptions are Clement (1983), Langley et al. (1987), McCloskey (1983), Qin and Simon (1990), and Wiser and Carey (1983).

6. An analysis of some of Faraday's explorations of thought experiments is to be found in Gooding (1990).

7. See Gooding (1990) for a fuller discussion of this point with respect to Faraday's experiments.

8. Berkson (1974), Bromberg (1968), Nersessian (1984, 1986), and Siegel (1986) present arguments in favor of the centrality of the analogical model, while Chalmers (1986), Duhem (1902, 1914), and Heimann (1970) are among those who deny its importance.

References

Berkson, W. 1974. *Fields of Force: The Development of a World View from Faraday to Einstein*. New York: John Wiley and Sons.

Bromberg, J. 1968. "Maxwell's Displacement Current and His Theory of Light." *Archive for the History of the Exact Sciences* 4:218–34.

Brown, H. 1986. "Sellars, Concepts, and Conceptual Change." *Synthese* 68:275–307.

Brown, J. R. 1986. "Thought Experiments since the Scientific Revolution." *International Studies in the Philosophy of Science* 1:1–15.

Campbell, N. R. 1920. *Physics, the Elements*. Cambridge: Cambridge University Press.

Carbonell, J. 1986. "Derivational Analogy: A Theory of Reconstructive Problem Solving and Expertise Acquisition." In *Machine Learning: An Artificial Intelligence Approach*, ed. R. Michalski, J. Carbonell, and T. Mitchell, pp. 371–92. Los Altos, Calif.: Morgan Kaufmann.

Carey, S. 1985. *Conceptual Change in Childhood.* Cambridge, Mass.: MIT Press.

Chalmers, A. F. 1986. "The Heuristic Role of Maxwell's Mechanical Model of Electromagnetic Phenomena." *Studies in the History and Philosophy of Science* 17:415–27.

Clavelin, M. 1968. *The Natural Philosophy of Galileo: Essay on the Origins and Formation of Classical Mechanics.* Cambridge, Mass.: MIT Press.

Clement, J. 1983. "A Conceptual Model Discussed by Galileo and Used Intuitively by Physics Students." In *Mental Models,* ed. D. Gentner and A. Stevens, pp. 325–40. Hillsdale, N.J.: Lawrence Erlbaum.

Cohen, H. F. 1990. "The Banquet of Truth: An Historiographical Inquiry into the Nature and Causes of the 17th Century Scientific Revolution." Unpublished manuscript.

Drake, S. 1973. "Galileo's Experimental Confirmation of Horizontal Inertia: Unpublished Manuscripts." *Isis* 64:291–305.

Duhem, P. 1902. *Les théories électriques de J. Clerk Maxwell: Étude historique et critique.* Paris: A. Hermann and Cie.

———. 1914. *The Aim and Structure of Physical Theory.* Reprint, New York: Atheneum, 1962.

Einstein, A. 1905. "On the Electrodynamics of Moving Bodies." In *The Theory of Relativity.* Reprint, New York: Dover, 1952.

———. 1917. *Relativity: The Special and the General Theory.* Reprint, London: Methuen, 1977.

Faraday, M. 1839–55. *Experimental Researches in Electricity.* Reprint, New York: Dover, 1965.

Franklin, N., and B. Tversky. 1990. "Searching Imagined Environments." *Journal of Experimental Psychology* 119:63–76.

Galilei, G. 1638. *Two New Sciences.* Trans. S. Drake. Reprint, Madison: University of Wisconsin Press, 1974.

Gentner, D. 1989. "The Mechanisms of Analogical Learning." In *Similarity and Analogical Reasoning,* ed. S. Vosniadou and A. Ortony, pp. 200–241. Cambridge: Cambridge University Press.

Gentner, D., and D. R. Gentner. 1983. "Flowing Waters and Teeming Crowds: Mental Models of Electricity." In *Mental Models,* ed. D. Gentner and A. Stevens, pp. 99–130. Hillsdale, N.J.: Lawrence Erlbaum.

Giere, R. N. 1988. *Explaining Science: A Cognitive Approach.* Chicago: University of Chicago Press.

Goldman, A. I. 1986. *Epistemology and Cognition.* Cambridge, Mass.: Harvard University Press.

Gooding, D. 1990. *Experiment and the Making of Meaning: Human Agency in Scientific Observation and Experiment.* Dordrecht: Kluwer Academic Publishers.

Gorman, M. E., and W. B. Carlson. 1989. "Interpreting Invention as a Cognitive Process: The Case of Alexander Graham Bell, Thomas Edison, and the Telephone." Forthcoming in *Science, Technology and Human Values.*

Heimann, P. M. 1970. "Maxwell and the Modes of Consistent Representation." *Archive for the History of the Exact Sciences* 6:171–213.

Hesse, M. 1966. *Models and Analogies in Science.* Notre Dame, Ind.: University of Notre Dame Press.

Holmes, F. L. 1985. *Lavoisier and the Chemistry of Life: An Exploration of Scientific Creativity.* Madison: University of Wisconsin Press.

Holyoak, K., and P. Thagard. 1989. "Analogical Mapping by Constraint Satisfaction: A Computational Theory." *Cognitive Science* 13:295–356.

Johnson-Laird, P. N. 1983. *Mental Models.* Cambridge, Mass.: Harvard University Press.

Keil, F. C. 1989. *Concepts, Kinds, and Conceptual Development.* Cambridge, Mass.: MIT Press.

Koyré, A. 1939. *Galileo Studies.* Atlantic Highlands, N.J.: Humanities Press, 1979.

———. 1968. *Metaphysics and Measurement.* Cambridge, Mass.: Harvard University Press.

Kuhn, T. S. 1964. "A Function for Thought Experiments." In *The Essential Tension: Selected Studies in Scientific Tradition and Change.* Chicago: University of Chicago Press.

———. 1965. *The Structure of Scientific Revolutions.* Chicago: University of Chicago Press.

———. 1986. "Possible Worlds in History of Science: Nobel Symposium — August 1986." Unpublished manuscript.

———. 1987. "What Are Scientific Revolutions?" In *The Probabilistic Revolution*, vol. 1 of *Ideas in History*, ed. L. Kruger, L. J. Daston, and M. Heidelberger, pp. 7–22. Cambridge, Mass.: MIT Press.

Langley, P., et al. 1987. *Scientific Discovery: Computational Explorations of the Creative Processes.* Cambridge, Mass.: MIT Press.

Larkin, J. H., and H. A. Simon. 1987. "Why a Diagram Is Sometimes Worth Ten Thousand Words." *Cognitive Science* 11:65–100.

Laudan, L. 1977. *Progress and Its Problems: Towards a Theory of Scientific Growth.* Berkeley: University of California Press.

McCloskey, M. 1983. "Naive Theories of Motion." In *Mental Models*, ed. D. Gentner and A. L. Stevens, pp. 299–324. Hillsdale, N.J.: Lawrence Erlbaum.

McNamara, T. P., and R. J. Sternberg. 1983. "Mental Models of Word Meaning." *Journal of Verbal Learning and Verbal Behavior* 22:449–74.

Mani, K., and P. N. Johnson-Laird. 1982. "The Mental Representation of Spatial Descriptions." *Memory and Cognition* 10:181–87.

Manukian, E. 1987. "Galilean vs. Aristotelian Models of Free Fall and Some Modern Concerns in Artificial Intelligence." Unpublished manuscript.

Maxwell, J. C. [1855] 1991. "On Faraday's Lines of Force." In *The Scientific Papers of James Clerk Maxwell*, ed. P. M. Harman. Cambridge: Cambridge University Press.

———. 1890. *The Scientific Papers of J. C. Maxwell*, ed. W. D. Niven. Cambridge: Cambridge University Press.

Morrow, D. G., G. H. Bower, and S. L. Greenspan. 1989. "Updating Situation Models during Narrative Comprehension." *Journal of Memory and Language* 28:292–312.

Murphy, G. L., and B. Cohen. 1984. "Models of Concepts." *Cognitive Science* 8:27–58.

Naylor, R. 1976. "Galileo: Real Experiment and Didactic Demonstration." *Isis* 67:398–419.

Nersessian, N. J. 1984. *Faraday to Einstein: Constructing Meaning in Scientific Theories.* Dordrecht: Martinus Nijhoff.

———. 1985. "Faraday's Field Concept." In *Faraday Rediscovered*, ed. D. Gooding and F. James, pp. 175–87. London: Macmillan.

———. 1986. "A Cognitive-Historical Approach to Meaning in Scientific Theories." In *The Process of Science: Contemporary Philosophical Approaches to Understanding Scientific Practice*, ed. N. J. Nersessian, pp. 161–78. Dordrecht: Martinus Nijhoff.

———. 1989. "Conceptual Change in Science and in Science Education." *Synthese* 80:163–84.

———. In press(a). "Discussion: The 'Method' to Meaning: A Reply to Leplin." *Philosophy of Science.*

philosophers sometimes talk of the importance of practice, most philosophical practice allows the scientist fewer resources to investigate the perceptual world than nonscientists have.[4] As Ronald Giere points out, psychologists allow rats greater ability to map their environment than some philosophers recognize scientists as having.[5]

It will be objected that scientists' own accounts of experimental work support the disembodied view that philosophers tacitly endorse: published papers and texts present their work as a highly deliberative and methodical activity in which theory leads and experiment merely responds. This ignores what we know about how and why scientists construct narrative accounts of their works. I deal with it in sections 2 and 3, below. A second difficulty is that the local, situated procedures of experimental science are beyond the reach of philosophical practice, which is, after all, representation-bound. The remaining sections of this chapter present a method of mapping experimental processes that can display procedures and the structures in which they occur.

2. Exemplary Science

Kuhn argued in 1961 that the image of science held by most nonscientists is too dependent on highly reconstructed accounts found in published papers or textbooks. The purpose of such texts is to make the argumentative weight of evidence as strong as possible and to disseminate easily understood practices that support the production of the evidence. They are written to be as logically transparent as possible and to conform to methodological standards.[6] This means editing out much of what went into achieving the results. The reduction of complex perceptual experience to visual experience shows that a reconstructive process is even at work in writing informal experimental narratives. As I show below, diaries and notebooks show that complex series of actions enabled seeing (and thinking), yet comparison with published accounts shows that there, thinking is juxtaposed more directly to visual perception, and theory to observation, than either had been during the discovery process. The most transparent (or least problematic) results are selected and highlighted so as to create a good evidential argument. "The manner in which science pedagogy entangles discussion of a theory with remarks on its exemplary applications has helped to reinforce a confirmation theory drawn predominantly from other sources."[7] The result is that cognitive aspects of the construction of new representations and new arguments are generally overlooked or relegated to the logically impenetrable area of "discovery."

I contend that the image of science predominant in analytical philosophy of science is still too dependent upon highly reconstructed accounts of published papers and texts. If we are interested in empirical access and empirical constraint, then we cannot ignore two facts. The first is the historical fact that the textbook image has reinforced a consequentialist approach to theory confirmation that became popular during the last century.[8] Again, the insight is an old one: "Given the slightest reason for doing so, the man who reads a science text can easily take the [exemplary] applications to be the evidence for the theory."[9]

A linear, "one-pass" view of discovery made it easy for philosophers to foreground the logical norms found in the methodological canons of a scientific field and to ignore the amount of qualitative, constructive work that enables quantitative precision.[10] The apparent preeminence of quantitative over qualitative work suited the idea that theory predominates and the best, if not the only really significant, use of experiment is to make quantitative tests.[11] Are these unassailable grounds for treating applications as evidence? Or is the neglect of qualitative, premensurative practice but a reflection of the linguistic turn in philosophy — itself an expression of more widely held assumptions about the priority of head over hands? Ryle criticized this "intellectualist legend."[12] Yet, by continuing to project an exclusively intellectual view of science, philosophies of science tacitly endorse views that philosophers might reasonably be expected to criticize. (Even if the critical responsibilities are not accepted, there is still the fact that neglect of practice makes a false mystery of how thought and talk successfully engage a material world.)[13]

Philosophical presuppositions are a contingent fact about philosophy, not science. Recognizing them makes the second difficulty — the fact that scientists manipulate material things as well as representations — much less daunting. Taking a broader view of philosophical practice is more than a matter of reckoning with the implications of historical and sociological studies of science. To situate cognitive processes alongside larger historical and social ones involves grasping the nettle of reconstruction. We need to remove the layers of narrative reconstruction that conceal important features of observation and experiment, namely: (1) their nonlinearity — reflexiveness, recursiveness, and multiple pathways between goals and solutions, (2) the importance of human agency in the manipulation and transformation of real and imaginary objects, (3) the interaction of concepts, percepts, and objects ("head and hands"), and (4) the creative possibilities opened up by uncertainty.[14] Like social aspects of experiment, these cognitive features are edited out of published accounts as new skills are developed and disseminated, as rules of thumb become accepted practice, and as enabling assumptions are articulated into arguments.

The inability of computational approaches to deliver discovery programs that can use data that are as "raw" as the stuff scientists work on is largely due to the fact that most work with the impoverished "one-pass" notion of discovery favored by analytical philosophy. But with representations powerful enough to deal with practical reasoning processes, students of discovery could investigate the dynamics of scientific research more realistically.[15] Extended to artificial intelligence, a naturalistic approach to discovery would recover experimental strategies and interpret situated problem solving through maps of 'real-time' processes; this approach would thus display the variety of strategies that enables creative use of observation and experiment. Analysis of these pathways would show many nonformal structures in early stages of experimentation but increasing conformity to standardized strategies and forms of argument as research programs develop.[16]

3. Recovering Reconstruction

Thinking and its interaction with the material world have a dynamical quality that rarely survives translation into the linear form of the narratives in which we learn about experiments. Published papers reconstruct discovery so that situated, active learning processes are rarely even glimpsed through them; they do not convey the procedural, skilled aspects of theoretical and experimental work. After all, their purpose is to insure that the contingencies and messiness of learning are precisely situated, if they appear at all. This commonplace is as important to understanding thought experiments as it is to recovering real ones. Again, Kuhn's early, practice-oriented work signals a warning for philosophies of science. He pointed out that "large amounts of qualitative work have usually been prerequisite to fruitful quantification in the physical sciences" and that the textbook image of science conceals this historical fact. He concluded that "it is only because significant quantitative comparison of theories with nature comes at such a late stage in the development of science that theory has seemed to have so decisive a lead."[17] The invention of qualitative procedures and the skills needed to use them has been no less important to the development of natural science than, say, calculus as a method of representing physical process has been. So far, there is little history and even less philosophy of how maker's knowledge has enabled scientific knowledge.[18]

From the standpoint of historical, sociological, and cognitive studies, the weakness of mainstream philosophy of science is that it imposes a sequential structure — the linear form of propositional argument — upon all reasoning. The promise of a logically proper methodology of science has faded; that promise cannot justify continued neglect of natural reasoning processes.

3.1. Reticularity and Reasoning

All accounts involve an element of reconstruction. Some of this is deliberate and articulate; some is prearticulate and outside the time of ordinary consciousness. In Arber's image, thought itself is reticular, folded like the cortex.[19] I shall identify six kinds of reconstruction, beginning with the two least studied by — but of most interest to — cognitive science and philosophy. Reconstruction is needed to produce an account ordered enough to enable further action, the communication of what is being done, and the redefinition of problems. This narrative ordering is already well advanced in the accounts we find in manuscript notes, records, correspondence, and drafts of papers. This sort of accounting involves what I call *cognitive reconstruction* (see Table 1). Roger Penrose speaks directly to this cognitive process when he writes that "the temporal ordering of what we appear to perceive is something we impose on our perceptions in order to make sense of them, in relation to the uniform forward time progression of an external physical reality."[20]

The time of conscious perception need not flow linearly; it may be multidirectional. Cognitive reconstruction is inherent in the processes of making sense — qua narrative order — of situated or context-dependent behavior, and of generating construals that communicate the sense of that behavior, however tentative these construals may be.[21] Such reconstruction makes experimenters' behavior intelligible to the actors involved; its purpose is not to argue or defend interpretations of the events, images, objects, or models it produces. This distinguishes cognitive from *demonstrative reconstruction*, which generates two sorts of things: lists of instructions on how to carry out observations or experiments — essential to the literary technology of vicarious or virtual witnessing of phenomena[22] — and arguments based on these explicit accounts of experimental procedure. Cognitive and demonstrative reconstruction interact. For example, J. B. Biot's narrative account of his resolution of the unruly behavior of magnetized needles near currents shows how, by means of learned manipulative skills, he arrived at the image of a 'circular contour' that organized both his perception and enabled the reader's vicarious participation in his experience.[23] The same narrative provided grounds for construing the needle behavior in terms of circles, which became an important heuristic for exploratory work by Faraday, examined below. Similarly, Lavoisier's drafts of his papers on respiration show concepts being articulated alongside arguments.[24] Historical studies can recover research processes back to a point where

Table 1. Six Types of Reconstruction

	Activity	*Narrative*	*Enables*
1. Cognitive (real-time, nonlinear)	constructive, creative, reasoning	notebooks, sketches, letters	representation, communication, argument
2. Demonstrative (real-time, nonlinear)	reasoning, argument	drafts of papers and letters	ordering, description, demonstration
3. Methodological (retrospective, linear)	demonstration	research papers, monographs	communication, criticism, persuasion, reconstructions, 4, 5, 6
4. Rhetorical (prospective, linear)	demonstration	papers, treatises	persuasion, dissemination
5. Didactic (prospective, linear)	exposition	textbooks, treatises	dissemination of exemplars
6. Normative (linear)	reconstruction	–	logical idealization

the need to develop an argument is becoming as important as the need to describe and order.

We can distinguish this development (though not too sharply) from *methodological reconstruction* of an informal record or working draft, which transforms it into a good piece of public argumentation. Here the main concern is to fashion an evidential argument that conforms to the methodological canons of a particular discipline. An example is Millikan's creative selection of data in his oil-drop experiments.[25] Medawar's well-known complaint about the fraudulence of scientific papers identifies other reconstructions of this kind, though his objection to the inductive form of research papers reflects his own preference for a deductively structured research process.[26] Ampère's reconstruction of the history of his electrodynamic experiments shows, however, that reconstruction can further rhetorical purpose without favoring an inductive methodology. Ampère's account also illustrates the unreliability of published accounts if treated as discovery accounts.[27] An example of more overtly *rhetorical reconstruction* would be Galileo's exaggeration of numerical values in his account of the leaning tower experiment. Again, methodological and rhetorical reconstruction are closely related.

Another type of reconstruction streamlines the discovery process to make every step as transparent and self-evident as possible. As Kuhn saw, the dissemination of new and established science depends on the transfer of "exemplary practices."[28] This facilitates the demonstrations of science texts, which are concerned neither with the complexities of research nor, therefore, with justifying the selection of particular methods. Thus, texts typically introduce physical quantities by defining them in terms of a set of procedures, but rarely mention alternatives or debate the grounds for selecting one set of practices rather than another. Yet this tacit consensus consisting of a web of implicit decisions is always open to challenge; it will be questioned whenever there is a challenge to the facticity of, say, a set of measurements. Writing transparency into examples by editing contingencies out of the practices that created them has a pedagogical as well as a demonstrative role: I call it *didactic reconstruction*.

3.2. Generation, Discovery, and Justification

This is a loose classification of reconstructions, not a hierarchy of distinct processes. All are motivated by the need to impose meaning and structure, though of course these are used to different ends (exploration, communication, invention, consensus-seeking, argumentation, dissemination). They make a closely woven texture of activity. I have unraveled them a little in order to show the unreality of standard philosophical views compared to naturalistic approaches to science. Consider that textbooks and monographs were (and to some extent still are) the main source of the image of experimentation that dominates analytical philosophy of science.[29] This made it easy to formalize scientists' deliberations in line with philosophical theories of perception and reasoning. This reconstruction is normative and antinaturalistic, so I call it *normative reconstruction*. Historically and *cognitively*, it presupposes the other five kinds of reconstruction in Table 1.[30] Yet it fails to recognize them. This failure has forced philosophy of science into two largely divergent paths: historical and formal.

Reichenbach's hallowed distinction between the contexts of discovery and justification was part of a highly selective view of what scientists do and what they produce; it included little if anything outside the realm of declarative knowledge. The justification of a claim to know something can of course be separated from its 'generation', but that separability is made possible by a lot of work: it reflects the conclusion of at least the first three reconstructive processes in Table 1. Moreover, however stable the claim becomes, this enabling work never loses its importance. As studies of controversies show, it remains peripheral only as long as the knowledge-claim is not being challenged. As soon as a challenge arises, experimental procedures and competences come under scrutiny alongside logical, computational, and conceptual ones.[31] Scientific arguments are not constructed once and for all as a simple, formal argument might be. Recent sociological studies here support Kuhn's historical point about the importance of qualitative practices to more formal and quantitative ones. Was the so-called naturalistic fallacy anything more than antinaturalistic dogma?

3.3. Recognizing Reconstruction

Recognizing reconstruction opens the door to cognitive work behind the narrative and formal structures conferred on learning and discovery as these processes are recorded and written up. If such structures are made rather than given, then they can be recovered empirically. That calls for a more cognitive approach in philosophy of science. The problem is that the processes of most interest — cognitive and demonstrative reconstruction — are also the least accessible to retrospective analysis of the sort that historians and philosophers undertake. The discipline that can study constructive processes in something like 'real-time' — sociology (or social anthropology of the laboratory) — tends to 'black-box' cognitive processes in favor of social ones. But where constructivist sociology of science argues that the bottom line in any explanation of innovation and the closure of controversy must invoke social relations, I argue that there is an important cognitive element as well: all *construction* involves *reconstruction*.

We need to study the emergence of structure from process. Here a cognitive-philosophical approach has something to offer both science studies and science education. Study of the construction of narrative structures and formal structures requires some bridge-building between formal and historical approaches in philosophy of science. Neither philosophy nor cognitive science, however, can say much about the dynamics of experimental processes while they lack modes of representing informal reasoning processes in real discovery situations. The remainder of this chapter addresses this practical problem.

4. The Procedural Turn

Much of what is done — especially at the frontiers of observation — involves activity that is unpremeditated as well as preverbal. Can we represent material and verbal practice in a manner that characterizes both the situated nature of procedural knowledge and the unpremeditated parts of discovery processes? I want to avoid prejudging the intentional status of certain experimental "acts" because, as my earlier discussion of reconstruction suggests, rationales for actions often emerge as understanding develops or as an account unfolds.[32] Particular acts are part of activity motivated by larger goals; however, even when they are verbalized, particular bits of agency need not be rationalized explicitly in terms of such goals. Although the term *procedures* is not without its problems, I shall use it to denote a sequences of acts or operations whose inferential structure we do not yet know. The term *procedural* connotes know-how. Such knowledge may be fluent in that it is skilled, and yet be nondeclarative, either because it has not yet been represented and articulated as, say, a list of instructions or because it has largely been packaged into a piece of apparatus (as we see below), a mathematical operation (such as integration), or a computer program.[33] Thus "procedure" connotes two complementary, interactive aspects of a single process: on the one hand, there is the manipulation of objects, instruments, and experience; on the other hand, there is the manipulation of concepts, models, propositions, and formalisms. Procedures are developed and refined over a period of time. During that time explicit rules or protocols may be established (e.g., as in calibrating a measuring instrument or cleaning a sample for carbon dating).[34] Many procedures can be machine based: sorting, mathematical integration, comparing.

4.1. Cognitive Regress

When we drop the assumption that ratiocination with ready-made representations is the only sort of agency worth recognizing, then we can look at how reasoning interacts with other activities. We get a very different picture of the relationship between the world and our representations of it. That relationship is based upon agency in a world of things, ideas, and other people. To draw attention to the context of action from which scientists' talk and thought about the world emerge, I shall represent agency in experimentation directly. I do this with diagrams that map experimental procedures. I shall use them to retrace a path from a finished piece of apparatus and the simple instructions needed to produce a new phenomenon with it, to the published narrative that juxtaposed these and other phenomena to theoretical issues, and finally to the situated learning and procedural knowledge developed to invent the apparatus. The procedural approach enables us to pursue a cognitive regress, moving from explicit, transparent, declarative knowledge — the stuff of which computational discovery programs are made — to implicit, opaque, and unpremeditated learning and invention.

4.2. Discovery Paths

We begin with something far removed from traditional philosophical concerns. Figure 1A shows an electric motor invented by Faraday on September 3, 1821. This was a refined version of the first device in the history of science to produce continuous mechanical effect from a non-mechanical source.[35] Faraday made it to send to other scientists to insure that they, too, could produce the new "electromagnetic rotations." Place some mercury in the glass tube, connect top and bottom to a battery, bring a magnet to the soft-iron rod in the base, and the wire will rotate about the rod. The device was designed so that the instructions could be simple. No new observational procedures needed to be learned. This is a typical example of a ubiquitous but largely invisible process — the dissemination of phenomena by the packaging of skilled procedures into devices. Just how much procedural knowledge is embodied in this little device will emerge as we unpackage it by retracing the path of practice that produced it. Neither the path nor its outcome was unique. The device illustrated in Figure 1B, Barlow's rotating "star," also produces continuous motion from a chemical source of energy, and is reached from the same initial procedures that led Faraday to the rotation motor.

Figure 2 shows a number of pathways leading from Oersted's discovery in 1820 that an electric current affects a magnetized needle (electromagnetism), to Faraday's exploration of this effect in the summer of 1821, to a number of new phenomena and the devices that exhibit them, ending finally in a new problem for theory: Do the new phenomena indicate a fundamentally new, non-Newtonian force and the prospect of a new sort of physics (as Barlow, Faraday, Davy, among others, believed) or can they be reduced to familiar ponderomotive, inverse-square forces (as Biot, Ampère, and others argued)?[36] The paths shown in Figure 2 involve different contingencies and implicate different experimental technologies: some feature contemporary accounts; others were recovered by repeating Faraday's experiments, which showed them to be missing from the manuscript notes of his day's work.[37]

The usual discovery story goes from Oersted's result, to its exploration by Davy (assisted by Faraday), to Faraday's learning of a conjecture by Wollaston. When Faraday tried to test this he got another effect that was, nonetheless, mistakenly construed as Wollaston's predicted effect. That is sequence 1-2-9-2-3-4 in Figure 2, developed as 5 and elaborated by Barlow as 6, 7, and 8. Result 10, Ampère's later success with 9, made clear that Faraday's result 4 realized something other than what Wollaston had suggested. Study of Faraday's notebook, together with actual repetition of his experiments, suggests other and more likely routes. For

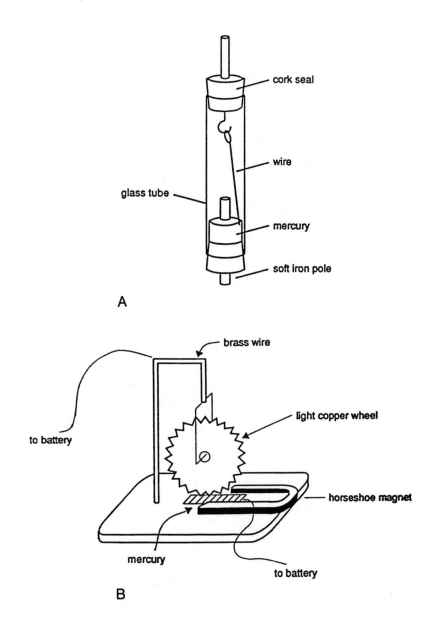

Figure 1. **A** *The simple apparatus Faraday used to display the newly discovered rotation of a current-carrying wire about a fixed magnet;* **B** *Barlow's rotating "star" of March 1822, in which the copper star rotates when the apparatus is connected to a battery.*

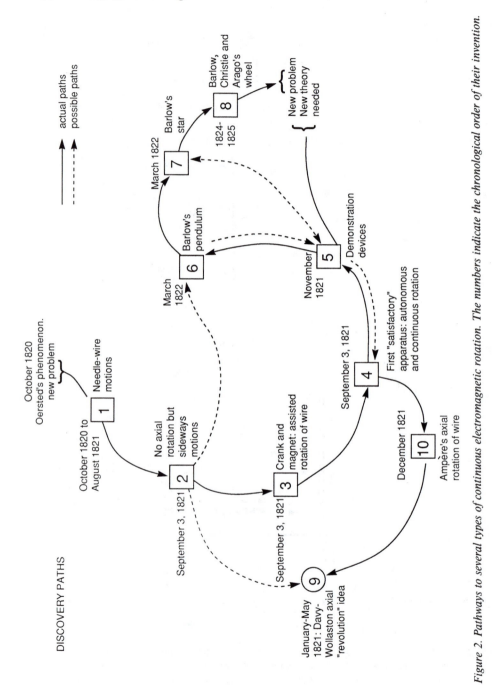

Figure 2. Pathways to several types of continuous electromagnetic rotation. The numbers indicate the chronological order of their invention.

example, though Faraday considered some sort of axial rotation early in his experiments of September 3, he soon abandoned this in favor of lateral motions of the wire, which he eventually fashioned into continuous circular motion around a magnetic pole. Repetition of Faraday's procedures shows that even with the correct configuration the density of the mercury makes it difficult to get continuous motion of any kind but very easy to get lateral motions: the wire behaves like a pendulum. Faraday's notes do not record his having followed lateral motions to an "electric pendulum" effect (pathway 1-2-6). This was demonstrated by Barlow's pendulum, from which Barlow developed the rotating star by March 1822 (Figure 1B). So we have three possible routes to continuous motion: one via Faraday's early sideways motions (1-2-[failed 9]-3-4-5); a second via the lateral motions and Barlow's pendulum effect (1-2-6-7) to the star, from which other forms of continuous motion could be derived (e.g., 5, 8); and a third via Wollaston's hypothesis (1-2-9-10-4-5).

This shows a lack of necessity about the discovery path Faraday recorded in his notebook. Faraday's first *published* account differs from all three paths. In his paper on the rotations Faraday presented the sequence 1-2-3-4-(failed 9)-5 as the actual discovery path. Within a month he could present lay observers directly with 5: the device shown in Figure 1A. This needs only operating instructions, not a discovery narrative.

5. Representing Experimental Paths

In this section I introduce a notation for mapping such pathways. I introduce this by mapping Faraday's instructions first, then his published narrative, and, finally, the discovery path recorded in his notebook. This adapts and extends a notation suggested by Ryan Tweney's "problem-behavior graphs" of know-how that has been articulated into verbal protocols.[38] My discussion of the pathways in Figure 2 shows that the notion of "the actual discovery path" is a chimera, a possibility easily stated in words but one made irretrievable by the reticular, reconstructive nature of actors' thought. We know that alternative paths were either not perceived, perceived but not pursued, or pursued to a dead end and subsequently forgotten. Thus, the maps are *interpretations* of processes. The plausibility of a given map depends on how well it interprets the information available to us.[39]

5.1. Representing Objects

The objects in an experimentalist ontology include material things, images, ideas, mental models, algorithms, etc. I use circles to denote concepts (ideas, or mentally represented things) while squares denote things taken to be in the material world (bits of apparatus, observable phenomena). The familiar theory-observation relationship of the hypothetico-deductive model would be mapped as in Figure 3. Here a hypothesis (say, that electromagnetic force circulating around a wire will push it around a fixed magnet) H_1 is derived from theory T_1 (incorporating a model of the structure of electromagnetic forces circulating around a current), where H_1 implies observation O_H (wire goes around magnet). A real-world possibility is imagined in which O_H occurs in a material situation realized by setup A. When this is realized, a result O_1 is observed. Comparison of result O_1 to imagined O_H shows whether the result supports the original theory (via the hypothesis). This simplifies the interaction of thought and action into a linear, one-way movement. As I show elsewhere, actual hypothesis-testing involves more frequent movement between conceptual and material domains.[40]

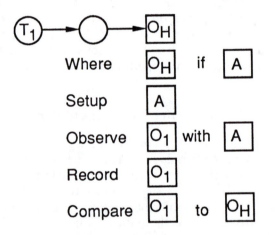

Figure 3. Relationship of theory, hypothesis, and observation on the standard hypothetico-deductive view.

Toward the end of the afternoon of September 3, Faraday began to reason in this way. We need to recover the work that enabled him to articulate his ideas and his procedures to a point where material practice could realize expectations. To do this we need a more subtle ontology and a less formal notion of structure. Many statements in theories are about entities that might exist in some world — say, of mathematical or formally defined objects, of theoretically imaginable objects, or of physical objects. They are based in turn on other claims containing information about such worlds. We can combine the squares and circles of Figure 3 to represent the ontological ambiguity of the entities in play. For example, there are mental representations of things taken to be in the real world, but which enter discourse only as interpreted through a complex of theories. A schematic drawing of an optical microscope represents our understanding of a class of instruments; such a drawing would appear in the map as a square (material artifact) inside a circle (concept or model of artifact); see Figure 4. This is what Latour calls an "inscription" and Hacking a "mark."[41] A model of a hypothetical, possible, but as yet unconstructed instrument would also be shown as a square inside a circle. Uninterpreted traces on a bubble-chamber photo would appear as a square (an event in the world). However, after interpretation as the tracks of elementary particles — a learning process with its own history — the square representing these traces would appear inscribed within a circle, indicating its theory-ladenness. Similarly, a representation or model of a putatively real entity or mechanism (e.g., a bacterium or an electron) would have the same composite form until such time as it is realized (or made directly observable), when it could be represented simply by a square. "Directness" of observation is a function of the mastery, dissemination, and acceptance of observational practices.[42]

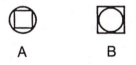

A   B

Figure 4. **A** *An idea or representation of a thing;*
B *An object or event taken to be in the world.*

In empirical science (as distinct from philosophy or pure mathematics) most of the manipulated objects will be 'composite'; they will be either mental models of putatively material entities (a model of an atom) or material embodiments of possibilities (iconic models and instruments, mimetic experimental setups). The composite symbols of Figure 4 represent ideas (representations of actual or possible states of affairs) and things (events, records, apparatus). As we shall see, this notation can represent the changing status of the objects in play. Ian Hacking has made a full inventory of these objects, but he does not discuss what enables observers to transform them as they move from one state to the next.[43]

5.2. Representing Agency

So far we have represented the things, ideas, and other resources that experimenters manipulate. Agency enables scientists to move from one state or object to the next. I represent human agency directly, by a line against which an active verb is printed. Each square or circle identifies what we might call a cognitive state of a scientist or group of scientists, and each connecting line represents an action upon or arising out of that state. Combining these symbols gives a sequence of thoughts and actions, including thoughts about objects, about actions and their outcomes, and about actions leading to new thoughts or objects. Figure 5 shows how Faraday's instructions to operate the simple electric motor look in this notation.

The orientation of the action-lines conveys interpretative judgments about whether any particular bit of agency results in change, (e.g., novel experience, information, artifacts, or images). A horizontal orientation indicates a judgment that something appears that is new to the actor; a vertical orientation indicates that nothing new results. Such judgments are a part of the interpretative process: the notation simply makes them explicit.

Repeated trials with a given setup may accumulate similar results (thereby increasing inductive support). The notation allows us to represent this quite differently from repetitions made to implement variations, needed, say, to deal with anomalies by revising the theory of a piece of apparatus or the apparatus itself. A 45-degree orientation of a line allows us to represent repeated trials that accumulate experimental skill or the fine-tuning of apparatus. A sequence including repetitions of an observation would appear as the vertical segment in Figure 6.

5.3. Resources

The brackets in Figure 6 denote resources used to enable operations or processes. Such resources include mathematical or logical procedures (used in deriving H_1 from T_1); technological precedents for a proposed device or component of a system; a theory of the instrumentation or design work to generate a viable piece of apparatus A_1; manipulative skills and specialized techniques; verbal and pictorial images; computational or other representational procedures that enable, say, comparison of numerical output or of observed phenomena O_{1-n} to what was predicted by H_1; and so on. The brackets merely hint at more complex links between the processes shown and the context of practices, resources, and

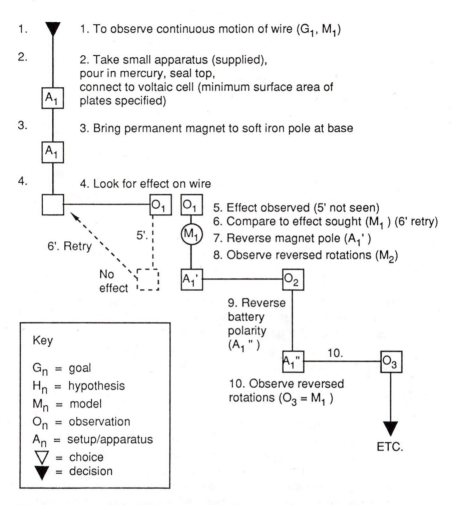

1. 1. To observe continuous motion of wire (G_1, M_1)

2. 2. Take small apparatus (supplied),
pour in mercury, seal top,
connect to voltaic cell (minimum surface area of
plates specified)

3. 3. Bring permanent magnet to soft iron pole at base

4. 4. Look for effect on wire

5. Effect observed (5' not seen)
6. Compare to effect sought (M_1) (6' retry)
7. Reverse magnet pole (A_1')
8. Observe reversed rotations (M_2)

6'. Retry   5'.

No
effect

9. Reverse
battery
polarity
(A_1'')

10.

10. Observe reversed
rotations ($O_3 = M_1$)

Key

G_n = goal
H_n = hypothesis
M_n = model
O_n = observation
A_n = setup/apparatus
▽ = choice
▼ = decision

ETC.

*Figure 5. A map of Faraday's instructions on how to produce electromagnetic motions
with the apparatus shown in Figure 1A.*

beliefs in which any piece of theoretical or experimental work is situated. The selection of experimental technologies and the theoretical
appraisal of instruments, for example, are often lengthy processes involving many decisions.[44] Though the maps are necessarily printed in
a two-dimensional field, they range over the space of concrete manipulations; mental spaces in which exploratory imaging and modeling take
place; a resource space that includes archives, databanks, and invento-

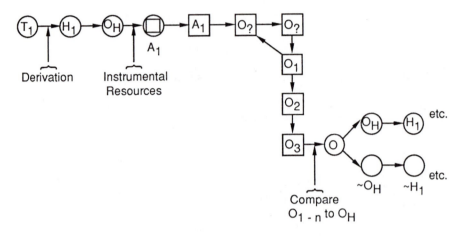

*Figure 6. A sequence representing the repetition of observations
and the use of resources.*

ries of techniques (often "stored" as human expertise); computational spaces; the social space in which observers negotiate interpretations of each others' actions; the physical (laboratory) space in which observations are fashioned; and the rhetorical and literary space in which they are reported and put to work in arguments.[45]

6. Experimental Reasoning

So far we have represented agency and its objects. What structure might such processes have? Answering this question requires a final piece of preparatory work: to represent choices or decisions that determine the path taken at any particular point. Comparison of a relatively unreconstructed laboratory record with a published account of the same experiment should show that the actual sequence of procedures and outcomes has been reordered and also that certain (contingent) choices have been promoted to decisions. I distinguish choices (made on the basis of incomplete information or partial understanding of a situation) from decisions (made on the basis of a rationale thought to be reasonably complete at the time a decision was made). Let choices be represented by white triangles; decisions by black ones, as in Figure 7. There are many more levels of embeddedness of choices and decisions than the two introduced here, but we have enough complexity to indicate how any particular decision is situated as a response to earlier outcomes.[46]

Because getting experiments to work involves exploring unexpected

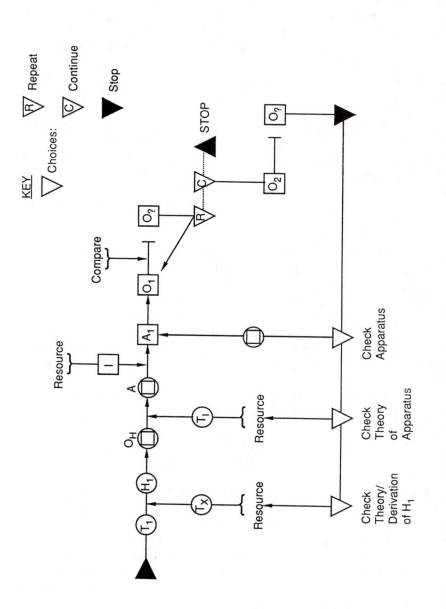

Figure 7. A sequence showing the distinction between choices and decisions and the extent of the repetition or recycling required by each sort.

results, we need to show the recalcitrance of instrumentation as well as of nature. A sequence can be broken to indicate anomalies, where procedures break down or where thoughts cannot be implemented. These discontinuities are important to understanding how reconstruction enables experiments (including thought experiments) to *work*. An experiment works when it shows that something is the case. The failure indicated by a discontinuity can be conceptual (e.g., a derivation by argument of a hypothesis from a theory) or it can be practical: a setup proves impossible to construct or to operate, fails to behave as expected, or produces unexpected output; an experimentalist lacks the dexterity needed to carry out certain material manipulations; or a theoretician lacks the mental agility to carry out certain formal procedures. Unexpected outcomes, whether the anomalies of real experiments or the paradoxes of thought experiments, enable experimentation to constrain and change theories. I indicate unanticipated outcomes by terminating an action-line with a T.

Retrospective narrative accounts of experiments reconstruct experimental pathways as decision trees. These structures recall forgotten, alternative paths and they illustrate the difference between contingent choices (which may generate branches) and well-grounded decisions (which can be seen as having encouraged growth of the main trunk). The tree image, however, is misleading. It implies inexorable growth in a particular direction (the trunk or main branches). It fails to capture the nonlinear, reconstructive character of scientific processes (bypasses, recycling, repetition, reinterpretation of earlier results, reformulation of expectations, refinement of concepts). This is because it fails to recognize that directionality emerges from reconstruction; in short, it benefits a great deal from hindsight. Comparing maps of different narratives concerning the same phenomenon shows this.

7. Comparing Narratives

I remarked earlier that Faraday's published account of his discovery of electromagnetic rotations presents the sequence 1-2-3-4-(failed 9)-5. Figure 8 is a map of the major part of this narrative. Figure 8 shows a great deal more activity than that required simply to produce the phenomenon with the ready-made device shown in Figure 1A. There is also more movement between images and models on the one hand and material manipulations on the other; however, this account shows far less activity than the laboratory record.

Figure 9 maps the day's work in which Faraday moved from exploratory observation of Oersted's electromagnetic effects, through the generation of new phenomenal possibilities, to the construction first of devices that failed to realize any of them, and on to the device that produced continuous, unassisted motion of a wire around a magnet. This map traces one of several possible paths, the sequence 1-2-3-4-5 in Figure 2.[47]

Analyzing the experiment into a sequence of actions on concepts and objects enables comparison of later narratives to early records. Visual comparison shows that the map of the published account displays much more horizontal movement (or new results) than the more vertically oriented map of the laboratory record. This displays an effect of cognitive and demonstrative reconstruction: individual moves are understood in the light of their outcomes and how these contributed to an overall result, so fewer moves are needed to make up a convincing account. Another difference is a change in the frequency of composite or ontologically ambivalent entities. The greatest difference here is between Figure 5 and Figure 9.

This comparison shows how reconstruction makes both agency and the contingency of choices disappear. Table 2 compares the laboratory record mapped in Figure 9 to Faraday's published narrative (center column, and Figure 8) and to his instructions on how to produce electromagnetic rotations (right-hand column, and Figure 5). The changing frequency (1) of different types of acts, (2) of successful to neutral or unsuccessful moves, and (3) of conceptual to material operations, illustrates the complementary packaging of phenomena (and the skills that enable them) into narratives and into devices. Systematic analysis of aspects of the process — acts, changes, choices, and the changing ratios of conceptual and material objects of manipulation — displays how the construction of narratives reduces the complexity of an experiment. This transformation enhances the self-evidence of its outcomes. For example, by November 1821, when Faraday produced the simple demonstration instructions, the rotations could appear as natural phenomena because they were produced with little skill or human intervention.

7.1. Ontological Ambivalence

Several points emerge from this comparison. First, the changing frequency of the composite or ontologically ambivalent entities illustrates how the status of what is manipulated changes. Ambivalence early on is crucial to the creative development of experiments because it enables free movement between possible and actual entities, that is, between ideas and things. Such ambivalence is therefore as essential to thought experiments as it is to real ones, but it would undermine the force of a demonstrative argument drawn prematurely from an experiment. As

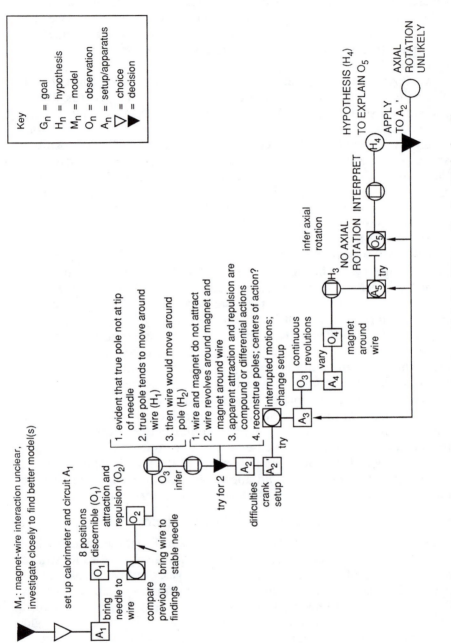

Figure 8. A map of the narrative Faraday published in 1821 (see Gooding 1990b, chap. 7).

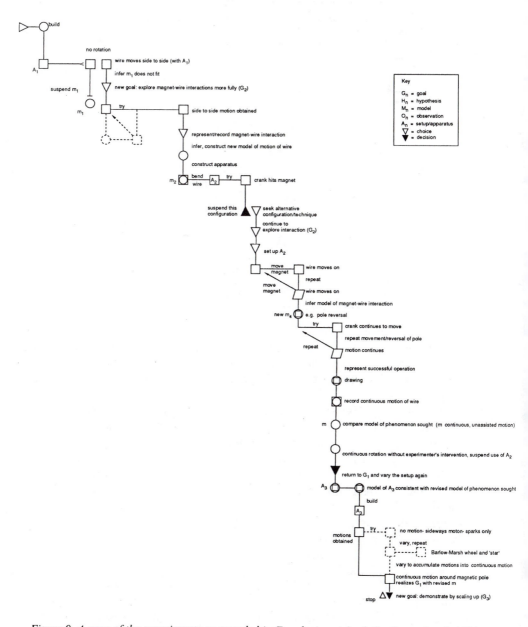

Figure 9. A map of the experiments as recorded in Faraday's notebook for September 3, 1821 (see Gooding 1990b, chaps. 5 and 6).

Table 2. Comparison of the narratives mapped in Figures 5, 8, and 9, indicating the increased complexity recovered as we move from the latest to the earliest version of the experiment.

	Notebook for September 3	October 1821 paper	Apparatus instructions
Decision points	12	4	2
Outcomes			
squares	16	9	10
circles	6	2	1
composite	6	8	0
Subtotals	28	19	11
Actions			
vertical	25	11	8
diagonal	3	0	(1)
horizontal	10	11	3
Subtotals	38	22	12
Totals	78	45	25

phenomena are elicited, as observational techniques are mastered, and as experimenters work out what the experiment can show, putatively idealike or thinglike entities are hypostatized as having *always* been one or the other.

Recognizing how reconstruction effects ontological disambiguation is an important step toward dispelling philosophical puzzlement about empirical access and realism. Judgments about the reality or necessity of an entity and about the directness of observations of it are inherently retrospective. In other words, their status reflects confidence based on certain representations being made and tried, on distinctions being made, on skilled practices established, and so on. Thus, the second point is that the *self-evidence* of the mental (conceptual) or material (real-world) status accorded to the entities to which language refers is conferred through reconstructive processes and the *concealment* of this aspect of their histories.

It follows that this cognitive and historical aspect of scientific practice will not appear in published texts because the ambiguous, composite status of objects and the enabling role of skill both disappear from retrospective accounts of experiments. Comparing maps of successive narratives of an experiment shows that this happens progressively, as possibilities are realized (as instruments, observable phenomena, etc.) and as decisions are made about the reality of what human agency has produced. Recognizing the existence of these processes takes some of the mystery out of empirical access, local consensus, and convergence.[48]

7.2. Thought Experiments

According to mainstream philosophy of science, natural phenomena are bounded by theory. I have argued that natural phenomena are bounded by human activity. Reasoning with ready-made representations is necessary to this, but it is not sufficient. The changing character displayed by the maps of Faraday's three narratives illustrates Quine's point (quoted earlier) about the power of verbal representation to describe and order worlds of experience. It also illustrates a complementary point about the power of techniques and technologies: Faraday's simple demonstration device is the technological counterpart of a successful semantic ascent. He needed fewer words because he had learned how to represent what he learned to do in the world. Introducing new representations involves introducing new practices and instruments that embody them.

I have argued the importance of engaging the material world beyond the world of representations because in science, as in most learning, language and action work together. Language appears sufficient for scientific purposes only if we *stay* in the world of monographs and textbooks. To deny this is to deny the ubiquity of human agency that enables Quineian semantic ascent to any of the languages of science, past and present.

Though it is a representation-making activity, science is not representation bound. I conclude by suggesting that embodiment and situated reasoning are just as important to the demonstrative use of thought experiments as they are to the evidential use of real- or material-world experiments. Thought experiment becomes possible when semantic ascent is largely complete, that is, when a world is sufficiently well represented that experiment can be conducted within it. The power of thought experiments to generate real-world assertions without apparently engaging the material world has been used to argue that representations are self-sufficient and to defend a rationalist view of science.[49] Koyré took the intellectualist legend for granted, playing down the empirical and material aspect of science in favor of the march of ideas and arguments. He was right about the importance of thought experiments in generating new science, but wrong to conclude that all new science is done a priori, by thought experiment. The alleged priority of thought experimentation is an artifact of a selective, overliterary reading of the history of science. Intellectual historians and philosophers sometimes overlook the vicious circularity of the relationship between their rationalism and their neglect of experimental practice.[50]

8. Why Do Thought Experiments Work?

How do scientists go from the actual to the possible, on to the impossible, and return to an actual world altered by that journey? The short answer is that thought experimentation and real experimentation have much in common. The demonstrative power of either sort of experiment depends on borrowing elements and strategies from the other kind. This follows from the close interaction of thinking and doing (illustrated by the map in Figure 9) and its counterpart in the mutual refinement of theoretical and experimental practice, which I detail elsewhere. Here I shall identify six similarities between thought experimentation and real experimentation in frontier or research science.

1. Thought experiments involve the manipulation of stable, well-defined objects, often using idealized versions of tools that enable interaction with aspects of an environment (measuring rods, clocks, detectors, etc.).

2. The demonstrative power of thought experiment depends on creating situations in which most moves or manipulations are wholly transparent but in which certain moves are impracticable, just as real-world experiments encounter anomalies and recalcitrances. Impracticability in thought is analogous to recalcitrance in material practice. It does much of the demonstrative work that methodologists prefer to attribute to formal contradiction. The fallibilist ideal is that theories are corrigible because it is possible to generate contradictions between statements derived from hypotheses and statements reporting observations. The complementary formalist ideal is that decisions about whether to accept statements deduced from theory are governed by logical rules.[51] No doubt Scriven spoke for many when he remarked that anything less than this has always seemed to abandon rigor "in favor of a kind of sloppy intuition."[52] There are, however, nonlogical yet rigorous alternatives. What is involved is the demonstration, using a paradox, of a contradiction in practice. Thought experimenters criticize a rival theory by showing the impracticability of doing something in the way required by that theory. In short, a theory is criticized through the practice that links it to those aspects of the world that it purports to be about. As Kuhn shows in his discussion of Galileo's dialogue on the inclined plane experiments, the paradox generated need not be a logical one such as squaring the circle, which could not be realized in any possible world.[53]

There are material constraints in thought experiments just as there are conceptual ones in real experiments.

3. To bring the impracticability of certain actions to bear in argument, the author of a thought experiment selects and isolates just those features of a phenomenon, the environmental framework, and the conceptual scheme that are mutually problematic. The author sets the stage for thought experimentation by readers, just as narratives of a real experiment enable vicarious observation through participation in exemplary practices. Galileo and Boyle well knew that witnessing must be participatory: a good narrative enables the lay-experimenter to follow the process in the manner that students of geometry learn to follow a geometrical proof. Such narratives are far easier to follow than realistic descriptions of fully contingent experiments like the one mapped in Figure 9.

4. Observations and operations carried out in this framework are performed by perfect observers: all the necessary skills and capacities are assumed. In this respect a thought experiment is the conceptual analog of Faraday's compact rotation device: few if any skills are needed to make it reproduce the intended experience in another observer. In virtual-witnessing of real experiments this experience is usually of the representation of a phenomenon; in thought experiments it is usually a paradox. Sometimes a threshold of ambiguity or inconsistency must be reached before thought is compelled to experiment in a manner that is open to conceptual reform. As I remarked earlier, "consistency" here is a practical matter, that is, a matter of practicability. Through cognitive and demonstrative reconstruction it becomes a conceptual matter and eventually, through thought experimentation, a logical one. But the force of the experiment as an argument depends on the impracticability of doing what is described. As an argument the experiment involves a subtle mix of material- and mental-world manipulations. Galileo did not attempt to escape the constraints of the actual world (say, by altering his inclined plane experiment to exclude motion of one sort or other): his thought experiments involve the world as his interlocutors knew it. No special type of knowing is presupposed. For thought experiment, the essential skill is the ability to follow a narrative describing the steps of a procedure. Because we acquire such abilities through general education, we tend to forget how important they are to the cogency and transparency of the experiments we mentally perform.

5. A thought experiment does what all formal, demonstrative procedures do: it transcends the particularities of practice in the material world. This gives the experiment both universality and transparency, properties it can have in a world of representations. The experimenter's

"actual" world is of course represented in thought, but thought highly constrained by the fact that it must be similar to a world that would-be observers already know. Real-world experiments achieve this transcendence only in their most reconstructed forms (see Tables 1 and 2). That the worlds investigated by real experiment appear to be contrived out of a different, "material" sort of stuff is explained by the process of ontological disambiguation (or reification) described earlier and illustrated by the experimental maps.

6. Finally, like real experiments in their demonstrative or textbook form, thought experiments always work. Their narratives have the inexorable character of geometrical demonstrations. No wonder philosophers prefer them to real ones!

Thought experiments work because they are distillations of practice, including material-world experience. Experimenters move readily between representations and their objects. The maps show how frequently experimentation shifts between these domains. To revise Quine's metaphor, semantic ascent is not a unidirectional, once and for all process: there are as many descents into the world of concrete particulars as there are ascents into the world of representations. The moral for philosophy of science of the interdependence of thought and action (or of theory and experiment) is that when you ignore one, you end up with a false view of the other and, therefore, with false problems of empirical access, representation, meaning, and realism.

Insofar as it is achievable, the formal consistency of theory and observation is an idealization of the practical consistency of the thinkable and the doable. Thought experiments work when they idealize the key features of real experiment, including manipulative skills and the ability to execute procedures. The converse is also true. Real experiments work as demonstrations when they implement key features of thought experiment. But, as my analysis has shown, real experiments only do this in their most reconstructed form. To many historians and philosophers, the most visible and epistemologically interesting function of any experiment is the demonstrative one. This points to a seventh similarity: if we could recover the private, "mental" prehistory of thought experiments, this would show that, like real ones, they work only after a lot of reconstruction has been done.[54]

Notes

Parts of this chapter were read to the 14th Annual Conference on Philosophy of Science at the Inter-University Centre of Postgraduate Studies, Dubrovnik; to a meeting of the British Society for the Philosophy of Science; and to the Conference on the Implications of Cognitive Sciences for Philosophy of Science at the Minnesota Center for the

Philosophy of Science. I would like to thank participants, particularly Ryan Tweney and Nancy Nersessian, for a number of constructive criticisms.

1. Quine (1957), p. 5 and (1960), pp. 270ff.

2. Sachs (1986), chap. 3.

3. Gregory (1986), pp. 28–30.

4. See Putnam (1974); Rescher (1980).

5. Giere (1988), pp. 109–10.

6. Kuhn (1961) and (1962), both reprinted in Kuhn (1977).

7. Kuhn (1962), in Kuhn (1977), at p. 80.

8. On generative justification see Nickles (1989), pp. 304–9.

9. Kuhn (1962), in Kuhn (1977), at p. 80.

10. On reconstruction and the one-pass fallacy see Nickles (1988).

11. See Kuhn (1961) and Gooding (1990b), chaps. 7 and 8.

12. Ryle (1949), pp. 27ff.

13. See the editors' introduction to Gooding, Pinch, and Schaffer (1989) and Gooding (1990b), chap. 1.

14. On the creative use of uncertainty see Gooding (1989a).

15. See Tweney (1990).

16. See Tweney and Gooding (1990).

17. Kuhn (1961), in Kuhn (1977), at p. 201.

18. A recent example is Perez-Ramos (1989).

19. Arber ([1954] 1985), chap. 1.

20. Penrose (1989), pp. 443–44.

21. For construals see Gooding (1986) and (1990b), chaps. 1–3.

22. On Boyle's literary technology of virtual witnessing see Shapin (1984).

23. For details of Biot's work see Gooding (1990b), chap. 2.

24. Holmes (1987).

25. See Franklin (1986).

26. Medawar (1963).

27. On Ampère see Hofmann (1987) and Gooding (1990b), chap. 2.

28. Kuhn (1962), in Kuhn (1977), at p. 80.

29. For more on this see Kuhn (1961).

30. Lakatos was ardently normative in his reconstruction of historical episodes (Lakatos 1970), but formalist philosophies either ignore or else reconstruct cognitive and historical processes to make them amenable to analysis. The tension between normative (methodological) and empirical (historical) concerns is evident in Laudan (1988).

31. For examples see Pinch (1985) and Collins (1985).

32. For further discussion of this issue see Gooding (1990a; 1990b, chaps. 5 and 6).

33. See Tweney and Gooding (1990).

34. See Collins (1985) and Laverdière (1989).

35. See Gooding (1990a).

36. For details of this divergence of interpretations, see Gooding (1990b, chap. 2).

37. Repetition of Faraday's experiments is discussed in Gooding (1989b; 1990b, chaps. 5 and 6).

38. See Tweney (1986) and Tweney and Hoffner (1987).

39. This includes the repetition of experiments (Gooding 1989b), study of extant instruments, and searches through laboratory manuals and records to recover shared practices, supplemented by published and other retrospective accounts.

40. See Gooding (1990b), chaps. 6–8.

41. Hacking (in press); for inscription devices see Latour (1987).

42. In a manuscript in preparation (Tweney, in preparation) Ryan Tweney argues that Hooke's drawing of a louse — an example of a situation in which apparently direct observations yielded representations now accepted as accurate — involved many manipulations of the object under his microscope to buildup a composite model from successive, discrete "scans." Hooke's engraving — more accurate than we would expect from the simple technology he had — depicts an image derived from this model. For other examples of how observations are made accurate and direct see Galison and Asmuss (1989) and Hacking (1983).

43. Hacking (in press).

44. See Gooding (in press).

45. For examples of this work see Bazerman (1988) and Shapin and Schaffer (1985).

46. I discussed this further in Gooding (1990b), chap. 6.

47. For detailed treatments of Faraday's notebook record see Gooding (1990b), chaps. 5 and 6.

48. Mystery arguments for realism are canvassed by Boyd (1973), Brown (1982), and Jardine (1978).

49. According to Koyré (1968), pp. 75ff., the significant innovation of the seventeenth century was not practical experimentation but a priori experimentation.

50. See, e.g., Brown (1986).

51. See Gooding (1990b), chaps. 7 and 8.

52. Scriven (1961), p. 219.

53. Kuhn (1962), in Kuhn (1977), at p. 253.

54. An example is the thought experimentation that led Faraday to build the first Faraday cage; see Gooding (1985) and for the prehistory of this experiment, Gooding (1990b), chap. 9.

References

Arber, A. [1954] 1985. *The Mind and the Eye.* Cambridge: Cambridge University Press.

Bazerman, C. 1988. *Shaping Written Knowledge.* Madison: University of Wisconsin Press.

Boyd, R. 1973. "Realism, underdetermination and a causal theory of evidence." *Noûs* 7:1–12.

Brown, J. R. 1982. "The miracle of science." *Philosophical Quarterly* 32:232–44.

———. 1986. "Thought experiments since the scientific revolution." *International Studies in the Philosophy of Science* 1:1–15.

Collins, H. M. 1985. *Changing Order: Replication and Induction in Scientific Practice.* Beverly Hills, Calif., and London: Sage.

Franklin, A. 1986. *The Neglect of Experiment.* Cambridge: Cambridge University Press.

Galison, P., and A. Asmuss. 1989. "Artificial clouds, real particles." In Gooding, Pinch, and Schaffer, eds. (1989), pp. 224–74.

Giere, R. 1988. *Explaining Science: A Cognitive Approach.* Chicago: University of Chicago Press.

Gooding, D. 1985. "'In nature's school': Faraday as an experimentalist." In D. Gooding and F. A. J. L. James, eds., *Faraday Rediscovered: Essays on the Life and Work of Michael Faraday, 1791–1867.* London: Macmillan; reprinted 1989, Macmillan/American Institute of Physics, pp. 105–35.

———. 1986. "How do scientists reach agreement about novel observations?" *Studies in History and Philosophy of Science* 17:205–30.

————. 1989a. "Thought in action: making sense of uncertainty in the laboratory." In M. Shortland and A. Warwick, eds., *Teaching the History of Science*. Oxford: Blackwells/ BSHS, pp. 126–41.

————. 1989b. "History in the laboratory: can we tell what really went on?" In F. James, ed., *The Development of the Laboratory: Essays on the Place of Experiment in Industrial Civilization*. London: Macmillan; New York: American Institute of Physics, pp. 63– 82.

————. 1990a. "Mapping experiment as a learning process." *Science, Technology and Human Values* 15:165–201.

————. 1990b. *Experiment and the Making of Meaning*. Dordrecht, Boston, and London: Kluwer Academic.

————. In press. "Putting agency back into experiment." In A. Pickering, ed., *Science as Practice and Culture*. Chicago: University of Chicago Press.

Gooding, D., and T. Addis. 1990. "Towards a dynamical representation of experimental procedures." In *Bath 3: Rediscovering Skill in Science, Technology and Medicine, Abstracts*. Bath: University of Bath, Science Studies Centre, pp. 61–68.

Gooding, D., T. J. Pinch, and S. Schaffer, eds. 1989. *The Uses of Experiment: Studies in the Natural Sciences*. Cambridge: Cambridge University Press.

Gregory, R. 1986. *Hands on Science*. London: Duckworth.

Hacking, I. 1983. *Representing and Intervening: Introductory Topics in the Philosophy of Natural Science*. Cambridge: Cambridge University Press.

————. In press. "The self-vindication of the laboratory sciences." In A. Pickering, ed., *Science as Practice and Culture*. Chicago: University of Chicago Press.

Hacking, I., ed. 1981. *Scientific Revolutions*. Oxford: Oxford University Press.

Hofmann, J. 1987. "Ampère, electrodynamics and experimental evidence." *Osiris* 3:45– 76.

Holmes, F. L. 1987. "Scientific writing and scientific discovery." *Isis* 78:220–35.

Jardine, N. 1978. "Realistic realism and the philosophy of science." In C. Hookway and P. Pettit, eds., *Action and Interpretation*. Cambridge: Cambridge University Press, pp. 107–25.

Koyré, A. 1968. *Metaphysics and Measurement*. London: Chapman and Hall.

Kuhn, T. S. 1961. "The function of measurement in modern physical science." *Isis* 52:161– 90; reprinted in Kuhn (1977).

————. 1962. "A function for thought experiments." In *L'aventure de la science, melanges Alexandre Koyré*. Paris: Hermann, vol. 2, 307–34; reprinted in Kuhn (1977).

————. 1977. *The Essential Tension*. Chicago: University of Chicago Press.

Lakatos, I. 1970. "Falsification and the methodology of scientific research programmes." In I. Lakatos and A. Musgrave, eds., *Criticism and the Growth of Knowledge*. Cambridge: Cambridge University Press, pp. 91–196.

Latour, B. 1987. *Science in Action*. Milton Keynes, Eng.: Open University Press.

Laudan, R. 1988. "Testing theories of scientific change." In A. Donovan, L. Laudan, and R. Laudan, eds., *Scrutinizing Science: Empirical Studies of Scientific Change*. Dordrecht, Boston, London: Kluwer Academic, pp. 3–44.

Laverdière, H. 1989. "The Sociopolitics of a Relic: Carbon Dating the Turin Shroud." Ph.D. thesis, Science Studies Centre, University of Bath.

Medawar, P. 1963. "Is the scientific paper a fraud?" *The Listener* (September 12, 1963): 377–78; reprinted in D. Edge, ed., *Experiment: A Series of Scientific Case Histories*. London: BBC, 1964.

Naylor, R. 1989. "Galileo's experimental discourse." In Gooding, Pinch, and Schaffer, eds. (1989), pp. 117–34.

Nickles, T. 1988. "Reconstructing science: discovery and experiment." In D. Batens and J. P. van Bendegem, eds., *Theory and Experiment*. Dordrecht and Boston: Reidel, pp. 33–53.

————. 1989. "Justification and experiment." In Gooding, Pinch, and Schaffer, eds. (1989), pp. 299–333.

Penrose, R. 1989. *The Emperor's New Mind: Concerning Computers, Minds, and the Laws of Physics*. Oxford: Oxford University Press.

Perez-Ramos, A. 1989. *Francis Bacon's Idea of Science and the Maker's Knowledge Tradition*. Oxford: Oxford University Press.

Pickering, A. 1989. "Living in the material world." In Gooding, Pinch, and Schaffer, eds. (1989), pp. 275–97.

Pinch, T. 1985. "Theory-testing in science — the case of solar neutrinos. Do crucial experiments test theories or theorists?" *Philosophy of the Social Sciences* 15:167–87.

Putnam, H. 1974. "The 'corroboration' of theories." In P. A. Schilpp, ed., *The Philosophy of Karl Popper*. La Salle, Il.: Open Court, vol. 1, pp. 221–40; reprinted in Hacking, ed. (1981), pp. 60–79.

Quine, W. V. O. 1957. "The scope and language of science." *British Journal for the Philosophy of Science* 8:1–17.

————. 1960. *Word and Object*. Cambridge, Mass.: MIT Press.

Rescher, N. 1980. "Scientific truth and the arbitrament of praxis." *Noûs* 14:59–74.

Ryle, G. 1949. *The Concept of Mind*. London: Hutchinson.

Sachs, O. 1986. *The Man Who Mistook His Wife for a Hat*. London: Picador.

Scriven, M. 1961. "Explanations, predictions and laws." In H. Feigl and G. Maxwell, eds., *Scientific Explanation: Space and Time*. Minneapolis: University of Minnesota Press, pp. 170–230.

Shapin, S. 1984. "Pump and circumstance: Robert Boyle's literary technology." *Social Studies of Science* 14:481–520.

Shapin, S., and S. Schaffer. 1985. *Leviathan and the Air-pump: Hobbes, Boyle, and the Experimental Life*. Princeton, N.J.: Princeton University Press.

Tweney, R. 1986. "Procedural representation in Michael Faraday's scientific thought." In A. Fine and P. Machamer, eds., *PSA 1986*. Ann Arbor, Mich.: Philosophy of Science Association, vol. 2, pp. 336–44.

————. 1990. "Five questions for computationalists." In J. Shrager and P. Langley, eds., *Computational Models of Discovery and Theory Formation*. Palo Alto: Morgan Kaufmann, pp. 471–84.

————. In preparation. "On doing science," unpublished.

Tweney, R., and D. Gooding. 1990. "Qualitative skills in quantitative thinking: Faraday as a mathematical philosopher." In *Bath 3: Rediscovering Skill in Science, Technology and Medicine, Abstracts*. Bath: University of Bath, Science Studies Centre, pp. 152–55.

Tweney, R., and C. Hoffner. 1987. "Understanding the microstructure of science: an example." Abstracted in *Proceedings of the 9th Annual Conference of the Cognitive Science Society*. Hillsdale, N.J.: Erlbaum.

Serial and Parallel Processing
in Scientific Discovery

The very existence of this book suggests that the "cognitive turn" in the philosophy of science is already at an advanced stage. Indeed, not since the seminal work of Kuhn (1962) has there been more activity among philosophers directed at what we can loosely characterize as the "processes" of science, rather than at its "products." In the present essay, I will seek to clarify some aspects of the cognitive turn. In particular, I hope to show that the oft-cited distinction between parallel and serial processes needs considerably more careful attention in its application to science than has so far been the case. It is not an "either-or" issue, but a "both-and" issue, one where cognitive scientists are not yet of one voice on the right way to proceed.

To set the stage, I will first briefly summarize my understanding of the nature of the distinction, describe an empirical domain (based on my earlier work on Michael Faraday) where the distinction has some utility, briefly discuss a well-known formal application of parallel processing models to scientific cognition (namely, Paul Thagard's), together with a replication of that model, and conclude by trying to draw out some specifically philosophical implications that may have consequences for future investigations.

1. Serial and Parallel Processing

Until recently, most formal models in cognitive science were based upon a serial analog in which unitary processing stages were strung together in daisy-chain fashion, each stage awaiting the output of the previous stage before beginning to process its bit or byte or whatever of information. Such models are based on the analogy between a Von Neumann computing device and human information processing, and they have proven themselves to be of great utility in both artificial intelligence (AI) applications and as formal tools in the analysis of human cognition. In particular, since a good deal of human problem solving is phenomenologically serial (e.g., any mental arithmetic problem that exceeds whatever multiplication table we have memorized), serial simulations of problem solving could be developed that matched quite closely the "introspective" analysis of self-reflection. The threatened subjectivity of such matches can be overcome by using the proper methodology (see Ericsson and Simon 1984), provided of course that one is willing to accept a few basic assumptions about the size and role of short-term memory. Furthermore, though seemingly limited in efficiency by the very seriality of processing that is their defining characteristic, the growing speed of modern digital computers has kept even very extensive serial models within comfortable real-time windows. This itself becomes a problem for the modeling of human cognition, however, since the cycle time of a modern computer far exceeds the cycle time of the processing units found in the brain. One sometimes needs so many production rules applied so often in such a simulation that no brain could possibly carry out the computation in that fashion. To get the phenomenological veridicality of a serial simulation, then, we have to simply abandon hope that the model can be tied to any conceivable neuroscientific model.

Parallel models of information processing seem to offer a way out of the impasse. They are not new models, of course. William James suggested something that sounds very much like parallel processing when he discussed associationist models of memory (James 1890, pp. 566–94), and Hebb's "cell-assembly" notion is an even better known precursor of the modern efforts (Hebb 1949). But the last decade has seen a resurgence of such models; parallelism is the new holy grail of cognitive science, it seems, fueled by the promise of truly parallel hardware (still only a promise) and, more importantly for us, by the promise of better models of human cognition, models that are conceivably implementable in brains. Promises, promises, to be sure, but intriguing ones nonetheless.

The really intriguing thing for me, however, is not the hoped-for tie to neuroscience. Instead, it is the promise of a better ability to capture at all (never mind how closely to the neural substrate) some of the processes of cognition that so far have had to be simply "black boxed," taken for granted as components but otherwise unanalyzed. In this category, I place "imaging," perceptual functioning, and certain aspects of memory and memory retrieval. With parallel models, we have the hope of capturing some very elusive aspects of cognition within the computational part of our formal models. This is important to my interest in modeling scientific cognition in particular, because perception, imagery, and memory play very central roles in such cognition, roles that have not been adequately captured in prior serial models.

So far, I have pitched the discussion of serial and parallel models in ways that are fully consistent with what we might loosely call the "Pittsburgh–San Diego Axis," that is, in terms that suggest the major dividing line is between Newell and Simon's serial models (as manifested, say, in their 1972 *Human Problem Solving*) and parallel models such as those described by McClelland, Rumelhart, and the PDP Research Group (1986). There is a third pole in cognitive science, however, one that, though actually closer to the concerns of those seeking to model scientific cognition, has received far less attention. I am referring to the "Natural Cognition" school (e.g., Neisser 1976; 1982), which insists that studies of cognition rely too much on laboratory-based studies and on formal computerized models. You cannot understand cognition, says this view, without paying careful attention to its manifestations in an ecologically valid natural context. For Neisser and his followers and fellow travelers, cognition is contextually dependent and must be described in that context before it is understood at all. I share this view, and part of my fascination with parallel models is that they seem to me to be a step in the right direction in this very important respect.

Using this third-pole notion, I can justify one aspect of my fascination with parallel models that otherwise might strike the reader as inappropriately motivated, namely, the fact that watching a network run can be enormously entertaining! Though I suspect that the suspicion with which many regard network models stems directly from this feature, I believe it is actually relevant. Parallel models work in ways that manifest *qualitative* fits to the world of natural cognition that far exceed the persuasiveness of the more usual sort of *quantitative* fits to data that we use to justify our models. And are they fun to watch! Something as simple as the McClelland and Rumelhart (1989) Interactive Activation Model (IAC) demonstration program "Jets and Sharks" is almost endlessly fascinating — it generalizes, makes goofy errors, retrieves information, is surprisingly fault-tolerant, and forms "concepts" in ways that have all the right phenomenological "feels" for a model of memory. Something this much fun just has to have something going for it.

The fun has a serious side, of course. A simple model of memory that can generalize, make humanlike errors, and retrieve information in the wacky content-addressable way that people do is an important achievement, demonstration though it be. It is compelling partly because it is so simple (IAC networks are the simplest type discussed in McClelland and Rumelhart 1989) and partly because the most compelling properties of its behavior were not designed in — they are emergent properties. No one had to write a "generalize memory" routine, or a "repair fault" algorithm. The properties are natural consequences of the architecture of the model. Furthermore, the dynamic quality of the models, the fact that they operate in a recognizably human "real time" (and within the so-called Feldman Limit of one hundred or so processing cycles [see Feldman 1985]) and that such things as change and even *rate* of change become meaningful variables, also gives credence (and adds to the fun). Thus parallel models promise a better capturing of what Nersessian (see her essay in this volume) refers to as the "cognitive dynamics" of problem solving in science (see also Tweney 1990 and Gooding and Tweney, in preparation).

Parallel models of cognition have so far been applied with something approaching testable validity only to visual processes at a fairly low level (e.g., Marr and Poggio 1976). They have been tried for higher-level processes also (e.g., language learning by McClelland and Rumelhart 1981, and to scientific explanation by Thagard 1989), but these must really count as demonstrations only. Thus, Pinker and Prince (1988) have shown that the language learning model simply is not up to the real data of language acquisition, and I will argue below that Thagard's model is not up to real scientific cognition. But each is a hopeful start nonetheless. To suggest that they are not entirely false starts, I will describe some aspects of a specific case study that are usefully thought about in terms of the parallel versus serial distinction.

2. A Specific Case

As I noted earlier, one of the gains to be hoped for is a better understanding of the role of memory in scientific thinking. To illustrate the point, I would like to consider a few examples of how the serial/parallel issue can shed light on the uses to which Michael Faraday (1791–1867) put his very extensive notebooks and memory aids. This is a good place to start because notebooks are, after all, a form of permanent memory; we can learn, from the uses to which they are put, something about the natural propensities of people who need to deal with large amounts of scientific information. One caveat: I am *not* presenting a parallel model as such; instead, I am trying to see if thinking about the problem in terms of the serial/parallel dichotomy can be of assistance. I am also not going to present the case study in all of its interesting detail, for which see Tweney (1991, in press). A bird's eye view will suffice here.

Faraday is an appropriate case in part because of the sheer volume of surviving notes, diaries, and notebooks, covering virtually his entire scientific career. Part of the reason for this resides in a strongly held conviction that Faraday derived first from the Lockean tradition of the commonplace book, which he learned about in 1809 by reading Isaac

Watts ([1741] 1809). From 1809 on, we have a variety of notebooks and diaries, which grew progressively more complex as Faraday struggled to find a storage and retrieval format that would efficiently work for him. To capsulize a much longer story, he moved from simple diarylike notebooks, to content-divided books in nonchronological order (e.g., his 1822 "chemical hints" notebook), to strictly chronological diaries for his laboratory work (begun in 1820 and continued till 1831), to strictly chronological diaries in which each entry was numbered in a strict, never varying sequence (from 1832 to the 1860s). The following table shows the progression of types:

Approx. Dates	*Notebook Type*
1809–1815	Ordinary diary, miscellaneous subjects
1816–1821	Lockean commonplace book, serial diaries
1821–1831	Nonchronological subject-specific notebooks; subject-specific laboratory diaries
1832–1860s	Strictly chronological laboratory diary and notebook, entries numbered in strict order

Each type of diary was accompanied by progressively more complex retrieval aids. At first, there is evidence that Faraday used ordinary alphabetical indexes, prepared after the notebook was complete. In the Lockean commonplace book, there is a mnemonic alphabetical index at the beginning, prepared as the notebook was kept. This works just like an alphabetical index, except that it can be kept up-to-date while the notebook is in use. In 1827, in the form of his only published book, *Chemical Manipulation*, Faraday experimented with a numbered address system. Each paragraph in the book is numbered, and cross references to other paragraph numbers are used freely, especially at the end, when Faraday provided an "inductive course" of exercises designed for self-study in chemistry. While such schemes were not invented by Faraday (see, e.g., Parkes 1822), he slowly began to use them in his own notebooks. After several tentative starts, he began in 1832 to use a strict numbering system for his laboratory diary, which began also to include nonlaboratory records, ideas to try, miscellaneous observations, and the like.

The numbered entry addresses are the key to retrieval in Faraday's mature system. The numbers provide each entry with a unique address, of course, and we can see how these were used from the huge variety of special-topic indexes, loose slips with numbered addresses, sheets that provide keys to other sheets of index numbers, and so on. He clearly did a lot of cut-and-paste work in which small slips with numbered addresses were pasted up onto larger sheets, presumably with a good bit of prior rearrangement. Faraday actively used these retrieval devices throughout his career, and there is even a surviving unfinished manuscript in which we can see that the index system was used in the composition of his papers. In effect, Faraday had something very much like a modern database system at his disposal, one that bears certain resemblances to a non-electronic HyperCard!

For the point at hand, the complete system, diary + retrieval aids, can be conceptualized as a way to bridge the gap between the serial organization of a huge long-term memory base and the need to retrieve selected aspects of the database into "parallel" classes of immediately available material. Here, the numbering of the entries provides a random-access address of enormous flexibility, though of course not, by our standards, of very great speed. Further, it would be a mistake to think that the surviving diary + retrieval aids amount to the *entire* record at Faraday's disposal; we cannot ignore the fact that a diary also is a means of accessing the diarist's own long-term memory.

The last point was brought home to me in vivid fashion by a failed "experiment" of my own. Thinking that it might be helpful to reassemble the complete text of the diary corresponding to one of the special retrieval lists, I set several undergraduate assistants to work cutting and pasting. Whenever Faraday had made a numbered reference to a paragraph in the diary, I had the students photocopy the relevant paragraph, and paste the entries, in the order they were referred to, onto larger sheets. I then tried to read the sheets, thinking that I would have a bird's eye view of an entire block of material relevant to one content area. In retrospect, perhaps it is no surprise to learn that the pasted-up sheets are almost unreadable for the simple reason that the paragraphs have been taken out of context. I was able to do much better by reading the sequence of numbers in the complete diary, flipping back and forth in search of the next number; the eye can then pick up the relevant surrounding context, reading more deeply when needed and skimming when not. For Faraday, there would be an even richer record surrounding each entry since each could then cue specific memories of the experiment itself, images associated with its running, the fact that he was mad at his assistant that day, and so forth. In short, what gets parallelized for Faraday is more than just the text of the diary itself.

My use of the term *parallel* in this context may puzzle the reader. How can we know that these are truly parallel processes? Is not the claim at the end of my last paragraph simply presuming the point? But note that the term *parallel* is relative in this context. The diary Faraday used is, strictly speaking, a serial record; to read it, you have to start at some beginning point and process the text more or less serially. But having an index that lets you move back and forth renders a large part of the text available with *near* simultaneity. The effect is much like one widely regarded as separating expert knowledge from novice knowledge. As Chi (see the essay in this volume) has argued, one of the distinguishing characteristics of the expert is the ability to "chunk" information, i.e., the ability to group what, for the novice, are disparate pieces of information into larger groups, hence reducing the load on short-term memory. Faraday's retrieval system allows him to do something similar, and this is an important sense in which he parallelizes his notes.

The point is of great importance for the "third pole" view of natural cognition, and I do not think we would be quite so prepared to acknowledge it in the absence of the tension between parallel and serial models that now characterizes cognitive science. It is also of potentially great relevance to the philosophical analysis of scientific *text*. My emphasis of the word is deliberate, to highlight the claim made so forcefully by Gooding (see the essay in this volume) that science does not proceed by text alone. Instead, each text has associated with it a memory or image or perceptual phenomenon that is not captured when we regard the text in the form of a translation into logical propositions. This too is a serial versus parallel issue, insofar as the propositional representation of text is something that can, by and large, be regarded as strictly serial. But that leaves out all of the nonserial surrounding context, be it in the environment or in the mind of the scientist (for a similar argument, see Gooding 1990).

3. ECHO in the Dark

The above does not touch on what to some will be the more interesting issue, namely, whether, natural cognition aside, formal network models are better models of scientific cognition than serial models. Two of the leading protagonists within the philosophy of science of the network approach are the Churchlands (see Paul Churchland's essay in this book) and Paul Thagard (1989). I would like to focus on the latter because, like my fruitless attempt to paste-up the Faraday diary from Faraday's index sheets, I think it is an instructive failure that illuminates some important issues.

Thagard's paper is an attempt to verify a specific model of explanation using a network instantiation. I will not address the validity of the model as such, since the philosophical adequacy of his work goes beyond my expertise. I do want to focus on the specifics of his network, however, and report a replication that we have carried out that sheds further light on just how we are going to be able to make use of formal network models.

Thagard's proposal is that explanation can be approached via a notion of "explanatory coherence" among propositions and evidence. To oversimplify, he regards two propositions (where a proposition can be either a bit of evidence or a hypothesis) as coherent if one can explain the other. He then provides seven principles or axioms that describe the coherence or incoherence of relations among a set of propositions. Some of these refer to more than just the binary or pairwise relations among propositions, for example, the Principle of System Coherence, which says that the global explanatory coherence of a set of propositions is a function of the individual local coherence of the pairwise propositions, and the Principle of Analogy, which says, in effect, that correspondences (or inconsistencies) among analogous relations result in increases (or decreases) of the strength of the pairwise coherence of the propositions. In all of these, of course, "explain" is an undefined, primitive term. Thagard is proposing that we can understand explanation by modeling the coherence of explanatory relations rather than by defining the essence of explanation directly.

To test his model, Thagard used a connectionist network called ECHO, in which propositions were represented by nodes in the network and the links between the nodes were given weights corresponding to the degree of explanatory coherence or incoherence. The weights between the links were generated by using a simple set of algorithms that instantiate the seven axioms of coherence. Once all the weights were loaded for a specific network, the network was activated for a number of cycles until a stable state was reached. Nodes could activate each other using a simple linear activation rule. In effect, the activation of each node was transmitted (via the weighted links) to each other node in the network on each cycle, and the process was repeated cycle by cycle until no further updating of node activation occurred. Thagard tried his networks with several historically motivated cases in the history of science, one based on Lavoisier's acceptance of the oxygen hypothesis over phlogiston and the other based on the arguments in Darwin's *The Origin of Species* (1859). In each case, Thagard found that the network correctly settled on the appropriate hypotheses, giving, for example, high positive activation to the specific subhypotheses of the oxygen theory and high negative activations to the subhypotheses of the phlogiston theory.

In our replication (Chitwood and Tweney 1990), we were able to use the demonstration software provided by McClelland and Rumelhart (1989) for the Jets and Sharks example. By altering the network template and matrix of weights between nodes (in accordance with the instructions given by McClelland and Rumelhart), we obtained simple networks for the Lavoisier and Darwin examples that work exactly like Thagard's. Since we were using a simpler software system than was available to Thagard, we had to make several changes. First, unlike Thagard, we incorporated two nodes in each network that corresponded to the overall theory and its alternate, oxygen versus phlogiston and evolution versus creation, respectively. Each such theory node was connected to its own explanatory hypotheses via excitatory links and to its competing theory via an inhibitory link. This simplified our network, in effect using two upper-level nodes to summarize, as it were, the overall activation of the relevant hypotheses. In addition, we were able to dispense with the differential weights among the links that Thagard had used. Because of the higher-level theory nodes, each link in our network was coded as either -1, 0, or $+1$. (We had intended to calculate the node weights directly using Thagard's rules, but found the description of them to be ambiguous in a number of key instances.) Given the changes, it is doubly remarkable that our networks behaved exactly as did Thagard's. By looking closely at what our networks did, and comparing our activation curves (i.e., the plot of activation versus cycle number) to Thagard's, we were able to see why. In effect, the outcome of the network activation in his networks and in ours depends on one — and we think *only* one — factor, namely, which theory has the larger number of explanatory links in its favor. In both of the examples tested, the historically right answer (oxygen or evolution) simply has a larger number of explained facts than the alternate answer. The networks amount to a complicated way to compute a box score! In fact, it is not even surprising that the box scores favor the right answer; in each case, Thagard based his networks on descriptions written by the winners. Both Darwin and Lavoisier described more evidence favoring their hypothesis than evidence favoring the alternate.

For the formal modeling of scientific cognition, we think the failure of Thagard's model to provide more than a simple box score is an instructive failure. One of the problems that arises inevitably in the application of new methods to a subject is the danger that we will get so carried away with the method that we will do very simple things in very complex ways and not notice that that is what we have done. Networks are deceptive in that regard; so much goes into building one that it is easy to miss the forest for all the trees (if a weak pun can be permitted). That does not mean that networks have no utility for science studies, but it does mean that their use needs to be attended with much care and with much careful attention to alternate ways of achieving the same effect.

4. Some Moral Lessons

Why should the philosophy of science care about the serial versus parallel debate within cognitive science? I would like to suggest several morals that can be drawn from the above two cases. The first has to do with the goal of constructing a naturalized epistemology (see Callebaut, in press) and the second has to do with what kind of desiderata a "consumer" ought to demand of cognitive science before accepting its wares.

On the first point, if you accept the view that at least some degree of naturalized epistemology is needed in the philosophy of science (as, for example, Giere [1989] has, as well as most of the authors in the present book), then it is important that some of the results of cognitive science be used to guide the endeavor. In the Faraday example given above, I have suggested that this and similar studies of natural cognition can go a long way toward addressing the need to incorporate nonverbal, nonlinear, and nonserial processes into our models of science.

As for the formal theories, the best and most workable ones now available are, by and large, serial at heart. They can go a long way toward capturing the regularity inherent in scientific thinking as is clearly shown by, for example, Kulkarni and Simon's (1988) successful simulation of the discovery of the Krebs Cycle in biochemistry. Parallel models hold great promise, but I do not believe that they will supplant the best of the serial models just yet, as I hope the replication of Thagard's work shows. One problem is that we have a long and difficult road ahead before the *truly* parallel aspects of thought — memory, imagery, and perception — are captured in network models. I think they will be eventually but the time is not yet. Once we do, then we will be able to use networks in formal models that generate something that really looks like a concept or an image, rather than black-boxing them as we do now or trying to simulate them in ways that are really not adequate (see, for example, the bit-maps used by Shrager [1990] to simulate images). Suppose, for example, that we could write a simulation like Kulkarni and Simon's in which the logic of experimentation is captured by production rules (as now) and the data inputs (now provided via symbols input from a keyboard) are simulated by a perceptual network that starts with features as input and provides a conceptual class as output to the serial production rules. Such a model would permit us to capture the dynamics of science in a way that is so far only hinted at.

As things stand now, the parallel models of higher processes are for the most part based on "grandmother cells," single nodes that stand for entire large concepts. This negates completely any hope of a neuroscientific tie (hence the ironic term "grandmother cell," as if a single cell represented the concept of grandmother). Perhaps even more seriously, such models fail to place the power of a parallel network at the place where it is most needed, at the interface between the serial flow of propositions and all of the nonverbal inputs to that flow. From this point of view, we are a long way from doing anything really useful with simulations like ECHO or its IAC replication.

References

Note: Getting familiar with the details of the various parallel models now so widely discussed is no mean feat, as the reader may have noticed. A number of general introductions have been written, of which Thagard (1989) is one exemplar, but there is really no substitute for hands-on experience with a network program. Thus, I highly recommend the inexpensive software package of McClelland and Rumelhart (1989), which is accompanied by an excellent manual designed to be used in conjunction with the two-volume McClelland, Rumelhart, and the PDP Research Group (1986). Beware of the fact that both the two-volume set and the later demonstration manual seem to exemplify what they are discussing; i.e., each seems to have been written in parallel fashion with no distinct beginning or end. The best approach, I believe, is to begin with the manual. Skim chapter 1, then skip to p. 24 and get the program running. Skip again to p. 38, run the "Jets and Sharks" and go back to the skipped material as needed. After this, chapters 1 and 2 in the two-volume set will make a great deal more sense. For those interested in the historical emergence of network models, the best single source is the set of introductions to the papers in Anderson and Rosenfeld (1988).

Anderson, J. A., and E. Rosenfeld, eds. 1988. *Neurocomputing: Foundations of research.* Cambridge, Mass.: MIT Press.

Callebaut, W. In press. *How to take the naturalistic turn.* Chicago: University of Chicago Press.

Chitwood, S., and R. Tweney. 1990. Explanatory coherence or interactive activation and competition? Unpublished manuscript.

Darwin, C. 1859. *The origin of species.* London: John Murray.

Ericsson, K. A., and H. A. Simon. 1984. *Protocol analysis: Verbal reports as data.* Cambridge, Mass.: MIT Press.

Faraday, M. [1822] In press. *Michael Faraday's "Chemical notes, hints, suggestions, and objects of pursuit,"* ed. R. D. Tweney and D. Gooding. London: Peregrinus Press.

———. 1827. *Chemical manipulation: Being instructions to students in chemistry.* London: Murray.

Feldman, J. A. 1985. Connectionist models and their applications: Introduction. *Cognitive Science* 6:205–54.

Giere, R. N. 1989. *Explaining science: A cognitive approach.* Chicago: University of Chicago Press.

Gooding, D. 1990. *Experiment and the making of meaning: Human agency in scientific observation and experiment.* Dordrecht: Kluwer Academic Publishers.

Gooding, D., and R. Tweney. In preparation. Mathematical thinking about experimental matters: Faraday as a mathematical philosopher. Unpublished manuscript.

Hebb, D. O. 1949. *The organization of behavior.* New York: Wiley.

James, W. 1890. *The principles of psychology.* New York: Henry Holt and Company.

Kuhn, T. 1962. *The structure of scientific revolutions.* Chicago: University of Chicago Press.

Kulkarni, D., and H. Simon. 1988. The processes of scientific discovery: The strategy of experimentation. *Cognitive Science* 12:139–76.

McClelland, J. L., and D. E. Rumelhart. 1981. An interactive activation model of context effects in letter perception. Part 1: An account of basic findings. *Psychological Review* 88:375–407.

———. 1989. *Explorations in parallel distributed processing: A handbook of models, programs, and exercises.* Cambridge, Mass.: MIT Press.

McClelland, J. L., D. E. Rumelhart, and the PDP Research Group, eds. 1986. *Parallel distributed processing: Explorations in the microstructure of cognition.* Vol. 2. Cambridge, Mass.: MIT Press.

Marr, D., and T. Poggio. 1976. Cooperative computation of stereo disparity. *Science* 194:283–87.

Neisser, U. 1976. *Cognition and reality.* San Francisco: W. H. Freeman.

Neisser, U., ed. 1982. *Memory observed: Remembering in natural contexts.* San Francisco: W. H. Freeman.

Newell, A., and H. Simon. 1972. *Human problem solving.* Englewood Cliffs, N.J.: Prentice-Hall.

Parkes, S. 1822. *The chemical catechism.* 10th ed. London: Baldwin, Craddock, and Joy.

Pinker, S., and A. Prince. 1988. On language and connectionism: Analysis of a parallel distributed processing model of language acquisition. *Cognition* 28:73–193.

Rumelhart, D. E., J. L. McClelland, and the PDP Research Group, eds. 1986. *Parallel distributed processing: Explorations in the microstructure of cognition.* Vol. 1. Cambridge, Mass.: MIT Press.

Shrager, J. 1990. Commonsense perception and the psychology of theory formation. In *Computational models of scientific discovery and theory formation,* ed. J. Shrager and P. Langley, pp. 437–70. San Francisco: Morgan Kaufmann Publishers.

Thagard, P. 1989. Explanatory coherence. *Behavioral and Brain Sciences* 12:435–502.

Tweney, R. D. 1990. Five questions for computationalists. In *Computational models of scientific discovery and theory formation,* ed. J. Shrager and P. Langley, pp. 471–84. San Francisco: Morgan Kaufmann Publishers.

———. 1991, in press. Faraday's notebooks: The active organization of creative science. *Physics Education* 26.

Watts, I. [1741] 1809. *The improvement of the mind.* London: Maxwell and Skinner.

The Origin and Evolution
of Everyday Concepts

The contributors to this volume were charged to explore how research in cognitive science bears on issues discussed in the literature on the philosophy of science. Most took this as a challenge to show how results from cognitive psychology or artificial intelligence inform theories of the processes of theory development and theory choice. I focus on a different issue — the origin of scientific concepts.

Let me begin by settling some terminological matters. By *concept*, *belief*, and *theory*, I refer to aspects of mental representations. Concepts are units of mental representation, roughly the grain of single lexical items, such as *object, matter, weight*. Beliefs are mentally represented propositions taken by the believer to be true, such as *Air is not made of matter*. Concepts are the constituents of beliefs; that is, propositions are represented by structures of concepts. Theories are complex mental structures consisting of a mentally represented domain of phenomena and explanatory principles that account for them.

The theories of the origins of concepts I discuss in this chapter fall on the nativist side of the nativist/empiricist debate. But even on the same side of the debate, there is room for very different positions. For example, Fodor (1975) claims that all lexical concepts are innate. Others, whom we may call "mental chemists," hold that all concepts arise by combination from a small set of innate primitives (e.g., Jackendoff 1989; Wierzbicka 1980). Unlike classical empiricists, the modern mental chemists do not hold that the initial vocabulary consists of perceptual primitives alone. Rather, they posit such abstract concepts as *object, want,* and *good*. Wierzbicka (1980) stands at the opposite extreme from Fodor; she argues that all lexical concepts can be derived by combinations of twenty-three universal, innate primitives.

A variety of evidence supports the nativist position. Take the concept of *object* as a case in point. By objects, I mean bounded, coherent wholes that endure through time and move on spatio-temporally continuous paths. Two extremely convincing lines of argument show this concept to be largely innate. The first is direct empirical evidence demonstrating it in infants as young as two to four months. The second derives from learnability considerations. If one wants to argue that two-month-olds have constructed the concept of an object, one must show, in principle, how they could have done so. From what primitives, and on what evidence? Not for lack of trying, nobody has ever shown how this concept could be formed out of some prior set of primitives. What would lead an organism existing in a Quinean perceptual quality space, sensitive only to similarity among portions of experience (Quine 1960), to see the world in terms of enduring objects?

The infant experiments depend upon babies' differential interest in physically impossible events, compared to physically possible ones. I refer you to recent reviews by Spelke (1991) and Baillargeon (1990) for detailed discussion of the methodology, and to an elegant paper by Leslie (1988) for an argument that these phenomena reflect violations of expectancies at a conceptual level. Here I will give just the merest sketch from this literature. Let us consider one experiment that shows that four-month-olds expect objects to move on spatio-temporally continuous paths. Babies are shown two screens, side by side. In one condition they watch a ball go back and forth behind two screens. The ball moves behind one screen, to the next and behind the second, out the other side, and back again. They watch until they get bored. Adults see this as one object passing back and forth behind the two screens. In a second condition another group of babies watch an object go behind the first screen from the left; then nothing happens; then an object comes out from behind the second to the right and returns behind it; then nothing happens; then an object comes out from behind the first to the left; etc. The event the second group of babies watch is identical to that of the first group, except that they see no ball in the space between the screens. They too watch until they get bored. Adults, of course, see this event as involving two numerically distinct balls. After the babies are bored, the screens are removed, and either one ball or two balls are revealed. Those in the first group maintain their boredom if presented with one ball, but stare at two balls; those in the second group maintain their boredom if presented with two balls, but stare at one ball. That is, four-month-old babies see these events just as do adults. Four-month-olds track numerical identity, and know that a single object cannot move from one place to another without passing through the intervening space (Spelke 1988).

For the purposes of the present essay, I will take it that the existence of rich innate concepts is not in doubt. While the most detailed work concerns infants' concept of an object, there is also substantial evidence for innate concepts of causality (e.g., Leslie 1988), number (e.g., Baillargeon, Miller, and Constantino, in press), and person (e.g., Mandler 1988). Spelke (1991) defends a stronger thesis: the initial representations of physical objects that guide *infants'* object perception and *infants'* reasoning about objects remain the core of the *adult* conception of objects. Spelke's thesis is stronger because the existence of innate representations need not preclude subsequent change or replacement of these beginning points of development. Her argument involves demonstrating that the principles at the core of the infants' concept of objects, spatio-temporal continuity (see above), and solidity (one object cannot pass through the space occupied by another; Baillargeon, Spelke, and Wasserman 1985; Spelke 1991) are central to the adult concept of objects as well.

I do not (at least not yet) challenge Spelke's claim concerning the continuity of our concept of physical objects throughout human development. However, Spelke implies that the history of the concept of an object is typical of all concepts that are part of intuitive adult physical reasoning. Further, she states that in at least one crucial respect, the acquisition of common-sense physical knowledge differs from the acquisition of scientific knowledge: the development of *scientific* knowledge involves radical conceptual change. Intuitive concepts, in contrast, are constrained by innate principles that determine the entities of the mentally represented world, thus determining the entities about which we learn, leading to entrenchment of the initial concepts and principles. She suggests that going beyond these initial concepts requires the metaconceptually aware theory building of mature scientists. To the degree that Spelke is correct, normal cognitive development would involve minimal conceptual change and no major conceptual reorganizations.

Spelke's claim is implausible, on the widely held assumption of the continuity of science with common-sense explanation (see the essays in the present volume). Of course, Spelke rejects the continuity assumption. In this chapter I deny Spelke's conjecture that ordinary, intuitive, cognitive development consists only of enrichment of innate structural principles. The alternative that I favor is that conceptual change occurs during normal cognitive growth. In keeping with current theorizing in cognitive psychology, I take concepts to be *structured* mental representations (see E. Smith [1989] for a review). A theory of human concepts must explain many things, including concepts' referential and inferential roles. Concepts may differ along many dimensions, and no doubt there are many degrees of conceptual difference within each dimension. Some examples of how concepts change in the course of knowledge acquisition follow:

1. What is periphery becomes core, and vice versa (see Kitcher 1988). For example, what is originally seen to be the most fundamental property of an entity is realized to follow from even more fundamental properties. Example: In understanding reproduction, the child comes to see that smallness and helplessness are derivative properties of babies, rather than the essential properties (Carey 1985b, 1988).

2. Concepts are subsumed into newly created ontological categories, or reassigned to new branches of the ontological hierarchy. Example: Two classes of celestial bodies — stars and planets/moons — come to be conceptualized, with the sun and the earth as examples, respectively (Vosniadou and Brewer, in press).

3. Concepts are embedded in locally incommensurable theories. Example: the concepts of the phlogiston and the oxygen theories of burning (Kuhn 1982).

Knowledge acquisition involving all three sorts of conceptual change contrasts with knowledge acquisition involving only enrichment. Enrichment consists in forming new beliefs stated over concepts already available. Enrichment: New knowledge about entities is acquired, new beliefs represented. This knowledge then helps pick out entities in the world and provides structure to the known properties of the entities. Example: The child acquires the belief "unsupported objects fall" (Spelke 1991). This new belief influences decisions about object boundaries.

According to Spelke and Fodor, physical concepts undergo only enrichment in the course of knowledge acquisition during childhood. According to the mental chemists, new concepts may come into being, but only through definition in terms of already existing ones. In this chapter I explore the possibility of conceptual change of the most extreme sort. I suggest that in some cases the child's physical concepts may be incommensurable with that of the adult's, in Kuhn's (1982) sense of local incommensurability. It is to the notion of local incommensurability that I now turn.[1]

1. Local Incommensurability

1.1. Mismatch of Referential Potential

A good place to start is with Philip Kitcher's analysis of local incommensurability (Kitcher 1988). Kitcher outlines (and endorses) Kuhn's thesis that there are episodes in the history of science at the beginnings and ends of which practitioners of the same field of endeavor speak languages that are not mutually translatable. That is, the beliefs, laws, and explanations statable in the terminology at the beginning, in language 1 (L1), cannot be expressed in the terminology at the end, in language 2 (L2). As he explicates Kuhn's thesis, Kitcher focuses on the referential potential of terms. He points out that there are multiple methods for fixing the reference of any given term: definitions, descriptions, theory-relative similarity to particular exemplars, and so on. Each theory presupposes that for each term, its multiple methods of reference fixing pick out a single referent. Incommensurability arises when an L1 set of methods of reference fixing for some term is seen by L2 to pick out two or more distinct entities. In the most extreme cases, the perspective of L2 dictates that some of L1's methods fail to provide any referent for the term at all, whereas others provide different referents from each other. For example, the definition of "phlogiston" as "the principle given off during combustion" fails, by our lights, to provide any referent for "phlogiston" at all. However, as Kitcher point out, in other uses of "phlogiston," where reference is fixed by the description of the production of some chemical, it is perfectly possible for us to understand what chemicals are being talked about. In various descriptions of how to produce "dephlogisticated air," the referent of the phrase can be identified as either oxygen or oxygen enriched air.

Kitcher produces a hypothetical conversation between Priestley and Cavendish designed to show that even contemporaries who speak incommensurable languages can communicate. Kitcher argues that communication is possible between two parties if one can figure out what the other is referring to and if the two share *some* language. Even in cases of language change between L1 and L2, the methods of reference fixing for many terms that appear in both languages remain entirely constant. Further, even for the terms for which there is mismatch, there is still some overlap, so that in many contexts the terms will refer to the same entities. Also, agreement on reference is possible because the two speakers can learn each other's language, including mastering the other's methods of reference fixing.

The problem with Kitcher's argument is that it identifies communication with agreement on the referents of terms. But communication requires more than agreement on referents; it requires agreement on what is said about the referents. The problem of incommensurability goes beyond mismatch of referential potential.

1.2. Beyond Reference

If speakers of putatively incommensurable languages can, in some circumstances, understand each other, and if we can, for analogous reasons, understand texts written in a language putatively incommensurable with our own, why do we want to say that the two languages are incommensurable? In answering this question, Kuhn moves beyond the referential function of language. To figure out what a text is referring to is not the same as to provide a translation for the text. In a translation, we replace sentences in L1 with sentences in L2 that have the same meaning. Even if expressions in L1 can be replaced with co-referential expressions in L2, we are not guaranteed a translation. To use Frege's example: replacing "the morning star" with "the evening star" would preserve reference but would change the meaning of a text. In cases of incommensurability, this process will typically replace an L1 term with one L2 term in some contexts and other L2 terms in other contexts. But it matters to the meaning of the L1 text that a single L1 term was used. For example, it mattered to Priestley that all of the cases of dephlogisticated entities were so designated; his language expressed a theory in which all dephlogisticated substances shared an essential property that explained derivative properties. The process of replacing some uses of "dephlogisticated air" with "oxygen," others with "oxygen enriched," and still others with other phrases, yields what Kuhn calls a disjointed text. One can see no reason that these sentences are juxtaposed. A good translation not only preserves reference; a text makes sense in L1, and a good translation of it into L2 will make sense in L2.

That the history of science is possible is often offered as prima facie refutation of the doctrine of incommensurability. If earlier theories are expressed in languages incommensurable with our own, the argument goes, how can the historian understand those theories, and describe them to us so that we understand them? Part of the answer to this challenge has already been sketched above. While parts of L1 and L2 are incommensurable, much stays the same, enabling speakers of the two languages to figure out what the other must be saying. What one does in this process is not *translation*, but rather *interpretation* and *language learning*. Like the anthropologist, the historian of science interprets, and does not merely translate. Once the historian has learned L1, he or she can teach it to us, and then we can express the earlier theory as well.

On Kuhn's view, incommensurability arises because a language community learns a whole set of terms together, which together describe natural phenomena and express theories. Across different languages, these sets of terms can, and often do, cut up the world in incompatible ways. To continue with the phlogiston theory example, one reason that we cannot express claims about phlogiston in our language is that we do not share the phlogiston theory's concepts *principle* and *element*. The phlogiston theory's *element* encompassed many things we do not consider elements, and modern chemistry has no concept at all that corresponds to phlogiston theory's *principle*. But we cannot express the phlogiston theory's understanding of combustion, acids, airs, etc., without using the concepts *principle, element,* and *phlogiston,* for the concepts of combustion, acids, and airs are intertwined with the concepts of elements and principles (among others). We cannot translate sentences containing "phlogiston" into pure twentieth-century language, because when it comes to using words like "principle" and "element" we are forced to choose one of two options, neither of which leads to a real translation:

1. We use "principle" and "element," but provide a translator's gloss before the text. Rather than providing a translation, we are changing L2 for the purposes of rendering the text. The translator's gloss is the method for teaching L1 to the speakers of L2.

2. We replace each of these terms with different terms and phrases in different contexts, preserving reference but producing a disjointed text. Such a text is not a translation, because it does not make sense as a whole.

1.3. Conceptual Differentiation

As is clear from the above, incommensurability involves change at the level of individual concepts in the transition from one language to the other. There are several types of conceptual change, including:

1. Differentiations, as in Galileo's drawing the distinction between *average velocity* and *instantaneous velocity* (see Kuhn 1977).

2. Coalescences, as when Galileo saw that Aristotle's distinction between *natural* and *violent* motion was a distinction without a difference, and collapsed the two into a single notion.

3. Simple properties being reanalyzed as relations, as when Newton reanalyzed the concept *weight* as a relation between the earth and the object whose weight is in question.

Characterizing change at the level of individual concepts is no simple matter. We face problems both of analysis and evidence. To explore these problems, take just one type of conceptual change — conceptual differentiation. Developmental psychologists often appeal to differentiation when characterizing conceptual change, but not all cases in which distinctions undrawn come to be drawn imply incommensurability. The two-year-old may not distinguish collies, German shepherds, and poodles, and therefore may have an undifferentiated concept *dog* relative to adults, but the concept *dog* could well play roughly the same role in both the two-year-old's and the adult's conceptual system. The cases of differentiation involving incommensurability are those in which the undifferentiated parent concept from L1 is incoherent from the point of view of L2.

Consider McKie and Heathecoate's (1935) claim that before Black, *heat* and *temperature* were not differentiated. This would require that thermal theories before Black represented a single concept fusing our concepts *heat* and *temperature*. Note that in the language of our current theories, there is no superordinate term that encompasses both of these meanings — indeed any attempt to wrap heat and temperature together could produce a monster. Heat and temperature are two entirely different types of physical magnitudes; heat is an extensive quantity, while temperature is an intensive quantity. Extensive quantities, such as the amount of heat in a body (e.g., one cup of water), are additive — the total amount of heat in two cups of water is the sum of that in both. Intensive quantities are ratios and therefore not additive — if one cup of water at 80° F is added to one cup at 100° F, the resultant temperature is 90° F, not 180° F. Furthermore, *heat* and *temperature* are interdefined — e.g., a calorie is the amount of heat required to raise the temperature of one gram of water 1° C. Finally, the two play completely different roles in explaining physical phenomena such as heat flow. Every theory since Black includes a commitment to thermal equilibrium, which is the principle that temperature differences are the occasion of heat flow. This commitment cannot be expressed without distinct concepts of *heat* and *temperature*.

To make sense of McKie and Heathecoate's claim, then, we must be able to conceive how it might be possible for there to be a single undifferentiated concept fusing *heat* and *temperature* and we must understand what evidence would support the claim. Often purely linguistic evidence is offered: L1 contains only one term, where L2 contains two. However, more than one representational state of affairs could underlie any case of undifferentiated language. Lack of differentiation between *heat* and *temperature* is surely representationally different from mere absence of the concept *heat*, even though languages expressing either set of thermal concepts might have only one word, e.g., "hot." A second representational state that might mimic nondifferentiation is the false belief that two quantities are perfectly correlated. For example, before Black's discoveries of specific and latent heat, scientists might have believed that adding a fixed amount of heat to a fixed quantity of matter always leads to the same increase in temperature. Such a belief could lead scientists to use one quantity as a rough and ready stand-in for the other, which might produce texts that would suggest that the two were undifferentiated.

The only way to distinguish these two alternative representational states of affairs (false belief in perfect correlation, absence of one or the other concept) from conceptual nondifferentiation is to analyze the role the concepts played in the theories in which they were embedded. Wiser and Carey (1983) analyzed the concept *heat* in the thermal theory of the seventeenth-century Academy of Florence, the first group to systematically study thermal phenomena. We found evidence supporting McKie and Heathecoate's claim of nondifferentiation. The academy's *heat* had both causal strength (the greater the degree of heat, the more ice would be melted, for example) and qualitative intensity (the greater the degree of heat, the hotter an entity would feel) — that is, aspects of both modern *heat* and modern *temperature*. The Experimenters (their own self-designation) did not separately quantify heat and temperature, and unlike Black, did not seek to study the relations between the two. Furthermore, they *did* relate a single thermal variable, *degree of heat*, to mechanical phenomena, which by analyzing contexts we now see sometimes referred to temperature and sometimes to amount of heat. You may think of this thermal variable, as they did, as the *strength* of the heat, and relate it to the magnitude of the physical effects of heat. The Experimenters used thermometers to measure degree of heat, but they did so by noting the rate of change of level in the thermometer, the interval of change, and only rarely the final level attained by the alcohol in their thermometers (which were not calibrated to fixed points such as the freezing and boiling points of water). That is, they did not quantify either temperature or amount of heat, and certainly did not attempt to relate two distinct thermal variables. Finally, their theory provided a different account of heat exchange from that of the caloric theory or of modern thermodynamics. The Experimenters did not formulate the principle of thermal equilibrium; their account needed no distinct concepts of heat and temperature. For all these reasons, we can be confident in ascribing a single, undifferentiated concept that conflated *heat* and *temperature* to these seventeenth-century scientists. No such concept as the Experimenters' *degree of heat* plays any role in any theory after Black.

The Experimenters' concept, incoherent from our point of view, led them into contradictions that they recognized but could not resolve. For example, they noted that a chemical reaction contained in a metal box produced a degree of heat insufficient to melt paraffin, while putting a solid metal block of the same size on a fire induced a degree of heat in the block sufficient to melt paraffin. That is, the *latter* (the block) had a greater degree of heat. However, they also noted that if one put the box with the chemical reaction in ice water, it melted more ice than did the heated metal block, so the *former* (the box) had a greater degree of heat. While they recognized this as a contradiction, they threw up their hands at it. They could not resolve it without differentiating temperature from amount of heat. The chemical reaction generates more heat but attains a lower temperature than does the block; the melting point of paraffin is a function of temperature, whereas how much ice melts is a function of amount of heat generated.

1.4. Summary

When we ask whether the language of children (L1) and the conceptual system it expresses (C1) might sometimes be incommensurable with the language (L2) and conceptual system (C2) of adults, where C1 and C2 encompass the same domain of nature, we are asking whether there is a set of concepts at the core of C1 that cannot be expressed in terms of C2, and vice versa. We are asking whether L1 can be translated into L2 without a translator's gloss. Incommensurability arises when there are simultaneous differentiations or coalescences between C1 and C2, such that the undifferentiated concepts of C1 can no longer play any role in C2 and the coalesced concepts of C2 can play no role in C1.

2. Five Reasons to Doubt Incommensurability between Children and Adults

I have encountered five reasons to doubt that children's conceptual systems are incommensurable with adults':

1. Adults communicate with young children just fine.

2. Psychologists who study cognitive development depict children's conceptions in the adult language.

3. Where is the body? Granted, children cannot express all of the adult conceptual system in their language, but this is because L1 is a subset of L2, not because the two are incommensurable. Incommensurability requires that L2 not be able to express L1 as well as L1 not being able to express L2. Just as we cannot define "phlogiston" in our language, so holders of the phlogiston theory could not define "oxygen" in theirs. Where do children's conceptual systems provide any phenomena like those of the phlogiston theory? Where is a preschool child's "phlogiston" or "principle"?

4. There is no way incommensurability could arise (empiricist version). Children learn their language from the adult culture. How

could children establish sets of terms interrelated differently from adult interrelations?

5. There is no way incommensurability could arise (nativist version). Intuitive conceptions are constrained by innate principles that determine the objects of cognition and that become entrenched in the course of further learning.

Those who offer one or more of the above objections share the intuition that while the young child's conceptual system may not be able to express all that the adult's can, the adult can express the child's ideas, can translate the child's language into adult terms. Cognitive development, on this view, consists of enrichment of the child's conceptual system until it matches that of the adult.

2.1. Adults and Young Children Communicate

The answer to this objection should by now be familiar. Incommensurability does not require complete lack of communication. After all, the early oxygen theorists argued with the phlogiston theorists, who were often their colleagues or teachers. Locally incommensurable languages can share many terms that have the same meaning in both languages. This common ground can be used to fix referents for particular uses of nonshared terms, e.g., a use of "dephlogisticated air" to refer to oxygen enriched air. With much common language the two sides can have genuine disagreements about the nature of dephlogisticated air. Anyway, it is an empirical question just how well adults understand preschool children.

2.2. Developmental Psychologists Must Express Children's Beliefs in the Adult Language; Otherwise, How Is the Study of Cognitive Development Possible?

I discussed earlier how it is possible for the historian of science to express in today's language an earlier theory that was expressed in an incommensurable language. We understand the phlogiston theory, to the extent that we do, by *interpreting* the distinctive conceptual machinery and enriching our own language. To the extent that the child's language is incommensurable with the adult's, psychologists do not express the child's beliefs directly in the adult language. Rather, they interpret the child's language, learn it, and teach it to other adults. This is facilitated by the considerable overlap between the two, enabling the psychologist, like the historian, to be interpreter and language learner.

2.3. Where Is the Body?

As mentioned above, those who raise these objections believe that the child's concepts are a subset of the adult's; the child cannot express all adult concepts, but the adult can express all the child's. The body we seek, then, is a child's concept that cannot be expressed in the adult's language.

There are two cases of the subset relation that must be distinguished. If concept acquisition solely involves constructing new concepts out of existing ones, then the child's concepts will be a subset of the adult's and no incommensurability will be involved. However, in some cases in which one conceptual system is a subset of another, *one-way* incommensurability obtains. For example, Newtonian mechanics is a subset of the physics of Maxwell. Maxwell recognized forces Newton did not, but Maxwell did not reconceptualize mechanical phenomena. That is, Maxwell's physics could express Newton's. The reverse is not so. It is not possible to define electromagnetic concepts in terms of Newtonian concepts.

While I would certainly expect that there are cases of one-way incommensurability, full two-way incommensurability is the focus of the present analysis. In the most convincing cases of incommensurability from the history of science, some of the concepts of C1, such as "phlogiston" and "principle," have no descendants at all in C2. The body we seek is a case in which C1 contains concepts absent from C2, concepts that cannot be defined in C2. Note that *concepts* are at issue, not terms. Since children learn language from adults, we would not expect them to invent terms like "phlogiston" or "principle" that do not appear in the adult lexicon. However, two-way incommensurability does not require *terms* in L1 with no descendants in L2. Newtonian mechanics is incommensurable with Einsteinian mechanics, but Newton's system contains no bodies in this sense. Similarly, though the Florentine Experimenters' source-recipient theory of thermal phenomena is incommensurable with our thermal theory, there is no Florentine analog of "phlogiston." Their "degree of heat" is the ancestor of our "temperature" and "heat." In these cases, incommensurability arises from sets of core concepts being interrelated in different ways, and from several simultaneous differentiations and coalescences. Thus, while there may be no bodies such as "phlogiston" or "principle" in the child's language, it remains an open empirical question whether cases of two-way incommensurable conceptual systems between children and adults are to be found.

2.4. How Would Incommensurability Arise? Empiricist Version

The child learns language from adults; the language being spoken to the child is L2; why would the child construct an L1 incommensurable with L2? This is an empiricist objection to the possibility of incommensurability because it views the child as a blank slate, acquiring the adult language in an unproblematic manner. But although children learn language from adults, they are not blank slates as regards their conceptual system. As they learn the terms of their language, they must map these onto the concepts they have available to them. Their conceptual system provides the hypotheses they may entertain as to possible word meanings. Thus, the language they actually construct is constrained both by the language they are hearing and the conceptualization of the world they have already constructed. Incommensurability could arise when this conceptualization is incommensurable with the C2 that L2 expresses.

Presumably, there are no phlogiston-type bodies in L1 because the child learns language from adults. The child learning chemistry and the explanation for combustion would never learn words like "principle" or "phlogiston." However, it is an open empirical question whether the child assigns meanings to terms learned from adult language that are incommensurable with those of the adult.

2.5. How Would Incommensurability Arise? Nativist Version

Empiricists question why the child, learning L2 from adults, might ever construct an incommensurable L1. Nativists worry how the developing mind, constrained by innate principles and concepts, would ever construct an L2 incommensurable with L1. This is Spelke's challenge cited in the opening of the present essay. Spelke does not deny the phenomenon of conceptual change in the history of science. That is, Spelke grants that innate constraints do not preclude the shift from the phlogiston theory to the oxygen theory; nor does she deny that this shift involves incommensurable concepts. Innate constraints do not preclude incommensurability *unless* children are different from scientists. Thus, Spelke's nativist objection requires the noncontinuity position, which is why she speculates that conceptual change requires mature scientists' explicit scrutiny of their concepts, and their striving for consistency. Of course, merely positing noncontinuity begs the question.

In considering these speculations, we must remember that the child develops his or her conceptual system in collaboration with the adult culture. Important sources of information include the language of adults, the problems adults find worthy and solvable, and so on. This is most obvious in the case of explicit instruction in school, especially in math and science, but it is no less true of the common-sense theories of the social, biological, and physical worlds constructed by cultures. Not all common-sense knowledge of the physical, social, and biological worlds develops rapidly and effortlessly. One source of difficulty may be incommensurability between the child's conceptual system and that the culture has constructed. Again, it is an open empirical issue whether common-sense conceptual development is continuous with scientific conceptual development in the sense of implicating incommensurability.

In this section I have countered five arguments that we should not expect incommensurability between young children's and adults' conceptual systems. Of course, I have not shown that local incommensurability actually ever obtains. That is the task of the next section.

3. The Evidence

I have carried out case studies of children's conceptualization of two domains of nature, and in both cases some of the child's concepts are incommensurable with the adult's. One domain encompasses the child's concepts *animal, plant, alive, person, death, growth, baby, eat, breathe, sleep*, etc. (Carey 1985b, 1988). The other encompasses the child's concepts *matter, material kind, weight, density*, etc. (C. Smith, Carey, and Wiser 1985; Carey et al., in preparation; see also Piaget and Inhelder 1941). Here I will draw my examples from the latter case, for it includes physical concepts and thus bears more directly on Spelke's conjecture that common-sense physical concepts develop only through enrichment.

The central phenomenon suggesting developmental cases of incommensurability is the same as that suggesting historical cases as well. The child makes assertions that are inexplicable to the adult — for example, that a particular piece of styrofoam is weightless or that the weight of an object changes when the object is turned on its side. Of course, such assertions do not in themselves demonstrate incommensurability. They raise three possibilities as to the relations between the child's conceptual system and the adult's:

1. The child is expressing false beliefs represented in terms of the same concept of weight as the adult's.

2. The child is expressing beliefs in terms of a different concept of weight from the adult's, but the child's concept is definable in the adult vocabulary.

3. The child is expressing beliefs in terms of a different concept of weight from the adult's; the child's and adult's concepts are incommensurable.

The only way to decide among these three alternatives is to analyze the child's and the adult's concepts of weight in the context of related concepts and the intuitive theories in which they are embedded.

Spelke's work on infants' conceptions of objects tells us that from the earliest moment at which these conceptions have been probed, children represent objects as solid, in the sense that no part of one object can pass through the space occupied by any part of another (see Spelke 1991). Work by Estes, Wellman, and Woolley (1989) shows that three-year-olds draw a distinction between real physical objects, such as a real cookie, and mentally represented objects, such as an image of a cookie or a dream of a cookie. These very young children know that only the former can be seen and touched by both the child and other people, and only the latter can be changed by thought alone. The young child distinguishes physical objects from other entities in terms of properties that are at least precursors to those adults use in drawing the distinction between material and immaterial entities. We shall see, however, that the child does not draw a material/immaterial distinction on the same basis as does the adult. Furthermore, the child's conceptual system represents several concepts undifferentiated relative to the adult's, and the differentiations are of the type that implicate incommensurability, that is, are like the *heat/temperature* case rather than the *poodle/collie* case. One example is the undifferentiated concept *weight/density*. Like the concept *heat/temperature* before Black, an undifferentiated *weight/density* concept does not remain a useful superordinate concept in the conceptual systems of those who have drawn the distinction.[2] Like heat and temperature, weight and density are different sorts of physical magnitudes; weight is an extensive quantity and density an intensive quantity, and the two are interdefined. A single concept undifferentiated between the two is incoherent from the later point of view.

4. Weight, Density, Matter, Material Kind

4.1. Undifferentiated Concept: Weight/Density

We require evidence in two steps to support the claim that weight and density are not differentiated by young children. First, to rule out the possibility that young children simply lack the concept *density*, we must show that heaviness relativized to size plays some role in their judgments. Indeed, C. Smith, Carey, and Wiser (1985) found that many young children (three- to five-year-olds) appeared to lack the concept of density at all. Older children, in contrast, relativized weight to size in some of their judgments of heaviness. Second, once we have shown that *density* is not entirely absent, we must show that the child does not relate density to some physical phenomena and weight to others but rather accounts for all heaviness-related phenomena in terms of an undifferentiated weight/density concept. Of course, one can never establish this beyond doubt; it is always possible that tomorrow somebody will find some limited contexts in which the child has systematically distinguished the two. But C. Smith, Carey, and Wiser (1985) devised a series of tasks, both verbal and nonverbal, which probed for the distinction in the simplest ways we could think of. For example, we presented children with pairs of objects made of different metals, and asked, "Which is heavier?" or "Which is made of the heavier kind of metal?" Nonverbal versions of the same task involved the child predicting which objects would make a sponge bridge collapse (weight the relevant factor) and sorting objects into steel and aluminum families (density the relevant factor). In the steel and aluminum family task, for example, the child was first shown several pairs of identically sized cylinders, and it was pointed out that steel is a much heavier kind of stuff than is aluminum. Children with an undifferentiated concept showed intrusion of absolute weight on judgments we would base on density; in this case this meant sorting large aluminum cylinders into the steel family, because they were heavy.

C. Smith et al. (1988) corroborated these results with other simple tasks. They provided children with scales and with sets of objects that varied in volume, weight, and material kind, and asked them to order the objects by size, by absolute weight, and by density (explained in terms of heaviness of the kind of stuff). The ordering required no calculations of density; for instance, if one object is larger than another, but they weigh the same or the smaller is heavier, we can infer without calculation that the smaller is denser. Prior to instruction, few children as old as age twelve are able to correctly order the same set of items differently on the basis of absolute weight and density. Mistakes reveal intrusions of weight into the density orderings, and vice versa. These results are underscored when children are asked to depict in a visual model the size, weights, and densities of a set of such objects. Only children who show in other tasks that they have at least partially differentiated weight and density produce models that depict, in some way or another, all three physical magnitudes.

Just as the Experimenters' undifferentiated *heat/temperature* concept led them into contradictions, children's *weight/density* concept leads them into outright contradiction. C. Smith, Carey, and Wiser (1985) presented children in this conceptual state with two bricks, one of steel and one of aluminum. Though the steel brick was smaller, the two weighed the same and children were shown that they balanced exactly on a scale. Children were probed: "How come these weigh the same, since one is so much bigger?" They answered, "Because that one [the one made of steel] is made of a heavier kind of stuff," or "Because steel is heavier," or some equivalent response. They were then shown two bricks of steel and aluminum, now both the same size as each other, and asked to predict whether they would balance, or whether one would be heavier than the other. Now they answered that they would weigh the same, "Because the steel and aluminum weighed the same before" (Figure 1).

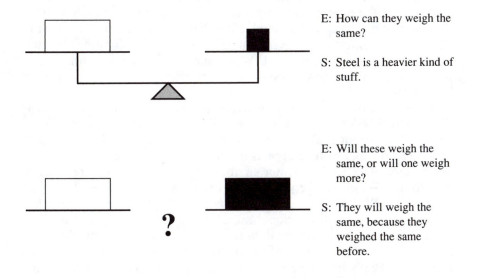

E: How can they weigh the same?

S: Steel is a heavier kind of stuff.

E: Will these weigh the same, or will one weigh more?

S: They will weigh the same, because they weighed the same before.

Figure 1. Concrete thought experiment.

Children give this pattern of responses because they do not realize that the claim that a given steel object weighs the same as a given aluminum object is not the same as that steel and aluminum weigh the same, even though they also understand that if a small steel object weighs the same as a large aluminum one, this is possible because steel is heavier than aluminum. It is not that children are unmoved by the contradiction in these assertions. They can be shown the contradiction, and since they, as well as adults, strive for consistency, they are upset by it. Drawing out contradictions inherent in current concepts is one of the functions of thought experiments (see Kuhn 1977; Nersessian, this volume). Here we have produced a concrete instantiation of a thought experiment for the child. Just as the Experimenters were unable to resolve the contradictions due to their undifferentiated *heat/temperature* concept, so too children cannot resolve the contradictions due to their undifferentiated *weight/density* concept.

4.2. How an Undifferentiated Weight/Density Concept Functions

The previous section outlined some of the evidence that six- to twelve-year-old children have a concept undifferentiated between weight and density. But how could such a concept function in any conceptual system, given the contradictions it leads the child into? The short answer is that the contexts in which the child deploys his or her weight/density concept do not, in general, elicit these contradictions. This is the same answer as for the Experimenters' *degree of heat* (undifferentiated between heat and temperature) (Wiser and Carey 1983), or for Aristotle's *speed* (undifferentiated between average and instantaneous velocity [Kuhn 1977]).

A sketch of the purposes for which children *do* use their concept provides a slightly longer answer. Like the Experimenters' *degree of heat*, the child's concept is *degree of heaviness*. Children appeal to heaviness of objects to explain some aspects of those objects' effects on themselves or on other objects. The greater an object's heaviness, the more difficult it is to lift, the more likely to hurt if dropped on one's toes, the more likely to break something else if dropped on it, and so on. Notice that "heavy," like other dimensional adjectives such as "big," is a relative term. Something is heavy relative to some standard, and the child can switch fluidly from one way of relativizing heaviness to another. An object can be heavy for objects of that type (e.g., a heavy book), heavy for the objects on the table, heavy for me but not my mother, and heavy for objects of that size. For the child with an undifferentiated weight/density concept, relativizing heaviness to a standard determined by size is no different from other ways of relativizing heaviness. Children differentiate *weight* and *density* as they realize that relativizing weight to size produces an independent physical magnitude, one related in systematic ways to distinct phenomena in the world.

The full answer to how children can have an undifferentiated weight/density concept that functions effectively within their conceptual system will require a description of their conceptual system. The claim that weight and density are not differentiated does not exhaust the differences between the child's concept and the adult's; indeed, it could not. Since an undifferentiated weight/density concept is incoherent from the adult's point of view, it must be embedded in a very different conceptual system to function coherently in the child's. We should expect, therefore, that the child's concept of heaviness differs from the adult's in many ways, beyond its being undifferentiated between weight and density.

4.3. The Material/Immaterial Distinction

The concepts of weight and density are embedded in an intuitive theory of matter. Weight is an extensive property of material entities; density an intensive property of material entities. Weight is proportional to quantity of matter; density is the ratio of quantity of matter to volume. The concepts of weight, density, matter, and quantity of matter have a long intellectual history (see Toulmin and Goodfield 1962; Jammer 1961, for comprehensive reviews). As Jammer (1961) tells the story, the late nineteenth century saw the flowering of the "substantial concept of matter," which identified matter and mass. The concept of inertial mass had been formulated by Kepler and systematized by Newton, who also fused it with the medieval concept "quantity of matter." A typical statement from the turn of the century: "If I should have to define matter, I would say: matter is all that has mass, or all that requires force in order to be set in motion" (Charles de Freycinet 1896, quoted in Jammer 1961, p. 86). On this view, mass is the essential property of matter and provides a measure of quantity of matter. In a given gravitational field, weight is an extensive quantity proportional to mass.

Clearly, prior to the formulation of the concept of mass, having mass could not be taken as the essence of material entities. And indeed, prior to the formulation of the concept of mass, weight was not seen as a candidate measure of quantity of matter; nor was having weight (even on earth) seen as necessary and sufficient for an entity's being material (Jammer 1961). The Greeks and the medieval Scholastics had different concepts of matter and of weight from post-Newtonian physicists. According to Jammer, Aristotle had no concept of quantity of matter, and saw weight as an accidental property of some material entities akin to odor. Even if the Greeks had a concept of quantity of matter, weight could not have served as its measure, since some entities, such as air, were thought to possess intrinsic levity. For the Greeks, weight was not an extensive quantity. There were no fixed units of weight; in practical uses, even within the same nation, different substances were weighed in terms of different standards. The weight of material particles was thought to depend upon the bulk of the object in which they were embedded. That is, Aristotle thought that a given lump of clay would itself weigh more when part of a large quantity of clay than when alone. Neither did the alchemists consider weight to reflect quantity of matter; they fully expected to be able to turn a few pounds of lead into hundreds of pounds of gold (Jammer 1961).

Density, like weight, was taken to be an irreducible intensive quality, like color, odor, and other accidents of matter. Density was not defined as mass/volume until Euler; what was actually quantified by the ancients was specific gravity (the ratio of a substance's density to that of water), not density itself. For example, Archimedes never used a term for density in his writings (Jammer 1961).

If weight was not seen as an essential property of material entities, what was? There were many proposals. Euclid proposed spatial extent — length, breadth, and depth. This was one dominant possibility throughout Greek and medieval times. Galileo listed shape, size, location, number, and motion as the essential properties of material entities — spatial, arithmetical, and dynamic properties. The spatial notions included impenetrability; that is, material entities were seen to occupy space uniquely. In another thread of thought, material entities were those that could physically interact with other material entities (Toulmin and Goodfield 1962). Again, weight was seen as irrelevant; on this view, heat, while weightless, is nonetheless material. Finally, another line of thought posited being inert, or passive, as the essence of matter. This was the precursor to the concept of mass; material entities are those that require forces for their movement (Kepler), or forms for their expression (Aristotle and the Scholastics).

The substantial conception of matter (the identification of matter with mass) occupied a brief moment in the history of science. Since Einstein, the distinction between entities with mass and those without is not taken to be absolute, since mass and energy are intraconvertible. It is not clear that the distinction between material and immaterial entities plays an important role in today's physics, given the existence of particles with no rest mass, such as photons, which are nevertheless subject to gravity, and, as Jammer (1961) points out, the concept of mass itself is far from unproblematic in modern physics.

Given the complex history of the concept of matter, what conception of matter should we probe for in the child? *Ours* would be a good bet, i.e., that of the nonscientific adult. What is the adult's intuitive conception of matter, and how is it related to the common-sense concepts of weight and density? While this is an empirical question that has not been systematically investigated, I shall make some assumptions. I assume that common-sense intuitive physics distinguishes between clearly material entities, such as solid objects, liquids, and powders, on the one hand, and clearly immaterial entities such as abstractions (height, value) and mental entities (ideas), on the other. I also assume adults conceptualize quantity of matter. Probably, the essential properties of matter are thought to include spatial extent, impenetrability, weight, and the potential for interaction with other material entities. Probably most adults do not realize that these four properties are not perfectly coextensive. Weight is probably seen as an extensive property of material entities, proportional to quantity of matter, while density is an intensive property, seen as a ratio of quantity of matter and size. This view is closely related to the substantial conception of matter achieved at the end of the nineteenth century, but differs from it in not being based on the Newtonian conception of mass, and being unclear about the status of many entities (e.g., gasses, heat, etc.).

There are two reasons why common-sense physics might be identified so closely with one moment in the history of science. First, common-sense science is close to the phenomena; it is not the grand metaphysical enterprise of the Greek philosophers. For example, in two distinct cases, common-sense science has been shown to accord with the concepts employed in the first systematic exploration of physical phenomena. Common-sense theories of motion share much with medieval impetus theories (e.g., McCloskey 1983), and common-sense thermal theories share much with the source-recipient theory of the Experimenters (see Wiser 1988). Both of these theories require a concept of quantity of matter. For example, the impetus theory posits a resistance to impetus proportional to quantity of matter, and the source-recipient theory of heat posits a resistance to heat proportional to quantity of matter. That untutored adults hold these theories is one reason I expect them to have a pre-Newtonian conception of quantity of matter. Second, the developments of theoretical physics find their way into common-sense physics, albeit at a time lag and in a watered down and distorted version. The mechanisms underlying this transmission include assimilating science instruction (however badly), making sense of the technological achievements made possible by formal science, and learning to use the measuring devices of science, such as scales and thermometers.

4.4. The Child's Material/Immaterial Distinction

We have four interrelated questions. Do young children draw a material/immaterial distinction? If yes, what is the essence of this distinction? Further, do they conceptualize "amount of matter"? If so, what is its measure?

Estes, Wellman, and Woolley (1989) claim that preschool children know that mental entities are immaterial; Piaget (1960) claims that until age eight or so, children consider shadows to be substantial, a claim endorsed by DeVries (1987). These works credit the young child with a material/immaterial distinction and with one true belief (ideas are immaterial) and one false belief (shadows are material) involving the concept of materiality. Assuming that children realize that shadows are weightless, this latter belief would indicate that, like Aristotle, they consider weight an accidental property of material entities. But is it true that they draw a material/immaterial distinction, and if so, on what grounds?

The claim of Estes, Wellman, and Woolley is based on the fact that children distinguish physical objects such as cookies from mental entities such as dreams and pictures in one's head. Estes, Wellman, and Woolley probed this distinction in terms of the properties of objective perceptual access (can be seen both by the child and others) and causal interaction with other material entities (cannot be moved or changed just by thinking about it). These clever studies certainly show that the child distinguishes objects from mental representations of objects in terms of features relevant to the material/immaterial distinction. But many distinctions will separate some material entities from some immaterial entities. Before we credit the child with a material/immaterial distinction, we must assess more fully the extension of the distinction, and we must attempt to probe the role the distinction plays in the child's conceptual system.

Shadows' materiality would be consistent with the essential properties of material entities being public perceptual access and immunity to change as a result of mental effort alone. Piaget's and DeVries's claim is based on children's statements like the following: "A shadow comes off you, so it's made of you. If you stand in the light, it can come off you. It's always there, but the darkness hides it"; or "The light causes the shadow to reflect, otherwise it is always on your body" (DeVries 1987). Such statements show that children talk as if shadows are made of some kind of substance and that they attribute to shadows some properties of objects, such as permanent existence. DeVries studied 223 children, ages two to nine, and only 5 percent of the eight- and nine-year-olds understood that shadows do not continue to exist at night, in the dark, or when another object blocks the light source causing the shadow. In discussing the question of the continued existence of shadows, virtually all spoke of one shadow being covered by another, or of the darkness of two shadows being mixed together, making it impossible to see the shadow, even though it was still there. In interpreting these data a problem arises that is similar to one that arises in the interpretation of the data of Estes, Wellman, and Woolley. These studies show that the child attributes to shadows some properties of material entities (i.e., independent existence and permanence), but what makes these properties tantamount to *substantiality?* It is not enough that these properties differentiate some entities we consider substantial, or material, from some we do not. Many properties do that.

We must assess whether the distinction between material and immaterial entities plays any role in the child's conceptual system. One reflection of such a role would be that children would find it useful to lexicalize the distinction. Preschool children surely do not know the word "matter" or "material," but they probably do know "stuff" and "kind of stuff." Have they mapped these words onto the distinction studied by Estes, Wellman, and Woolley? Do they consider shadows made of some kind of stuff, as Piaget and DeVries claim? In the context of an interview about kinds of stuff such as wood, metal, and plastic, C. Smith, Carey, and Wiser (1985) asked four- to nine-year-olds whether shadows are made of some kind of stuff. About three-fourths of the four- to seven-year-olds replied "Yes," and most volunteered, "Out of you and the sun." While this may reflect their considering shadows material, it seems more likely to reflect their understanding the question to be whether and how one can make a shadow.

In a recent study, my colleagues and I attempted to address directly whether the child distinguishes between entities made of some kind of stuff and entities not made of some kind of stuff, and if so, on what basis. We introduced children from the ages of four through twelve to the issue by telling them that some things in the world, such as stones and tables and animals, are made of some kind of stuff, are material, are made of molecules, while other things we can think of, like sadness and ideas, are not made of anything, are not material, are not made of molecules (Carey at al., in preparation). We encouraged children to reflect on this distinction and to repeat our examples of material and immaterial entities. We then asked them to sort a number of things into two piles: (1) material things like stones, tables, and animals; and (2) immaterial things like sadness and ideas. The things we asked them to sort were these: car, tree, sand, sugar, cow, worm, styrofoam, Coca Cola, water, dissolved sugar, steam, smoke, air, electricity, heat, light, shadow, echo, wish, and dream. We credited children with the distinction if they sorted objects, liquids, and powders in the material pile, and wish and dream in the immaterial pile. Where they placed the remaining items provided some information concerning the properties they considered central to the distinction.

As can be seen from Table 1, our instructions led to systematic sorting at all ages. At all ages, over 90 percent of the placements of the car, the

Table 1. Percent Judged Material

Age	4	6	10	12
car, tree, styrofoam	93%	96%	91%	100%
sand, sugar	65%	94%	95%	100%
cow, worm	55%	81%	95%	100%
Coca Cola	30%	88%	100%	100%
water	40%	25%	90%	100%
dissolved sugar	63%	63%	55%	88%
steam, smoke, air	20%	25%	30%	61%
electricity	40%	75%	73%	63%
heat, light	30%	38%	41%	31%
echo, shadow	25%	25%	9%	13%
wish, dream	5%	19%	5%	13%

tree, and styrofoam were into the material pile, and at all ages except age six, less than 5 percent of the placements of wish and dream were into this pile. Children understood something of the introductory instruction and certainly distinguished solid, inanimate objects from abstract entities and mental representations. Shadows were not considered material; at all ages except age four, shadows and echoes were patterned with wishes and dreams. These data do not support Piaget's and DeVries's claim that young children consider shadows substantial. Nonetheless, many of the younger children revealed very different bases for their sorts than did the older children. Around one-tenth of the four- and six-year-olds answered randomly. In addition, half of the preschool children took only solid, inanimate objects plus powders as material. That is, 50 percent of the four-year-olds denied animals and liquids are material, including a few who also denied sand and sugar are; 13 percent of the six-year-olds also showed this pattern; see Table 2. These data are striking, since the introduction of the material/immaterial distinction explicitly mentioned animals as examples of material entities. These children seemed to focus on the locution "made of some kind of stuff" and therefore answered affirmatively either if they could think of the material of which something is made (many commented trees are made of wood) or if they thought of the entities as constructed artifacts. Another reflection of this construal is seen in the six-year-olds' responses to Coke (88 percent sorted as material) compared to water (25 percent sorted as material). Children could think of ingredients of Coke (sugar and syrup), but saw water as a primitive ingredient, thus not made of any kind of stuff. This construal also contributed to the six-year-olds' affirmative judgments on wish and dream; some children commented that dreams are made of ideas. Thus, among the youngest children there were considerable problems understanding or holding on to what distinction was being probed. Sixty percent of the four-year-olds and 25 percent of the six-year-olds showed no evidence of a conception of matter that encompassed inanimate objects, animals, liquids, and powders. These children had not mapped the properties probed by Estes, Wellman, and Woolley onto their notion of "stuff."

Table 2. Percent of Subjects Showing Each Pattern

	Age 4	Age 6	Age 10	Age 12
adult, mass crucial	0	0	9%	0
mass crucial, gases judged not material	0	0	9%	38%
physical consequences — includes solids, liquids, powders, *gases, and* some of electricity, heat, light, shadow, echo	0	0	0	63%
physical consequences — excludes gases; includes solids, liquids, powders, and some of electricity, heat, etc.	40%	75%	82%	0
denies liquids, animals, gases, and immaterial entities	60%	13%	0	0
random	10%	13%	0	0

However, 40 percent of the four-year-olds, 75 percent of the six-year-olds, and 100 percent of the ten- to twelve-year-olds provided systematic sorts that clearly reflected a concept of matter. Nonetheless, *weighing something*, or having mass, was not coextensive with the entities even these children judged material. It was only the oldest children who sometimes claimed that all weightless entities were not material (38 percent of the oldest group, Table 2). As in Table 2, only one child in the whole sample had an adult pattern of judgments.

Three groups of entities were reflected in the sorts: solids, liquids, and powders on the one hand; echo, shadow, wish, and dream on the other; with all others firmly in between. For children under twelve, electricity, heat, and light were equally or more often judged material than were dissolved sugar, steam, smoke, and air (Table 1). Further, all children under twelve judged some immaterial entities (such as heat) material *and* some material entities (such as air) immaterial. In their justifications for their judgments, children mainly appealed to the perceptual effects of the entities — they mentioned that one can see and touch them. One child in a pilot study articulated the rule that one needs two or more perceptual effects for entities to be material. You can see shadows, but cannot smell, feel, or hear them; you can hear echoes but cannot see, smell, or touch them; therefore shadows and echoes are not material. Nor is air. But heat can be seen (heat waves) and felt, so heat is material.

To sum up the data from the sorting task: Of the youngest children (ages four to six), a significant portion did not know the meaning of "stuff" in which it is synonymous with "material." This leaves open the question of whether they draw a material/immaterial distinction, even though this task failed to tap it. However, about half of the younger children and all the older ones did interpret "stuff" in the sense intended, revealing a material/immaterial distinction. Up through age eleven, the distinction between material and immaterial entities was not made on the basis of weight. Only at ages eleven to twelve were there a few children who took all and only entities that weigh something as material.

4.5. Weight and Materiality, Continued

The sorting data show that early elementary children do not take an entity's weighing something as necessary for materiality (in the sense of being made of some kind of stuff). From ages four through eleven, virtually all children who deemed solids, liquids, and powders material also judged some weightless entities (electricity, heat, light, echoes, or shadows) material. However, they might hold a related belief. They may see weight as a property of all prototypical material entities (solids, liquids, and powders). C. Smith, Carey, and Wiser (1985) provide data that suggest that young children do not expect even this relation between materiality and weight. When given a choice between "weighs a lot, a tiny amount, or nothing at all," children judged that a single grain of rice, or a small piece of styrofoam, weighed nothing at all. Carey et al. (in preparation) probed for a similar judgment from those children who had participated in the material/immaterial sorting task. Virtually all had judged styrofoam to be material (Table 1). We began with a sheet of styrofoam twelve inches by twelve inches by a half inch and asked whether it weighed a lot, a little, a tiny amount, or nothing at all. If children judged that it weighed a little, we showed a piece half the size and asked again. If that was judged as weighing at least a tiny amount, a small piece the size of a finger tip was produced, and the question repeated. Finally, the child was asked to imagine the piece being cut again and again until we had a piece so small we could not see it with our eyes, and asked if that would weigh a lot, a little, or nothing at all — whether we could ever get a piece so small it would weigh nothing at all.

C. Smith, Carey, and Wiser's (1985) results were confirmed (Figure 2). More than half of the four-year-olds and fully half of the six-year-olds judged that the *large* piece of styrofoam weighed nothing at all, and all four- to six-year-olds judged that the small piece weighed nothing. Half of the ten- to eleven-year-olds judged that the small piece weighed nothing at all, and almost all judged that if one kept dividing the styrofoam, one would eventually obtain a piece that weighed nothing. Not until age twelve did half of the children maintain that however small the piece, even one so small one could no longer see it, it would weigh a tiny, tiny amount.

These data are important beyond showing that children consider an entity's weighing something as unrelated to its being material. They show that children, like the Greeks, do not take weight as a truly extensive property of substances. They do not conceive of the total weight of an object as the sum of weights of arbitrarily small portions of the substance from which it is made. This is one very important way in which the child's *degree of heaviness* differs from the adult's *weight*. The child's *degree of heaviness* is neither systematically intensive nor systematically extensive, as is required if the child's concept is undifferentiated between *weight* and *density*.

4.6. Occupation of Space by Physical Objects

We do not doubt that even four-year-olds know some properties that solids, liquids, and powders share, even if being "made of some kind of stuff" and having weight are not among those properties. Presumably, young children extend the properties of physical objects studied by Estes, Wellman, and Woolley (1989) to liquids and powders: public access, nonmanipulation by thought alone. Another place to look might be a generalization of the infants' solidity constraint (see Spelke 1991). Infants know that one physical object cannot pass through the space occupied by another; we would certainly expect four-year-olds to realize the related principle that no two objects can occupy the same space at the same time, and they might extend this principle to liquids and powders. We assessed this question by asking our subjects to imagine two pieces of material, one wood and one metal, cut to fill entirely the inside of a box. They were then asked whether we could put the wood and the metal in the box at the same time. No children had any doubts about this question; they answered that they both could not fit in at the same time (Table 3). When asked to imagine the box filled with water and then probed as to whether the steel piece and the water could be in the box at the same time they all (except one four-year-old who said that both could be in the box at the same time because the water would become compressed) again said no, that the water would be pushed out (Table 3).

Children are confident that solids and liquids (and, I am sure, although we did not probe it, materials such as sand as well) uniquely

**Table 3. Occupy Space:
Can Steel and x Fit in Box at Same Time?**

| | %NO | | |
	x = wood	*x = water*	*x = air*
Age 4	100%	90%	0%*
1st grade	100%	100%	25%
5th grade	100%	100%	55%
7th grade	100%	100%	62.5%

*n = 5. The remaining four-year-olds denied
there was air in the box.

occupy space. However, it is unlikely that this property defines a material/immaterial distinction for them. To assess that, we would have to see whether those who think electricity, heat, light, echoes, or shadows to be material all consider these to occupy space. Still, these data confirm our suspicion that children see physical objects, liquids, and powders as sharing properties relevant to the material/immaterial distinction. Having weight is simply not one of these properties.

4.7. A Digression: An Undifferentiated *Air/Nothing* Concept

The last questions about the box concerned air. Children were asked, of the apparently empty box, whether there was anything in it at the moment, and when they said no, we said, "What about air?" Except for half of the four-year-olds, who denied there was air in the box and insisted that there was nothing in it, all children agreed the box contained air. All who agreed were asked whether one could put the steel in the box at the same time as the air. If they said yes, they were further probed as to whether the steel and air would be in the box, then, at the same time. As can be seen from Table 3, the vast majority of the four- and six-year-olds thought that air and steel could be in the box at the same time, explaining, "Air doesn't take up any space"; "Air is all over the place"; "Air is just there — the metal goes in, and air is still there"; "Air isn't anything"; and so on. One child said baldly, "Air isn't matter." Almost half of the ten- to twelve-year-olds also provided this pattern of response.

The sorting task also suggests that young children consider air not material — air was judged made of some kind of stuff by none of the four-year-olds, 10 percent of the six-year-olds, and 36 percent of the ten- and eleven-year-olds. Only twelve-year-old subjects judged air made of some kind of stuff (75 percent) and also maintained that the steel would push the air out, just as it would the water (65 percent). While the characterization of the child as believing air to be immaterial is easy enough to write down, a moment's reflection reveals it to be bizarre. If air is not material, what is it? Perhaps children consider air to be an immaterial physical entity, like a shadow or an echo. But several children said outright, "Air is nothing; air isn't anything." However, "air" is not simply synonymous with "nothing" or "empty space," for children as young as six know that there is no air on the moon or in outer space, that one needs air to breathe, that wind is made of air, etc. Indeed, in a different interview in which we probed whether children of this age considered dreams and ideas made of some kind of stuff, an interview in which "air" was never mentioned, several different children spontaneously offered "air" as the stuff of which dreams and ideas are made of. This set of beliefs reflects another undifferentiated concept, *air/nothing* or *air/vacuum*, incommensurable with the concepts in the adult conceptualization of matter.

4.8. Interim Conclusions: The Material/Immaterial Distinction

Children distinguish solids, liquids, and powders, on the one hand, from entities such as wishes and dreams, on the other, in terms of properties related to the distinction between material and immaterial entities. These include uniquely occupying space, (probably) public perceptual access, and not being manipulable by thought alone. Not all four- to six-year-olds have related this distinction to the notion of "stuff," so the data available at this point provide no evidence that these properties determine a *material/immaterial* distinction, rather than, for example, an undifferentiated *real/unreal* distinction. Some children of these ages, and all children in our sample of ages ten and older, have related this distinction to the notion of "stuff," but do not yet see weight as one criterion for materiality.

4.9. Taking Up Space: Matter's Homogeneity

While young children may not draw a distinction between material and immaterial entities, they do conceptualize kinds of stuff such as plastic, glass, wood, sand, and water. They distinguish objects from the stuff of which they are made, realizing that the identity of an object does not survive cutting it into many small pieces, but the identity of the stuff is not affected. There is, however, some question as to their ability to preserve identity of stuff as it is broken into smaller and smaller pieces. C. Smith, Carey, and Wiser (1985) suggested that perhaps young children cannot conceive of substances as composed of *arbitrarily* small portions, each of which maintains the identity of the substance and some of its substance-relevant properties. In other words, they may not grasp that stuff is homogeneous. (Of course, matter is not homogeneous, but in many respects twelve-year-olds think it is, and the developmental progression toward that view will be described here.) This could underlie their lack of understanding that the total weight of an object is the sum of the weights of small portions. Alternatively, the problems young children have with conceptualizing the weight of tiny portions of matter could be independent of a conception of substance as homogeneous.

Children's commitment to solids and liquids occupying space led us to probe their understanding of homogeneity in this context (Carey et al., in preparation). Our first method of doing so drew on the weight probes described above. We asked children whether the big piece of styrofoam took up a lot of space, a little space, or no space at all. We then repeated that question concerning the small piece, the tiny piece, and imagined halves and halves again until we got a piece so small one could not see it with one's eyes.

Compare Figure 3 to Figure 2. At all ages children revealed a better understanding of homogeneity in the context of the question of whether a piece of styrofoam occupies space than they did in the context of the question of whether a piece of styrofoam weighs anything. Twelve-year-olds were virtually perfect on the task; only one said that one could arrive at a piece of styrofoam so small that it would not take up any space at all. More significantly, fully half of the six-year-olds and ten- to eleven-year-olds made these adult judgments. Only four-year-olds universally failed; all said that if one arrived, by cutting, at a piece too small to see with one's eyes, that piece would not take up any space. By this measure then, almost all twelve-year-olds, and half of the children between ages six and twelve, understand that solid substances are continuously divisible, and that an arbitrarily small piece of substance still occupies a tiny, tiny amount of space. These conceptualize substances as homogeneous. Equally important, by this measure, four-year-olds do not have this understanding.

Not all children understood the locution "take up space." As Nussbaum (1985) pointed out, children lack the Newtonian conception of space as a geometric construction that defines points that may or may not be occupied by material bodies. Because we could see that some children were not understanding what we were getting at, we devised another question to probe children's understanding of the homogeneity of matter. We presented an iron cylinder, told children that it was made of iron, and asked whether they could see *all* the iron in the bar. If children responded no, they were then shown a much smaller cylinder, and the question was repeated. Next they were shown an iron shaving, and

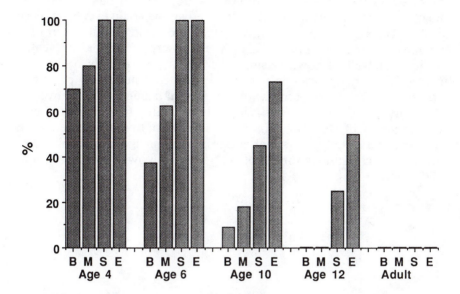

Figure 2. Weight of styrofoam. Percent judging piece of styrofoam weighs nothing at all as a function of the size of the piece. B, big; M, medium; S, small; E, ever, if one kept cutting it in half, repeatedly.

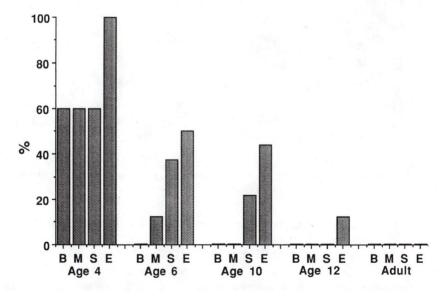

Figure 3. Styrofoam's taking up space. Percent judging piece of styrofoam takes up no space at all as a function of the size of the piece. B, big; M, medium; S, small; E, ever, if one kept cutting it in half, repeatedly.

the question was repeated; finally they were asked to imagine halving the iron repeatedly, were probed as to whether one could ever get a piece small enough so that (with a microscope) one could see all the iron. A commitment to the continuity and homogeneity of matter is revealed in the response that however small the piece, there will always be iron inside. Of course, matter is particulate, not continuous. In principle, one could arrive, by the process of dividing, at a single atom of iron, in which there would be no iron inside. Children are often taught the particulate theory of matter beginning in the seventh to ninth grades; work by science educators shows that children of these ages are deeply committed to a continuous theory of matter (e.g., Novick and Nussbaum 1978, 1981; Driver et al. 1987).

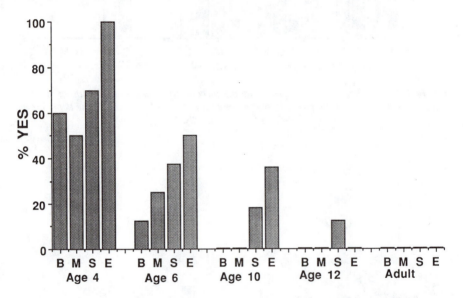

Figure 4. Visibility of all the iron. Percent judging one can see all the iron, as a function of the size of the piece of iron. B, big; M, medium; S, small; E, ever, if one kept cutting it in half, repeatedly.

There were two types of answers that showed children to be thinking about the iron as an object rather than as a continuous substance: (1) "Yes, you can see all the iron"; and (2) "No, because you can't see the bottom," or "No, because there is some rust on it." This probe for an understanding of homogeneity and continuity of matter reveals the same developmental pattern as did the questions of whether small pieces of matter occupy space (Figure 4; cf. with Figure 3). All of the twelve-year-olds said that one could never see all the iron, no matter how small the piece, because there would always be more iron inside. More than half of the six- to eleven-year-olds also gave this pattern of responses. Only four-year-olds universally failed. A majority of the preschool children claimed one could see all the iron in the two large cylinders, more said so for the shaving, and virtually all said one would eventually get to a speck small enough so one could see all the iron.

Figures 3 and 4 reveal nearly identical patterns. An analysis of consistency within individuals corroborates this result. Those children who revealed an understanding of continuity and homogeneity on the "see all the iron" task also did so on the "styrofoam occupies space" task, and those who failed on one, failed on the other. The relationship holds even when the four-year-olds (most all failing at both tasks) and the twelve-year-olds (most all succeeding at both tasks) are removed from the analysis (p. $< .05$, chi-square). The two tasks are really quite different from each other, so this consistency strengthens our conclusions that four-year-olds do not grasp the continuity and homogeneity of solid substances; that half of early elementary aged children do; and that by age twelve, virtually all children have constructed such an understanding of solid substance.

An understanding of substances as continuous and homogeneous may well be a conceptual prerequisite to an extensive understanding of weight. If children cannot think of a piece of iron as composed of arbitrarily small portions of iron, then they would not be able to think of the weight of an object as the sum of weights of arbitrary portions of the substance from which it is made. The data in Figures 3 and 4 show that all four-year-olds and half of the six- to eleven-year-olds lack this prerequisite for an extensive understanding of weight. But the comparisons between these data and those in Figure 2 show that more is required for a reconceptualization of *degree of heaviness* as true *weight*. What might that be?

My answer is speculative, going well beyond the data at hand. My guess is that an understanding of substance as continuous and homogeneous is a prerequisite for a concept of *quantity of substance* or *quantity of matter*. Even after one has formulated this concept of *quantity of matter*, the question whether heaviness is an accidental property of matter is open. In the course of differentiating *weight* and *density*, the child will see that volume cannot be a measure of quantity of matter, leading the child to be open to an extensive conception of weight as a measure of quantity of matter.

4.10. Mathematical Prerequisites

Like the Experimenters' *degree of heat* the child's *degree of heaviness* is not a fully quantitative concept. The child's *degree of heaviness* is certainly ordered. Children understand that one object (A) can be heavier than another (B), and they expect relative heaviness to be reflected in measurement of weight — if A weighs 250 grams, then B will weigh less than 250 grams. They take this relation to be transitive and asymmetric. However, the limits of children's quantification of degree of heaviness are revealed in their willingness to judge that a piece of substance weighing 250 grams could be broken into ten parts, each of which weighs nothing.

A true understanding of the extensivity of weight requires an understanding of division, a mathematical concept that is very difficult for most elementary school children (Gelman 1991). And a quantitative, extensive conception of weight is clearly required for a quantitative conception of density. This further requires an understanding of ratios and fractions, also conceptually difficult for children in these age ranges (Gelman 1991). Thus, as Piaget and Inhelder (1941) argued cogently, a quantitative understanding of density requires mathematical concepts that do not emerge in most children until early adolescence.

Black differentiated heat from temperature in the course of attempting to measure each independently from the other, and relating each quantified magnitude to distinct thermal phenomena. The full differentiation of weight and density is achieved by children during science instruction in the course of similar activities. Unlike Black, the young child in elementary school lacks the mathematical tools for this achievement. The Experimenters faced theory-specific conceptual barriers to differentiating heat and temperature. Similarly, the child faces theory-specific conceptual barriers to differentiating weight and density. But the child also lacks tools of wide application (Carey 1985a), here mathematical tools, important for the reconceptualization. In this sense there is a domain-general limitation on the young child's understanding of matter, just as Piaget and Inhelder (1941) argued.

5. Conclusions

Concepts change in the course of knowledge acquisition. The changes that occur can be placed on a continuum of types — from enrichment of beliefs involving concepts that maintain their core to evolution of one set of concepts into another incommensurable with the original. In this chapter I have explored Spelke's conjecture that spontaneous development of physical theories involves only enrichment. I argued, contra Spelke, that the child's intuitive theory of physical objects is incommensurable with the adult's intuitive theory of material entities.

As in cases of conceptual change in the history of science, this case from childhood includes differentiations where the undifferentiated concepts of C1 play no role in C2 and are even incoherent from the vantage point of C2. *Weight/density* and *air/nothing* were the examples sketched here. The child's language cannot be translated into the adult's without a gloss. One cannot simply state the child's beliefs in terms of adult concepts — the child believes air is not material, but the "air" in that sentence as it expresses the child's belief is not our "air" and the "material" is not our "material." Similarly, the child believes heavy objects sink, but the "heavy" in that sentence as it expresses the child's belief is not our "heavy." I can communicate the child's concepts to you, but I have provided a gloss in the course of presenting the patterns of judgments the child makes on the tasks I described. To communicate the child's concept *degree of heaviness*, I had to show its relation to the child's concept *density* and *substance*, for all these differ from the adult's concepts and are interrelated differently than in the adult conceptual system.

These are the hallmarks of incommensurable conceptual systems. Spelke might reply that the conceptual change described here was *originally* achieved by metaconceptually aware scientists and that children only achieve it, with difficulty, as a result of schooling. Thus it does not constitute a counterexample to her claim that spontaneous knowledge acquisition in childhood involves only enrichment. This (imaginary) reply misses the mark in two ways. First, even if the original development of the lay adult's conception of matter was achieved by metaconceptually sophisticated adults, and only gradually became part of the cultural repertoire of lay theorists, it is still possible that spontaneous (in the sense of unschooled) conceptual change occurs as children make sense of the lay theory expressed by the adults around them. Second, the construction of a continuous, homogeneous conception of substances occurs spontaneously between ages four and eleven, in at least half of the children in our sample. This is not taught in school; indeed, this theory is known to be false by science teachers. Similarly, in C. Smith, Carey, and Wiser (1985), roughly half of the children had differentiated weight from density by age nine, before they encountered the topic in the school curriculum. True, many children require intensive instruction to achieve this differentiation (see C. Smith et al. 1988). What we have here is analogous to Gelman's findings on fractions; some elementary aged children construct a conceptually deep understanding of fractions from minimal exposure to the topic while others do not (Gelman 1991).

Spelke's speculations concerning spontaneous knowledge acquisition includes two nested theses. She argues that conceptual change more extreme than enrichment (1) does not occur in the course of spontaneous development of intuitive concepts, in general, and (2) does not occur in the spontaneous development of the concept *physical object*, in particular. It is the first thesis I have denied in this chapter. Let us now turn to the second. True, babies and adults see the world as containing objects that obey the solidity and spatio-temporal continuity principles. But for adults, these principles follow from a more abstract characterization of objects as material, and in the adult version of the principles, liquids, powders, and even gasses obey the same principles. At the very least, conceptual change of the second and third degrees has occurred — what the baby takes as the core properties of objects are seen by the adult to be derived from more fundamental properties. And adults have constructed a fundamental theoretical distinction, material/immaterial, unrepresented by babies.

I would speculate that the conceptual evolution between the baby's concepts and the adult's passes over at least two major hurdles. Objects, for babies, are bounded, coherent wholes and, as such, are totally distinct from liquids, gels, powders, and other nonsolid substances. The distinction between objects and nonsolid substances is very salient to young children; it conditions hypotheses about word meanings, and relates to the quantificational distinction between entities quantified as individuals and entities not quantified as individuals (Soja, Carey, and Spelke 1991; Bloom 1990). It seems possible that young children believe objects can pass through the space occupied by liquids, since they experience their own bodies passing through water and objects sinking through water. The first hurdle is the discovery that in spite of these differences, physical objects and nonsolid substances share important properties, making liquids and powders *substantial* in the same sense as are objects. By age four, children apparently understand that liquids uniquely occupy space; it is not clear whether younger children do.

Liquids and powders are not quantified as individuals precisely because they have no intrinsic boundaries; they can be separated and recoalesced at will. The quantificational distinction between nonsolid substances and objects supports seeing nonsolid substances as homogeneous and continuous and not seeing objects in this light. The second hurdle involves extending this conception of nonsolid substances to solid substances. The data reviewed above show that by ages six to eleven, only half of the children in our sample had achieved this extension.

Changes of this sort go beyond mere enrichment. New ontological distinctions come into being (e.g., material/immaterial), and in terms of this distinction entities previously considered deeply distinct (e.g., objects and water) are seen to be fundamentally the same. The acquisition of knowledge about objects involves more than changes in beliefs about them. The adult can formulate the belief "Objects are material"; the infant cannot.

5.1. Origins of Concepts

If I am right, not all lexical concepts are innate; nor is it the case that all lexical concepts are definable in terms of a set of innate primitives. Rich innate concepts do not preclude subsequent conceptual change. How, then, do new concepts arise? The key to the solution of this problem is that sets of concepts can be learned together, so that some of their interpretation derives from their relations to each other, as they map as a whole onto the world. Insofar as aspects of the interpretation of concepts derive from that of those concepts from which they are descended, this answer avoids a vicious meaning holism. In this way, concepts are ultimately grounded in the innately interpreted primitives.

But how is this achieved in practice? My ideas have been developed in collaborative work with Marianne Wiser and Carol Smith, and are tested in the arena of science education. Wiser has developed curricula to effect change in high school students' thermal concepts, specifically, to induce differentiation of heat and temperature. Smith has developed curricula to effect change in junior high school students' concepts of matter, especially, to induce differentiation of weight and density. Both of these curricula are based on computer-implemented, interactive visual models that serve as analogies. For example, in one model, weight is represented by number of dots, volume by number of boxes of a fixed size, and density by dots per box. Students work with the models, and then work on mapping the models to the world. The latter is the hard part, of course, because without having differentiated weight and density, students do not always succeed in mapping number of dots per box to density, rather than, say, absolute weight.

As well as being involved in discussing the merits of the different models we have invented, students are engaged in making their own models, and in revising models as new phenomena to model are encountered. Note the relation between such curricular interventions and the uses of visual analogies discussed by Nersessian (this volume). Beyond the use of analogies, we also employ other techniques she discusses. We engage students in limiting case analyses and concrete thought experiments. To give just one example: students who maintain that a single grain of rice weighs nothing participate in the following activity. A playing card is balanced on a relatively wide stick, and the number of grains of rice placed on one edge necessary to tip it over is ascertained (approximately forty). Then it is balanced on a narrower stick, and students discover around ten will tip it over. Then it is balanced on a very thin edge, such that a single grain of rice tips it over. After each event, they are asked why the card tips over, and for the first two events, they typically appeal to the weight of the piles of rice. At the end, they are asked again whether they think a single grain of rice weighs anything at all. Lively discussions inevitably follow, for this demonstration is not enough to change every ten- to twelve-year-old's view on the matter. What is interesting are the arguments produced by those who *do* change their views. In every class in which I have witnessed this activity, students spontaneously come up with two arguments. First, that when they had thought that a single grain of rice weighed nothing at all, they were not thinking about the sensitivity of the measuring instrument. And second, that of course a single grain of rice weighs something. If it weighed nothing, how could ten grains or forty grains weigh something? In short, $0 + 0 + 0$ can never yield a nonzero result.

In the end, I would like to hold out an as yet unrealized promise. If the continuity hypothesis is correct, and conceptual changes in childhood are indeed of the same sorts as conceptual changes in the history of science, then interventions of the sort sketched here become the testing ground for ideas concerning the processes underlying conceptual change.

Notes

1. My explication of local incommensurability closely follows Carey (1988) and Carey (1991), though I work through different examples here.

2. The concept of density at issue here is a ratio of weight and volume, and is a property of material kinds. We are not probing the more general abstract concept of density expressing the ratio of any two extensive variables, such as population density (people per area).

References

Baillargeon, R. 1990. Young infants' physical knowledge. Paper presented at the American Psychological Association Convention, Boston.

Baillargeon, R., K. Miller, and J. Constantino. In press. 10.5-month-old-infants' intuitions about addition.

Baillargeon, R., E. S. Spelke, and S. Wasserman. 1985. Object permanence in 5-month-old infants. *Cognition* 20:191–208.

Bloom, P. 1990. Semantic structure and language development. Ph.D. diss., Massachusetts Institute of Technology.

Carey, S. 1985a. Are children fundamentally different thinkers and learners from adults? In S. F. Chipman, J. W. Segal, and R. Glaser, eds., *Thinking and Learning Skills*. Vol. 2, pp. 486–517. Hillsdale, N.J.: Lawrence Erlbaum.

———. 1985b. *Conceptual Change in Childhood*. Cambridge, Mass.: MIT Press.

————. 1988. Conceptual differences between children and adults. *Mind and Language* 3:167–81.

————. 1991. Knowledge acquisition: enrichment or conceptual change? In S. Carey and R. Gelman, eds., *The Epigenesis of Mind: Essays in Biology and Cognition*, pp. 257–91. Hillsdale, N.J.: Lawrence Erlbaum.

Carey, S., et al. In preparation. On some relations between children's conceptions of matter and weight.

DeVries, R. 1987. Children's conceptions of shadow phenomena. *Genetic Psychology Monographs* 112:479–530.

Driver, R., et al. 1987. *Approaches to Teaching the Particulate Theory of Matter.* Leeds: Leeds University, Children's Learning in Science Project.

Estes, D., H. M. Wellman, and J. D. Woolley. 1989. Children's understanding of mental phenomena. In H. Reese, ed., *Advances in Child Development and Behavior*, pp. 41–87. New York: Academic Press.

Feyerabend, P. 1962. Explanation, reduction, empiricism. In H. Feigl and G. Maxwell, eds., *Scientific Explanation, Space, and Time*, pp. 28–97. Minneapolis: University of Minnesota Press.

Fodor, J. 1975. *The Language of Thought.* New York: Thomas Y. Crowell.

Gelman, R. 1991. From natural to rational numbers. In S. Carey and R. Gelman, eds., *The Epigenesis of Mind: Essays in Biology and Cognition*, pp. 293–322. Hillsdale, N.J.: Lawrence Erlbaum.

Jackendoff, R. 1989. What is a concept, that a person may grasp it? *Mind and Language* 4:68–102.

Jammer, M. 1961. *Concepts of Mass.* Cambridge, Mass.: Harvard University Press.

Kitcher, Philip. 1988. The child as parent of the scientist. *Mind and Language* 3: 217–28.

Kuhn, T. S. 1977. A function for thought experiments. In T. S. Kuhn, ed., *The Essential Tension*, pp. 240–65. Chicago: University of Chicago Press.

————. 1982. Commensurability, comparability, communicability. In *PSA 1982*, vol. 3, pp. 669–88. East Lansing, Mich.: Philosophy of Science Association.

Leslie, A. 1988. The necessity of illusion. In L. Weiskrantz, ed., *Thought without Language*, pp. 185–210. Oxford: Oxford University Press.

McCloskey, M. 1983. Intuitive physics. *Scientific American* 4:122–30.

McKie, D., and N. H. V. Heathecoate. 1935. *The Discovery of Specific and Latent Heat.* London: Edward Arnold.

Mandler, J. 1988. How to build a baby: on the development of an accessible representational system. *Cognitive Development* 3:113–26.

Novick, S., and J. Nussbaum. 1978. Junior high school pupils' understanding of the particulate nature of matter: an interior study. *Scientific Education* 62:273–81.

————. 1981. Pupils' understanding of the particulate nature of matter: a cross-age study. *Science Education* 65, no. 2:187–96.

Nussbaum, J. 1985. The particulate nature of matter in the gaseous phase. In R. Driver, E. Guesner, and A. Tiberghien, eds., *Children's Ideas in Science*. Philadelphia: Milton Keynes.

Piaget, J. 1929. *The Child's Conception of the World.* London: Routledge and Kegan Paul.

————. 1960. *The Child's Conception of Physical Causality.* Patterson, N.J.: Littlefield, Adams, and Co.

Piaget, J., and B. Inhelder. 1941. *Le développement des quantites chez l'enfant.* Neuchatel: Delachaux et Niestlé.

Quine, W. V. O. 1960. *Word and Object.* Cambridge, Mass.: MIT Press.

Smith, C., S. Carey, and M. Wiser. 1985. On differentiation: a case study of the development of the concepts of size, weight, and density. *Cognition* 21:177–237.

Smith, C., et al. 1988. Using conceptual models to facilitate conceptual change: weight and density. Harvard University: Educational Technology Center, technical report.

Smith, E. 1989. Concepts and induction. In M. Posner, ed., *Foundations of Cognition Science*, pp. 501–26. Cambridge, Mass.: MIT Press.

Soja, N., S. Carey, and E. Spelke. 1991. Ontological constraints on early word meanings. *Cognition* 38:179–211.

Spelke, E.S. 1988. The origins of physical knowledge. In L. Weiskrantz, ed., *Thought without Language*. Oxford: Oxford University Press.

———. 1991. Physical knowledge in infancy: reflections on Piaget's theory. In S. Carey and R. Gelman, eds., *The Epigenesis of Mind: Essays in Biology and Cognition*, pp. 133–69. Hillsdale, N.J.: Lawrence Erlbaum.

Thagard, P. In press. Concepts and conceptual change. *Synthese*.

Toulmin, S., and J. Goodfield. 1962. *The Architecture of Matter*. Chicago: University of Chicago Press.

Vosniadou, S., and W. Brewer. In press. The construction of cosmologies in childhood. *Cognitive Psychology*.

Wierzbicka, A. 1980. *Lingua Mentalis*. New York: Academic Press.

Wiser, M. 1988. The differentiation of heat and temperature: history of science and novice-expert shift. In S. Strauss, ed., *Ontogeny, Philogeny, and Historical Development*, pp. 28–48. Norwood, N.J.: Ablex.

Wiser, M., and S. Carey. 1983. When heat and temperature were one. In D. Gentner and A. Sievens, eds., *Mental Models*, pp. 267–97. Hillsdale, N.J.: Lawrence Erlbaum.

Conceptual Change within and across Ontological Categories: Examples from Learning and Discovery in Science

1. Introduction

The simple working definition adopted in this essay for conceptual change is that it refers primarily to the notion of how a concept can change its meaning. Since a difference in meaning is difficult to define, one can think of it as a change in its categorical status: That is, changing the category to which the concept is assigned, since all concepts belong to a category. To assume that all concepts belong to a category is quite standard: For instance, A. R. White (1975) also assumed that a concept signifies a way of classifying something. Thus, this essay is not concerned with concept acquisition per se, but rather with how a concept can take on new meaning as a form of learning. This also presupposes that one already has some notion of what a particular concept is.

The term "conceptual change" is misleading in that it can refer either to the outcome of the change (or the resulting category to which a concept is assigned), or to the processes that achieve such changes. The literature on conceptual change is often ambiguous about this distinction. "Conceptual change" is used here to mean one of these ways and should be clear from the context; otherwise it is explicitly clarified.

The thesis to be put forth in this essay is that it may be important to discriminate between conceptual change that occurs within an ontological category and one that necessitates a change between ontological categories. The former I shall simply call conceptual change or conceptual change within an ontological category, and the latter *radical conceptual change* or conceptual change across ontological categories. Making such a discrimination may be critical for a better understanding of learning and development, as well as many other related phenomena, such as the occurrence of scientific discoveries.

1.1. The Nature of Ontological Categories

The nature of ontological knowledge has been investigated in the psychological literature primarily by Keil (1979). Ontology divides our knowledge into categories that are conceptually distinct in a way that is somewhat difficult to define, although the distinction can be easily captured. Consistent with theories provided by Sommers (1971) and Keil (1979), I propose that there exist a few major categories in the world that are ontologically distinct physically, and should be perceived by adults as ontologically distinct psychologically. Figure 1 shows three basic ontological categories: matter (or material substances), events, and abstractions. (There may be others as well.)

The intrinsic reality of ontological categories can be determined in two ways, as shown in the list below. First, a distinct set of physical laws or constraints governs the behavior and ontological attributes of each ontological category. For example, objects in the matter category must have a certain set of constraints that dictate their behavior and the kind of properties they can have. Matter (such as sand, paint, or human beings) has ontological attributes such as being containable and storable, having volume and mass, having color, and so forth. In contrast, events such as war do not have these ontological attributes and obey a different set of constraints. (An ontological attribute is defined by Sommers [1971] as a property that an entity has the potential to have, such as being colored, even though it does not have it. War or pain, for instance, does not have the ontological attribute of "being red," whereas squirrels can have such an ontological attribute even though squirrels are commonly not red.) Thus, events are governed by an alternative, distinct set of physical laws, such as varying with time, occurring over time, being unstorable, having a beginning and an end, and so on.

Corresponding Ways to Capture the Reality of Ontological Categories

Intrinsic Reality

1. A distinct set of constraints govern the behavior and properties of entities in each ontological category.

2. No physical operations (such as surgery, movement) can transform entities in one ontological category to entities in another ontological category.

Psychological Reality

1. A distinct set of predicates modify concepts in one ontological category versus another, based on sensibility judgment task.

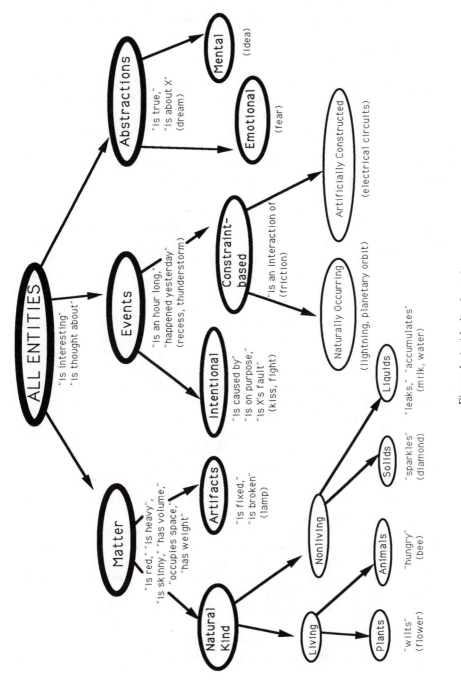

Figure 1. An idealized ontology.

2. No psychological mechanism (such as deletion or addition of features, analogy, generalization, specialization) can transform a concept in one ontological category to a concept in another ontological category.

A second way to capture the reality of ontological categories is to show that entities in distinct ontological categories cannot be transformed physically from one category to another. No physical operations (such as surgical operations, movement) can transform entities in one ontological category to entities in another. For example, "men" can participate in a "race" by moving and running (a physical operation), but "moving men" have not been transformed into a race: The men still preserve the identity of men.

Ontological categories can be determined empirically to be psychologically real and distinct if they are modified by distinct sets of predicates. A linguistic test commonly used by philosophers — the sensibility of the predicated term — can determine whether two categories are ontologically distinct or not. As illustrated in Figure 1, each of the major categories (matter, events, and abstractions) generates a "tree" of sub-categories. A predicate (indicated in quotes) that modifies one concept will sensibly modify all other concepts below it on the same (branch of a) tree (commonly known as "dominate"), even if the modification is false. Thus, a bee has the potential to be "heavy" even though it is false, whereas a bee cannot be "an hour long." Conversely, an event can be "an hour long" but not "skinny." The point is that predicates on the same tree can modify concepts below it sensibly even if it is false, because the truth or falsity of the sentence can be checked. However, when predicates from a different branch of the same tree or different trees are used to modify concepts on another branch or another tree, then the sentence does not make sense. For instance, it makes no sense to say "The thunderstorm is broken." Such statements are called category mistakes. "The thunderstorm is broken" is not merely a falsehood, for otherwise "The thunderstorm is unbroken" would be true. "The thunderstorm is broken" is a category mistake because "broken" is a predicate used to modify physical objects made of material substances, whereas thunderstorm is a type of event, thus belonging to the other ontological category. The psychological reality of some of the ontological categories depicted in Figure 1 has been tested by Gelman (1988) and Keil (1979; 1989). Besides the sensibility judgment task, Keil (1989) has also used a physical transformation task such as surgical operations to show that even young children will honor the distinction between natural kinds and artifacts.

In general, categories are ontologically distinct if one is not a superordinate of the other. Thus, branches on the same tree can presumably form distinct ontological categories as well, as in the case of the distinction of natural kind from artifacts. For purposes of discussion, the three major ontological categories in Figure 1 will be referred to as the "trees," and "branches" that do not occupy a subordinate-superordinate role may be considered the basic ontological categories.

Three caveats are necessary. First, Figure 1 can be a hierarchy reflecting intrinsically real ontology (in which the term "entities" will henceforth be used to refer to the category members). By intrinsically real, I mean an idealization based on certain scientific disciplinary standards. There can be another tree that corresponds to a psychological ontology (in which the category members will be referred to as "concepts"). The idealized and the psychologically represented ontological trees should be isomorphic, even though they may not be. (In fact, the thesis presented here addresses precisely the problem that arises for learning when there is a mismatch between the intrinsic and the psychological ontology.) There is no literature or psychological evidence about this isomorphism (or lack of it).

A second caveat is that the distinction between basic and major ontological categories is made here for the sake of discussion. Whether or not there is a greater physical and/or psychological distance among major ontological categories than among the basic ones within a tree is not known. This is an empirically tractable question though. A third caveat is that the argument put forth here assumes an intrinsic and a psychological distinction among the trees (or major ontological categories) only. The extent to which the branches are ontologically distinct (e.g., Are fish and mammals ontologically distinct? What about plants and animals?) remains an epistemological and psychological issue that is not to be addressed here. For instance, the criteria described in the above list can also apply to the distinction between classes and collections (Markman 1979). To what extent classes and collections are ontologically distinct needs to be explored.

1.2. Assertions of the Theory

The theory proposed in this essay makes two assertions. First, that conceptual change within ontological categories requires a different set of processes than conceptual change across ontological categories. In fact, the kind of learning processes that can foster conceptual change within an ontological category is inappropriate initially for achieving conceptual change across ontological categories. Moreover, it may be inappropriate to think of conceptual change across ontological categories as a change at all. It may be more proper to think of this kind of radical change as the development or acquisition of new conceptions, with the initial conceptions remaining more or less intact.

Figure 2 depicts the two types of conceptual change schematically. For conceptual change within a tree (depicted by the two trees *a* and *b* in the left column of Figure 2), the concepts themselves do not change their basic meaning; thus, the shapes of the nodes in Figure 2*a* and 2*b* are preserved as the original tree evolves. What can change is the location of the nodes in the context of the tree: The concepts have migrated. Such migration can result from the concepts having acquired more attributes, or certain attributes may become more or less salient, and so forth. Examples of migration will be presented in the latter half of this essay. Thus, another good way to characterize nonradical conceptual change is reorganization of the tree, or perhaps even restructuring (although the term *restructuring* tends to imply radical conceptual change, so I will refrain from using that term).

For radical conceptual change (depicted by the tree *c* in the second row of Figure 2), the nodes themselves change meaning (thus represented by different shapes in Figure 2*c* from the original tree). And there is actually no isomorphic mapping between the nodes of the two trees, even though there may be superficial correspondences. For example, Black's concept of heat (which was differentiated from temperature in the eighteenth century) corresponds only superficially to the Experimenters' undifferentiated concept of heat (more analogous to hotness). This is what philosophers refer to as incommensurability of concepts. Thus, it is proposed here that radical conceptual change refers to an *outcome* of change in which a concept's original assignment to a category has shifted to a new assignment. One might conceive of a concept as initially belonging to the original tree, and subsequently after radical conceptual change it then belongs to the tree depicted in Figure 2*c*. The question that has been puzzling researchers is what mechanism moves a concept from one tree to another totally distinct tree. To preview a bit, the thesis to be put forth in this essay is that it may be inaccurate to think of conceptual change as a *movement* or *evolution* of a concept from one tree to another in a gradual or abrupt manner, as in the case of conceptual change within a tree. Rather, such concepts on distinct trees may be better developed independently, so that there is really no *shifting* per se, although the resulting outcome represents a *shift*. Thus, new conceptions on an ontologically distinct tree can be developed *gradually*, and yet the final outcome of the development (the shift) may appear to occur *abruptly*. This separation of outcome and process, I think, explicates the confusion in the literature about the abruptness or gradualness of conceptual change.

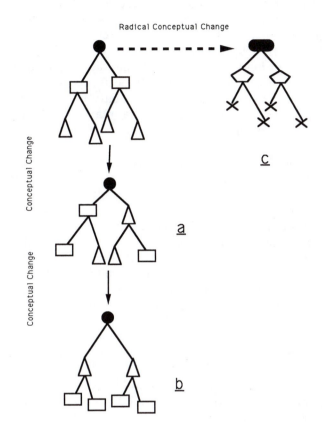

Figure 2. A schematic depiction of radical and nonradical conceptual change.

This view has two important implications. First, a concept may continue to remain on the original tree, even though its corresponding counterpart has developed on the new tree; and second, this view obviates the need to search for a mechanism that accounts for such a transition. Thus, this radical shifting (or lack of it) is depicted in Figure 2 by dotted lines.

Radical conceptual change (the outcome) must be more complicated than simply reassigning a concept to another category. Simple reassignment can occur on many occasions, such as when we are told that whales are mammals even though we initially thought that they were fish. This particular reassignment may be easy for adults to make for two reasons. The first possible explanation is that perhaps mammals and fish do not constitute distinct ontological categories for adults. That is, we know that mammals and fish are both kinds of animals; therefore our initial conception that whales are fish is merely false, and not a category mistake (Keil 1979). Category mistakes occur when we attribute a concept with a property that belongs to another ontological category. Thus the ease of reassigning whales to the class of mammals can take place precisely because no ontological boundary has to be crossed. An equally viable alternative explanation is that mammals and fish are distinct basic ontological categories, and the ease of reassigning whales to mammals is due to the fact that as adults, we are quite familiar with both the fish and mammals categories, so that we can easily attribute whales with all the characteristic and defining properties of mammals; and furthermore, we can verify that whales indeed do possess some of these properties.

By our definition, then, radical conceptual change requires that a concept be reassigned from one ontological category to another. If this definition prevails for radical conceptual change, then it is obvious that it must occur infrequently. So, for instance, we find it difficult to change our belief that an entity that belongs to one ontological category (such as a human being) can be changed to an entity in another ontological category (such as a spiritual being). Furthermore, a prerequisite for crossing such an ontological barrier is to have some understanding and knowledge of the target category to which the concept is to be changed. Our first assertion, then, is that different sets of processes are necessary for conceptual change within and across ontological categories. The exact nature of these processes will be elaborated later.

The second assertion is that learning certain science topics or domains, such as physics, requires conceptual change across ontological categories, and this is what causes difficulty. This is because the scientific meaning of physical science concepts belongs to a different ontological category than naive intuitive meaning. Some basic physical science concepts (those that are learned in an introductory physics course) are conceived by naive students as belonging to the ontological category of material substance, whereas scientists conceive of them as entities belonging to the ontological category of "constraint-based events" (a term I shall use to refer to entities such as current, see Figure 1 again).[1] Therefore, for physics entities, there is a mismatch between their intrinsic properties and how they are perceived psychologically. Students treat them as belonging to the same ontological category as other material substances, whereas in reality they are physically distinct. Thus, students perceive physics concepts to belong to the same ontological category as material substance, so that they use the behavior and properties of matter to interpret the behavior and properties of constraint-based events. This means that in learning physics, students need to differentiate their intuitive ontology of these concepts from their physics ontology.

The argument put forth in this essay, regarding the implication of a mismatch in the intrinsically real and the psychologically represented ontological categories for learning and discovering of science concepts, constitutes a preliminary sketch of a theory that implicates the psychological effect of ontological categories for learning and development. Although there is hardly any work in the literature about the psychological reality of ontological categories, my assertions are supported by the following three sets of general evidence. First, abundant data show that it has been extremely difficult to capture learning and understanding of basic physics in the laboratory or in the classroom, and similarly, it had been historically extremely difficult to discover the correct Newtonian views. This difficulty is consistent with the claim that such learning, understanding, and discovery require a radical conceptual change in students' and scientists' perception of individual physics entities from a type of substance to a type of constraint-based event. That is, to achieve radical conceptual change in this case is to learn to differentiate the two ontological categories, and not to borrow predicates and properties of the material substance category to interpret events in the alternative category. Second, there is an explicit similarity in the conceptions of physical science concepts by naive students and medieval scientists, and I interpret this correspondence to arise from both groups of people having adopted a material substance view. Third, the processes of discovery by medieval scientists and the processes of learning by naive students must be similar because they both require this radical conceptual shift. Details of this evidence will be presented later.

These assertions also have the following two implications: First, conceptual change across ontological categories is relatively difficult, not because the cognitive mechanism producing the change is complex and not well understood (as is often proposed for understanding transition, restructuring, and so forth), but because no cognitive processes (such as discrimination, differentiation, generalization, etc.) can *change* the meaning of a concept from one kind of a thing to another. For instance, adding or deleting features to the concept of "men" in a semantic network will not change "men" into the concept "war." "Men" can be a component of the concept "war," but the concept "men" will remain intact as an isolatable concept. Thus, the concept of "men" has not changed. More importantly, the concept "men" is embedded in a semantic network of human beings, as a kind of animal, which is then a kind of living thing, and so on. The concept "war" cannot possibly inherit these properties of "living things," so therefore the concept "war" cannot be modified from the concept "men" by an acquisition mechanism that we are familiar with. Instead, a new concept "war" can be built that consists of "men" as a necessary feature. Second, in order to have an ontological change as an outcome, we may have to first learn about the new category, before we can assign a concept from an original category to it. This suggests that the difficulty of radical conceptual change is not in the process of change itself (the process of reassignment), but the difficulty lies in two independent processes: (1) the new category of knowledge must first be learned and understood or induced; and (2) one often has to induce the realization (i.e., problem finding) that a concept does not belong to the category that the person had originally assigned it. For example, we may never realize that force is not a kind of impetus embodied in a moving object. To realize it on our own is clearly a nontrivial induction, comparable to a scientific discovery.

2. Conceptual Change across Ontological Categories

As mentioned earlier, radical conceptual change occurs infrequently in our everyday context. It is not often the case in our daily experiences that we have to change our conception of a physical object from being of one kind to another. Furthermore, to move a concept from one subtree to another appears counterintuitive, because it violates its physical meaning in terms of what kind of a thing it is. For example, even four-year-olds acknowledge that living things are fundamentally different from artifacts. Two pieces of evidence attest to this acknowledgment. First, four-year-olds never attribute artifacts with any animal or human properties (Carey 1985; Chi 1988). Keil's work (1989) shows that even at age five, children will agree that certain physical transformations are meaningful and others not, and the distinction is drawn between ontological categories such as natural kinds and artifacts. For example, they will agree that a skunk can be transformed to a raccoon with the appropriate operations, such as shaving off its fur, replacing the black fur with brown fur, replacing the tail with another, and so on. However, five-year-olds will not accept a transformation as possible that crosses boundaries of basic ontological categories, such as transforming a toy dog to a real dog. Thus, Keil's data provide evidence that even preschoolers will implicitly honor the boundaries between conceptual structures of "different" kinds. The following two sections discuss the extent to which physical science concepts belong to an ontologically distinct category from naive conceptions, and the processes of radical conceptual change.

2.1. Learning Science Concepts

One of the major occasions for needing radical conceptual change is in the learning of science concepts. This essay will argue that what creates difficulties for students in learning certain science domains is that the learning requires a fundamental change in the category to which one's initial conception of science concepts belongs. That is, naive conceptions about physics concepts belong to a different ontological category than scientific conceptions.

Understanding the nature of naive (or mis-) conceptions is a relatively recent inquiry. In the last decade or so, science educators have finally forgone Piaget's notion of viewing the learner as bringing to the task only logical-mathematical knowledge, and instead, have begun to investigate "the substance of the actual beliefs and concepts held by children" (Erickson 1979, p. 221). Thus, it has been recognized by science educators only within the last decade that students come to the classroom with naive conceptions or preconceptions of science concepts (as opposed to no conceptions), and these preconceptions are usually incorrect from the scientific point of view; thus they are referred to as *misconceptions.* A large number of studies (totaling more than fifteen hundred; see the bibliography of research on misconceptions in Pfundt and Duit 1988) in the past decade were devoted to analyzing children's and naive college students' ideas about forces, motion, light, heat, simple electrical circuits, and so forth (see Driver and Easley [1978] for a review). These misconceptions are very robust, and are very easily documented. For instance, the typical notion of force and motion is that force is an entity that can be possessed, transferred, and dissipated, such that there need to be agents for the causes and control of motion, as well as agents for the supply of force (Law and Ki 1981). Thus, force is a kind of impetuslike substance, and it is a property of material objects. This recognition was first brought to the attention of cognitive scientists by the publications of McCloskey and colleagues (McCloskey, Caramazza, and Green 1980).

What kind of conceptual change does learning science concepts require? Although the consensus among science educators is that a radical restructuring type of conceptual change is required, analogous to accommodation (Osborne and Wittrock 1983), no direct attempts have been made to explicate exactly what kind of changes are needed, and what mechanisms might produce such changes. Many instructional and laboratory attempts at producing changes have resulted in failures. The basic issues have then evolved into identifying whether or not students' naive beliefs are theorylike, and what kind of changes are required to modify the naive theory to a scientific theory.

The reason that learning science concepts is difficult and instruction has failed, according to the present theory, is because students treat scientific concepts as belonging to the same ontological category as matter. Therefore, in interpreting instruction about science concepts, they borrow ontological attributes of material substance to explain and predict behavior of scientific concepts. This assumption is supported by an extensive survey of the science education literature, in which Reiner, Chi, and Resnick (1988) and Reiner et al. (in preparation) have suggested that the fundamental conception that underlies most of the students' conception of physical science concepts is to treat them as a kind of substance. This conclusion was based on a survey of the literature examining students' misconceptions across four different concepts: A consistent pattern was found among students of all ages, in that they attribute heat, light, current, and forces with properties of substances.

Overall, four characteristics emerge from the data we have surveyed in the literature. First, there is an impressive amount of consensus among findings of different researchers on the same concept, even though the consensus is often implicit (for example, several researchers have suggested that students' naive conceptions of heat are substance-based: Albert 1978; Erickson 1979; 1980; Tiberghien 1980; Rogan 1988; Wiser 1987). Second, there is also an implicit consensus among the findings of different researchers on the different concepts. That is, we have argued that the evidence strongly suggests that all four concepts that we surveyed — heat, light, forces, current — are treated as substance-based entities (Reiner, Chi, and Resnick 1988). Third, there appears to be no obvious developmental improvement across age, nor instructional-based improvement across educational levels, suggesting that greater learning or experience with the concepts themselves do not seem adequate to promote conceptual change. In fact, McCloskey's study that identified the existence of these misconceptions was conducted with Johns Hopkins undergraduates who had completed a course in mechanics. And finally, students' naive conceptions are similar to medieval scientists' conceptions, inviting the inference that the medieval scientists' conceptions were substance-based as well. (This point will be developed further below.)

The conclusion I make is simply that the very nature of the general robustness of the results in terms of the consistency (1) across studies, (2) across concepts, (3) across ages, (4) across educational levels, and (5) across historical periods, suggests that substance is the ontological category to which students assign these physical science concepts. Thus, in order for students to really understand what forces, light, heat, and current are, they need to change their conception that these entities are substances, and conceive of them as a kind of constraint-based event (including fields), thereby requiring a change in ontology. That is, for a lack of better terms, let us consider these physics entities, such as fields, as a type of constraint-based event: They exist only as defined by relational constraints among several entities, and their existence depends on the values or status of these other variables. For example, current exists only when charges move in an electric field between two points. A field is what fills the space around a mass, an electrical charge, or a magnet, but it can be experienced only by another mass, electric charge, or magnet that is introduced into this region. Hence, fields are neither substances nor properties of substances. Viewed this way, it becomes apparent that learning science concepts in the classroom requires radical conceptual change, in which students' initial conception of certain entities as a kind of substance has to undergo a reassignment into the constraint-based event category. This kind of reassignment cannot be achieved through any kind of acquisition mechanism, such as deletion or addition, discrimination or generalization, since such operations cannot simply transform a substance-based type of entity to a constraint-based event entity (as illustrated earlier with the concepts of "men" and "race"). One of the most compelling pieces of evidence is provided by Law and Ki. They asked high school students to write down a set of Prolog rules to describe the trajectories and forces of an object's motion. When the students are confronted with motion and trajectories that violate their rules' predictions, they then modify their rules by adding, deleting, refining, or generalizing them — mechanisms that are well-defined and implemented in computational theories of learning. However, such local patching did not change the fundamental meaning of their concept of forces, thus supporting our conjecture that radical conceptual change cannot be achieved by *revising* existing conceptions with common acquisition processes.

2.2. Processes of Radical Conceptual Change

The assertion to be proposed below about the cognitive processes responsible for conceptual change across ontological categories (i.e., in learning physics) is isomorphic to those of the physical processes in that no cognitive processes can directly transform a concept from one ontological category to another. This means that the processes of learning that cognitive psychologists currently understand — addition, deletion, discrimination, generalization, concatenation, chunking, analogizing — cannot actually transform a concept originally stored as belonging to one ontological category to another. Instead, conceptual change that requires crossing ontological boundaries must take place via a three-step procedure, as shown in the list below. The sequencing of the first two steps depends on whether the conceptual change is achieved by induction or by instruction. In any case, one of the three steps (Step 1 in the list below) is that the students must induce or be told that physics entities belong to a different ontological category. We surmise that this knowledge is induced in the context of the current mode of physics instruction, and it should probably be told instead (to be elaborated in the end of this essay). To have students find out themselves would be comparable to the processes of discovery, thus too difficult for students to undertake. In any case, a necessary step in radical conceptual change is to know about the properties and behavior of events, and in particular about this specific constraint-based kind of event.

Processes of Radical Conceptual Change

1. Learn the new ontological category's properties via acquisition processes;
2. Learn the meaning of individual concepts within this ontological category via acquisition processes;
3. Reassign a concept to this new ontological category (there are three possible sets of processes to achieve this reassignment):
   a. actively abandon the concept's original meaning and replace it with the new meaning;
   b. allow both meanings to coexist and access both meanings depending on context;
   c. replace automatically via coherence and strength of new meaning.

A second step (Step 2 in the list) in the ontological shift is to learn the individual physics concepts (such as heat, current) in a slow and laborious manner, proceeding in much the same way as traditional instruction. The learning process for this second phase occurs by the common acquisition processes that we already understand, such as deletion, addition, discrimination, generalization, and so on. However, the accretion and assimilation should be done in the context of the properties of the overarching constraint-based ontological category, rather than in the context of naive substance-based conceptions.

After such learning has taken place, in which the acquisition is embedded in the context of this event-ontological category, then the final stage (Step 3 in the above list) is for the students to "reassign" these entities (displacement view) or "identify" these entities with the event category (coexisting view). This reassignment procedure can take place in three ways. An explicit way (3a of the above list) is for the student to actively abandon the concept's original meaning and replace it with the new meaning, assuming that the person now has dual meanings associated with a concept. This explicit way is the most commonly conceived of way, such as by historians when they discuss scientific revolutions, and implicitly by many psychologists. For a simple example, it means we have to consciously think of whales as a kind of mammal rather than as fish. A better example would be for students to no longer think of forces as a kind of impetus that can be emitted by objects, but as a kind of attraction between two objects. A second way (3b) that a reassignment can take place is for the two assignments to co-occur simultaneously. That is, a concept simply takes on two meanings or gets assigned to two different categories, and one does not replace the other. The two meanings are accessed under different circumstances. This coexisting view is probably what actually happens. Evidence shows that even physicists will occasionally revert back and use naive notions to make predictions of everyday events (McDermott 1984). The third way (3c) that a reassignment can take place is for it to happen automatically, without conscious effort, by the mere fact that the concept's new meaning is perhaps more coherent and robust, since this new category of knowledge is usually learned via instruction, whereas a concept's original meaning is usually induced from the environment in everyday context, in a haphazard fashion. Thagard's (1989) computational model of theory selection can be viewed as a mechanism to account for this kind of reassignment. All three reassignment procedures can take place. No empirical evidence exists to discriminate among them.

The totality of all the processes outlined in the above list constitutes "the processes of radical conceptual change." Notice that acquisition processes (such as deletion, addition, generalization, and so on) are embedded in the subprocesses of the total processes of radical conceptual change. The point is that radical conceptual change requires the learning of Steps 1 and 2 in the above list (either from induction or from being told). Acquisition processes are responsible for the learning at these stages.

To recapitulate, radical conceptual change (or a shift) has taken place after the processes of Steps 1, 2, and 3 have occurred. Traditional physics instruction focused primarily on Step 2. We surmise that students who have successfully learned physics have induced Step 1, and executed Step 3. Instruction may have to focus on teaching Step 1 and explaining Step 3.

2.3. Discovering Science Concepts

In the previous section, the argument put forth is that *learning* physical science concepts requires the student to undergo the processes of radical conceptual change as depicted in the list in the previous section. In this section, I explore the possibility that radical conceptual change is the critical factor that underlies scientific revolutions, scientific discoveries, as well as scientific thinking by lay adults and by children.

At the level of discussion that occurs in the literature, there is no question that there exists a superficial similarity in the processes of scientific revolutions and individual scientist's discoveries, as well as children's scientific reasoning. Revolutions, discoveries, and reasoning usually proceed in a broad five-step procedure, as shown in the accompanying list.

Five-step Procedure Leading to Scientific Discovery, Revolutions, or Thinking

1. a set of anomalies exists (or an unexplained phenomenon is encountered, or an impasse is reached);

2. their abundance reaches a crisis state (or the individual is overwhelmed by the conflicts or confrontations so that he/she realizes that there is a problem, sometimes referred to as the state of problem finding);

3. a new hypothesis is made or induced (i.e., abduction);

4. experiments are designed to test the new hypothesis, and a new theory is formulated on the basis of the results of the experiments;

5. the new theory is accepted (old theory is abandoned, overthrown, or replaced).

This five-step procedure varies only slightly depending on whether one is discussing a scientific revolution, an individual's discovery, or scientific thinking. In a revolution, there are some extraneous processes between Steps 4 and 5, such as the individual scientist has to convince a community of scientists to accept the new theory and abandon the old. Below, the argument presented proposes that a critical stage of discovery that has been overlooked may lie in a preceding step, one not depicted in the above list. This step is the stage at which a radical conceptual change is made, which then leads to the recognition that an anomaly presents a problem (Steps 1–2). But in order to recognize that a situation is anomalous, the scientist must undergo radical conceptual change prior to such recognition, assuming that the nature of medieval and naive conceptions are both substance-based. Such an assumption provides a reasonable interpretation to the somewhat confusing findings in the literature showing that sometimes anomalies are responsible for discoveries and other times not. (This will be elaborated below.) Unfortunately, the majority of the existing research on scientific discoveries does not focus on this preceding step of the discovery processes.

2.3.1. A Framework for Defining Scientific Revolution, Theory Change, and Conceptual Shift

Historians and philosophers of science have been discussing scientific revolutions for decades, as a form of change that comes about from processes that are beyond an incremental, piecemeal process (T. S. Kuhn 1970). They involve changes more extensive than at the level of individual theories; rather, such changes or revolutions involve the research traditions under which specific theories are constructed.

T. S. Kuhn (1970), Lakatos (1970), and Laudan (1977) all discuss theory change at the level of paradigms, or research programmes, or research traditions (these terms will be used interchangeably): that is, what constitutes a paradigm shift, or changes in world views. Paradigm shifts may be conceived of as analogous to crossing ontological boundaries. Perhaps a visual depiction of it will make this discussion clearer. In Figure 3, I have depicted the paradigms (X and Y) as the large circles, each encompassing individual theories (A, B, ... or E, F). Each theory within a paradigm includes a set of concepts (a_1, a_2, ... b_1, b_2, ...), and these concepts are shared by the different theories (A and B) in the same paradigm. Basically, a paradigm (corresponding to an ontological category) contains a small set of overarching, core assumptions (X_1, X_2, ...): assumptions about the kind of basic entities in the world, how these entities interact, and the proper methods to use for constructing and testing theories about these entities (Laudan 1977). All scientists working within the framework of a given paradigm adopt this implicit set of assumptions. For instance, "evolutionary theory" really refers to a family of doctrines in which the fundamental assumption is that organic species have common lines of descent. All variants of evolutionary theories would implicitly make that assumption. Atomic theories refer to a large set of theories all of which are predicated on the assumption that matter is discontinuous. Likewise connectionist theories in psychology would assume that mental processes occur in parallel, as opposed to serially (Schneider 1987).

Thus, two research paradigms (X and Y) differ if they differ in their ontology and methodology, meaning that (1) their core concepts (a_1,

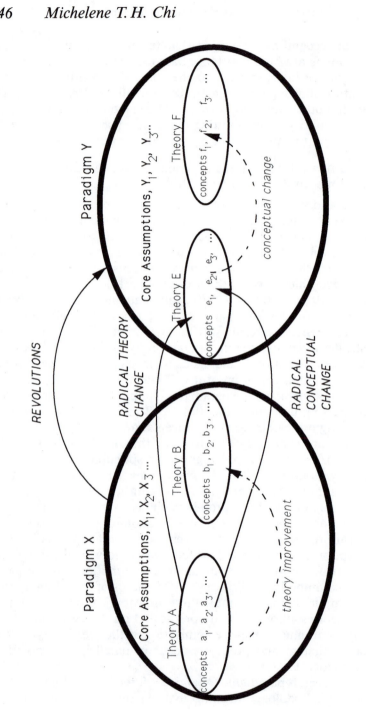

Figure 3. A schematic depiction of paradigm changes, theory changes, and conceptual changes.

$a_2, \ldots e_1, e_2, \ldots$) are ontologically different, (2) their explanations (based on those core concepts) must be different, and (3) the methodology by which the research is conducted is different. The idea is that because two paradigms are radically (or ontologically) different, it takes a revolution to overthrow one and replace it by another. By implication, the theories constructed from different paradigms (e.g., Theories A and E) would be radically different as well.

Different theories within a paradigm (A and B, or E and F), however, are not radically (or ontologically) different. They are merely modifications of one another, and they presumably assume the same set of concepts and scrutinize the same set of problems that are considered to be legitimate or admissible problems to be solved within the paradigm. Theory change within the same paradigm, on the other hand, is considered to be incremental, in the way normal science usually proceeds. Furthermore, scientists working within the same paradigm use the same set of methodologies or instruments to do their sciences. (All radical changes in Figure 3 are represented by solid arcs, and nonradical changes are represented by dotted arcs.)

Hence, in discussing paradigm shifts, historians and philosophers of science discuss the shifts predominantly at two levels: either at the level of the paradigm, such that the shift is referred to as a revolution, or at the level of individual theories embedded in *different* paradigms, referred to here as *"radical" theory change*, to be consistent with the terminology of radical conceptual change.

The proposal here is that one might want to interpret the occurrence and processes of paradigm and radical theory change to be predicated upon the existence of an underlying ontological conceptual shift. Thus, even though historians and philosophers of science discuss paradigm shift at the global level of revolutions, so that they have to worry about additional sociological and political processes involved in achieving the revolutions, this does not preclude the possibility that the impetus for such a revolution is the occurrence of an underlying radical conceptual change and a change in the explanations and principles based on these concepts. Of course, one often cannot discuss radical change at one level without simultaneously considering changes at other levels. So, for instance, although Nersessian's work (1989) discusses the historical shift at the level of a theory (from an impetus theory to a Newtonian inertial theory), because she is selecting theories that are embedded in different paradigms (thus radical theory change by definition), she is really addressing both the issues of paradigm change and radical conceptual change at the same time. However, because paradigm change is more global, we may gain greater understanding of paradigm shift if we first understand the radical conceptual change on which it is based, but the converse may not be true. Therefore, it seems critical that we understand conceptual change first.

To determine whether two theories *within* the same paradigm have undergone some significant change, two different sorts of analysis are probably required: one analysis considers the predictability and robustness of a theory in terms of how much evidence it can explain and how reliable the explanations are; the second analysis requires a determination of the structure and coherence of a theory. This is the sort of evaluation that all trained scientists are quite equipped to do. Thus, the issues that have to be considered for determining whether one theory is significantly better than another theory within a paradigm (let us call it *theory improvement*) are quite different from the issue of radical theory change (or how two theories from ontologically different paradigms differ). For theory improvement within a paradigm, one need not consider whether the two theories adopt the same concepts and/or the same methods. This is assumed. Instead, other considerations are needed, such as the criteria to be used for determining whether one theory is better than another. Thus, theory improvement is analogous to conceptual change within an ontological category, to be elaborated in the second half of this essay.

2.3.2. The Processes of Scientific Discovery, Reasoning, and Revolution

As stated earlier, by identifying the five-step procedure (see the list in sec. 2.3) as being responsible for scientific discovery, scientific reasoning, and scientific revolution, philosophers, psychologists, and developmentalists are not necessarily illuminating the dilemma that remains about how revolutions and discoveries occur. For example, this five-step process is often used to describe a child's adoption of a new theory. Karmiloff-Smith and Inhelder's data (1975) are a classic example, illustrating a discovery that is akin to the five-step process. Karmiloff-Smith and Inhelder observed what implicit theories children use to balance various beams whose center of gravity has been surreptitiously prearranged. Children began balancing by implicitly using the "theory" that the center of gravity necessarily coincides with the geometric center of the object. (Note that a more appropriate term to use here in describing their work might be "strategy" rather than "theory.") However, because the beams had been "fixed," Karmiloff-Smith and Inhelder were able to see when and how children would modify their original geometric-center theory. Children began by refusing to change their strategy of balancing the beam in its geometric center even though the feedback

5, Lavoisier thought that the gas obtained by Priestley was
elf except perhaps it was more pure;

77, Lavoisier finally concluded that the gas was one of the
main constituents of the atmosphere.

t the conclusion that Lavoisier came to was never reached by
, and in fact, oxygen was isolated much earlier by Scheele, in
y 1770s. Hence, what led one scientist to the proper recognition
others? Similarly, what led Faraday to recognize that the brief
le effect" of his galvanometer occurring when he connected the
the battery was a critical observation (Tweney 1985)? Or what
d Alexander Fleming to notice that the mold that killed the bacte-
at he was trying to grow on sterile agar plates was important? The
is that the knowledge embodied by some scientists is represented
way to prime them to recognize that a given phenomenon is an
ortant anomaly — a common finding in the perception literature;
rwise anomalies are rejected.

There are also many examples of assimilation. That is, anomalies are
en not treated as an unusual event, but instead are assimilated into
existing knowledge and explained away by it. Vosniadou and Brewer
989) have some great examples of assimilation. In their study of young
ildren's conception of the earth's shape, Vosniadou and Brewer found
hat children assimilate the new correct information that the experimenter provided in numerous interesting ways. For example, if children
think of the earth as square, and you tell them that it is round, they
will then think that it is round like a pancake, rather than round like a
sphere. If further instructed with the fact that the earth is round like a
sphere, children then reconcile their conception of a flat earth with the
information that the earth is a sphere by imagining a disklike flat surface
inside or on top of the sphere, and people residing on that disk. These
are clear examples of assimilation in that new information is incorporated into an existing knowledge structure. (Although young children
do not remove their misconception about the shape of the earth that
people stand on, this example of robustness illustrates a different cause
of resistance than the factor of ontological change [see Chi, Chiu, and
de Leeuw 1990].)

In those instances when new pieces of information cannot be assimilated, they become anomalies that can often then be explained away by
ad hoc explanations. Abundant evidence of ad hoc explanations can be
seen in the science education literature.

Besides the fact that anomalies can be rejected or assimilated or explained away, there is no reason why an individual or a scientist cannot

contradicted their predictions. Thus, tl
be a confrontation of anomalies. How
amples, children eventually developed t
their original geometric-center theory. N
Inhelder's use of the term "theory change
defined in this essay. (Such diverse uses ol
confusion to the literature.) In any case, bec
tific thinking and scientific discovery are ofte
"the metaphor of the lay adult — or the child
has gained wide acceptance in the last decad
probably inappropriately so.

At this global level, the only key difference b
lutions and individual scientist's and/or a child
the case of the individual, a scientist or a child
him/herself, whereas in the revolution case, the scie
a community of scientists, as well as the communi
the fact that such global mapping between historica
dividual analyses is too vague to be of any use, two
unanswered with respect to the first two steps of the li
First, empirically, there are myriads of countereviden
anomalies do not always foster discovery or conceptual
dividual. That is, Steps 1 and 2 (of the list in sec. 2.3) are
to produce Steps 3 and 4. This is particularly clear in em
using a training-type of paradigm. It is also evident in Piage
their replications. Instead, anomalies are often rejected or
or ad hoc explanations are provided to account for them. And
often rejected because they have not been recognized as suc
they are often rejected by theorists as inconsequential evidenc
is abundant evidence in both the developmental literature as w
cases of scientific discoveries of such failed recognition. For ex
the following sequence of events has been identified as having
Lavoisier's discovery of oxygen (taken from T. S. Kuhn's account [
pp. 53–56]):

1. In 1772, Lavoisier was convinced both that something was amiss
   with the phlogiston theory, and was already anticipating that
   burning bodies absorbed some part of the atmosphere;

2. In 1774, Priestley had succeeded in collecting the gas released by
   heated red oxide of mercury and identified it as nitrous oxide;

3. In 1775, Priestley further identified it as common air but with
   less than its usual amounts of phlogiston;

simultaneously operate under two theories, such that the anomalies can be explained in the context of one theory but not the other, as Laudan (1977) has also proposed. Therefore, there is no clear conclusion to be drawn about when and how anomalies can take the role of triggering a paradigm shift or conceptual change, unless we first understand the nature of the person's (or scientist's or child's) knowledge representation. This view is shared by Lakatos (1970), who has also claimed that paradigm shifts are not necessarily brought about by the accumulation of anomalies.

The second problem with such a global analysis of similarity in the five-step process of discovery, reasoning, and revolution is that the mechanism for what triggers discovery in both the historical and the individual cases is still not clear. It has been extremely difficult to capture the triggering event. For example, it may be nearly impossible to study the mechanism of discovery through historical analysis. Although the approach of diary analysis might be promising, the diary entries are often spaced too far apart to capture the processes of radical conceptual change. And numerous uncontrolled factors may play roles that are impossible to reconstruct historically.

Further, it has been equally difficult to capture the actual process of discovery in psychological laboratory studies. Most of the time, no overt observations can be made for inferring the internal mechanism (for example, in Karmiloff-Smith and Inhelder's data, the actual processes of shift could not be ascertained). At other times, the kind of evidence that has been captured for the presence of "invention" or "discovery" of a new strategy (which, as stated above, is not *akin* to discovering an ontologically new theory) is nothing more than a longer pause before a new strategy is adopted, or longer latencies. Siegler and Jenkins (1989), for instance, showed that the trial during which young children shift from using a simple counting strategy (the sum strategy) in which they add up two single-digit numbers (let us say $5 + 3$) by counting each hand (1,2,3,4,5; 1,2,3) then counting both hands together to get the sum (1,2,3,4,5,6,7,8), to a more sophisticated min strategy (in which they assert rather than count the quantity of one hand — 5, then increment the amount of the other hand, 6,7,8) was preceded by a trial with a very long latency, on the order of twenty seconds. It is not exactly clear what occurs during those twenty seconds of the preceding trial. Was the child testing the new strategy? In other words, perhaps the child simply adopted the min strategy, and double checked that it was accurate, thus requiring the extra twenty seconds. This length of time is consistent with this interpretation. However, the crucial question is: What made the child think of adopting this new strategy in the first place? Likewise,

in traditional verbal learning tasks (Postman and Jarrett 1952), subjects were either informed or uninformed about the principle underlying
the paired associations. However, uninformed subjects' performance improved dramatically on and after the trial at which they could state the
principle. There is no direct evidence to reveal exactly what triggered
the ability to finally state the principle, other than a noticeable improvement for a few trials preceding the first trial at which the principle was
stated. An interpretation similar to the one I proposed for the Siegler
and Jenkins result can be offered here as well, namely that in the trials preceding the principle-stating trial, subjects could have incidentally
made the correct response, and took notice of the rule governing the response. When they are certain that the rule applies for a few trials, they
then finally announced it. So, the trial at which they announced it simply reflects the point at which their confidence about the rule that they
have induced has reached a critical level. What triggered the induction
is still not captured.

Thus, it may actually be next to impossible to capture the very moment of transition, either during strategy shift or during some process
of discovery. The best one can hope for is to model computationally the
mechanism that might be taking place, and see whether the mechanism
proposed actually predicts the timing and location during which "discovery" or "strategy shift" actually occurs. This is precisely what has been
done by VanLehn (1991). (Note that a caveat is necessary: To repeat, the
aforementioned strategy-shift literature is not entirely analogous to radical conceptual change, which may not occur in an instantaneous way,
as the processes of the list in section 2.2 indicate. These works are cited
here to illustrate a methodological difficulty.)

An alternative research approach is to worry less about the actual
mechanism that may be responsible for the discovery or strategy shift,
but rather, be more concerned about the structure of the mental representation at the time that such a discovery was made, so that the
occurrence of an event that matched the representation could trigger
the process of recognition, and thereby the subsequent processes of discovery. In other words, perhaps the fruitful questions that should be
asked are those pertaining to Steps 1 and 2 of the discovery processes:
how anomalies are seen as such, and how they reach crisis proportion,
rather than Steps 3 and 4, in which one is concerned primarily with the
process of formulating the final hypothesis. "Discoveries" are made only
when the scientist's knowledge is represented in such a way that a specific recognition of the problem can be appropriately made. This is the
same rationale I have given to the early problem-solving work, in which
we focused on the initial representation that experts and novices bring

into the problem-solving situation, rather than focusing directly on the processes of problem solving as a search through a set of permissible states, thereby ignoring the role of the representation as it dictates the space from which the experts and novices search (Chi, Feltovich, and Glaser 1981). Thus, here, I propose the same line of analyses: unless we can clearly lay out what the scientist's initial representation is, we will not be able to fully uncover the mystery of the discovery process itself.

Because the nature of a scientist's representation in the process of discovery has not played a significant role in the research agenda of cognitive, developmental, and computational scientists, they have naturally focused their attention on the five different stages of this discovery process as depicted in the list in section 2.3. The majority of the research has focused on the processes of Steps 3–5 of the list (D. Kuhn 1989; Klahr and Dunbar 1988). Langley et al.'s work (1987), for example, addressed the problem of discovering laws given a set of empirical observations. For instance, BACON.5 can look for a linear relation among a set of variables by using a standard linear-regression technique. If the variables and their values fit a monotonically increasing relation, then the ratio between the two terms is found, if the signs of the values of the two terms are the same. If a monotonically decreasing trend occurs, then a product is defined, and so on. BACON.5's discovery is restricted to finding an appropriate fit of a mathematical expression to the empirical values that correspond to a prespecified set of variables. This process corresponds to the formulation of a new theory or principle (in the form of a mathematical expression) on the basis of experimental results (Step 4 of the list in sec. 2.3).

Much computational work is emerging concerning the other processes of discovery as well, such as the mechanism leading to the formulation of new hypotheses (Step 3 of the list in sec. 2.3, Darden and Rada 1988), and with designing experiments (Step 4, Falkenhainer and Rajamoney 1988; Dieterich and Buchanan 1983; Rajamoney 1988). Neither of these works focuses on the preceding stages of discovery — namely, the conditions of the knowledge representation that triggered "problem finding" or "recognizing a phenomenon as an anomaly," which is clearly viewed here as the most critical part of the discovery processes. In artistic productions, Getzels and Csikszentmihalyi (1976) also believe that problem finding is the most creative aspect.

In sum, in this section, I have argued that identifying what anomalies led to scientific discoveries, and/or inducing regularities in empirical data, and/or formulating the right hypotheses to test, may not be the loci of scientific discovery. Instead, scientific discoveries may be predicated upon the occurrence of radical conceptual change, which occurs prior to the five-step process. Radical conceptual change may have already occurred prior to the (1) recognition that a phenomenon is anomalous, (2) induction of regularities that exist in the empirical observations, and (3) formulation of hypotheses to test. Since these processes may occur after radical conceptual change has taken place, research efforts focused on these processes may not lead to an understanding of what triggered the processes of discovery, beginning with a recognition of anomalies.

2.3.3. Scientific Thinking

Because of the interest in the processes of discovery, there is a large body of literature concerning scientific thinking in the lay person. The idea behind this research is that the reason few discoveries are made (in either the everyday context or in scientific fields) may perhaps be due to the fact that children and the lay adults do not reason in a scientific way. Reasoning in a scientific way refers essentially to the methods that are used by scientists or what has been referred to as "the scientific method" (corresponding basically to the five-step procedure of the list in sec. 2.3). Hence, this line of research explores in what ways children and the lay adults are more naive in the way they reason: Is it because they do not weight all the available evidence? Is it because they cannot formulate all the hypotheses exhaustively? Is it because they do not search in a dual space? Is it because they do not engage in analogical reasoning? (See Klahr and Dunbar 1988; D. Kuhn, Amsel, and O'Loughlin 1988.) In short, most of the research in cognitive science, developmental psychology, and artificial intelligence concentrates on Steps 2–5 of the scientific method. Although the developmental findings offer interesting results, naturally showing that younger children are less efficient at scientific thinking than older children or adults (for example, younger children consider and/or generate fewer alternative hypotheses than older children), it is difficult to conclude from such results that younger children think less scientifically than older children or adults. An equally viable interpretation is that the form of reasoning most scientists can and do engage in is available to novices and perhaps children as well, but they merely lack the appropriate content knowledge to reason with. For example, Chi, Hutchinson, and Robin (1989) have shown that both experts (in the domain of dinosaurs) and novice children (of equal ability and age, about five to seven years old) are competent at using analogical comparisons to reason about novel dinosaurs (analogical reasoning is considered to be a significant part of the scientific method). The difference is that the expert children picked the exactly appropriate dinosaur family with which to incorporate and compare the novel dinosaur, whereas the novice children picked the closest animal family to make their analogical comparisons. The outcome of course is that the expert children made all the correct inferences about a novel dinosaur, whereas the novice children made many incorrect inferences, since they based their analogies on an inappropriate source. Likewise, Schauble et al. (1991) looked at poor and good solvers' predictions of the outcome of a simple circuit, and found that the solvers' understanding of the circuit does exhibit the use of analogy. Both the good and the poor solvers show the same frequency of analogy use. The success, then, seems to depend on picking the right analogy, one that shows deep understanding of both the analogy and the target domain. Thus, the crucial matter is not whether you are or are not able to use analogies (part of a scientific method), but whether your knowledge base allows you to pick the appropriate analogy and reason from it.

Although it is probably true that people who have not made discoveries, or children who use more naive methods or mental models to solve a problem, are more likely to be less systematic and use less sophisticated strategies, this may be an outcome, rather than a cause, of their poor performance. Thus, the conjecture proposed here is that the ability to recognize a problem (Steps 1–2) is the crucial process that determines how successfully one can subsequently formulate a good hypothesis, and perform an experiment, and so forth. This would suggest that being *told* what the problem is may not be as effective, since the knowledge structures required to be able to "recognize" the problem are not necessarily available to the scientist who is simply told the problem.

Conversely, even though the literature is full of evidence depicting the lay person's and the child's fallacious reasoning, it has also been shown that scientists can reason fallaciously as well. For example, it is not uncommon for scientists to show confirmation biases: that is, testing hypotheses that can confirm their beliefs rather than disconfirm them (Faust 1984; Greenwald et al. 1986). Thus, although research on scientific thinking may uncover interesting differences between the way children and adults reason, or between the lay adults and scientists, my interpretation of such differences is that they are the outcome, and not necessarily the cause, of scientific discoveries. That is, a person tends to think systematically, and/or generate hypotheses exhaustively, and/or use analogical reasoning appropriately, only if he/she has a well-organized representation of the problem space that he/she is working with. Therefore, analyzing the reasoning strategies per se will not tell us what representation the reasoner is acting on, and whether it is the representation that dictated the pattern of the reasoning processes.

In sum, the argument put forth in this section asserts that in order for scientific discoveries to occur, a scientist or an individual must first be able to recognize that there is a problem. This recognition in turn requires that the person's knowledge be represented in such a way as to permit such a recognition. This implies that some form of radical conceptual change (in the case of discovering in the physical sciences) has probably taken place so that the knowledge representation is primed for such recognition. Thus, radical conceptual change may underlie the entire (five-step) discovery process, rather than an efficient use of "the scientific method," for the kind of ontological changes discussed here.

2.4. Similarity between Medieval Theories and Naive Conceptions: What Constitutes a "Theory"?

There are basically two ways in which similarity between medieval beliefs and naive beliefs have been drawn. One of the ways is to point to similarity in the actual beliefs or in the underlying assumptions. Using the actual-belief approach, Nersessian and Resnick (1989), for example, pointed out three categories of beliefs that constrained both the medieval as well as the contemporary views about motion, and these are: (1) all motion requires a causal explanation; (2) motion is caused by a mover or "force"; and (3) continuing motion is sustained by impetus or "force." Similarly, using the underlying-assumption approach, McCloskey proposed that students' naive intuitions resemble the assumptions made by the medieval scientists. For instance, the naive notions about motion are like the impetus theory discussed by Buridan in the fourteenth century in that both the naive notions and the impetus theory use the following two assumptions:

1. that an object set in motion acquires an internal force, and this internal force (impetus) keeps the object in motion;

2. that a moving object's impetus gradually dissipates (either spontaneously or as a result of external influences) so that the object gradually slows down and comes to a stop.

Wiser (1987) also uses this kind of argument to make claims about the similarity between naive conceptions of heat and those held by medieval scientists holding the source-recipient model.

A second way to point to the similarity between intuitive beliefs and medieval beliefs is to appeal to the notion that they are both theorylike. Thus, science educators are concerned with the issue of whether naive theories or conceptions conform to the notion of a "theory." The problem is how to define what a theory is, and how one can tell whether one has a theory or not. This is a problem that philosophers have tackled extensively (e.g., see Suppe 1977). Science educators and cognitive scientists implicitly assume the definition of a theory to be one that applies to a scientific theory. Thus, in order to decide whether naive students' conceptions or beliefs are theorylike, many researchers point to the similarity between the students' thinking and that of medieval scientists, in a global sort of way. Besides drawing upon parallels between the contemporary students' views and the historical views in a global way (as illustrated in the foregoing paragraph), these researchers assume that naive students' beliefs are theorylike in the scientific or nomological sense of a theory, one in which the laws play a significant role in scientific explanation (Hempel 1966). Viennot (1979), for example, even tried to fit mathematical formulas to students' beliefs, thus raising the status of their beliefs to quantifiable laws. Viennot goes so far as to describe the students' conception in terms of mathematical equations, such as $F = \beta v$, and referred to these expressible pseudolinear relations as intuitive "laws." Clement (1982) also thought of students' conceptions as zeroth-order models that can be modified in order to achieve "greater precision and generality" (p. 15). A less principle-related way of claiming that students' naive conceptions are theorylike is to point to the general characteristics of their conceptions, which are (1) consistent or not capricious, and (2) robust and resistant to change.

It is easy to understand how the notions of a theory are borrowed from the scientific notion of a theory, since we are preoccupied with learning science domains, and all science domains are defined and bounded by their theories. Scientific theories tend to satisfy the classical positivists' view of a theory, in the sense that it explains a wide set of phenomena, is coherent (in the sense that there are no internal contradictions), and is often deductively closed. However, implicitly adopting the scientific definition of a theory may be the wrong assumption for determining the psychological nature of a "theory"; in other words, determining whether a body of knowledge from which one can generate explanations constitutes a "theory" or not is equivalent to determining what underlies a coherent theoretical structure for naive explanations. That is, does the knowledge from which naive explanations are generated have a coherent internal structure? Assessing the psychological coherence of such a knowledge structure should determine whether naive explanations are "theorylike." Furthermore, whether a set of naive explanations is theorylike cannot be determined by a single criterion, such as whether or not it is analogous to medieval conceptions. At the very minimum, one should determine whether naive explanations are theorylike on the basis of at least four criteria: consistency across studies, across concepts, across ages and/or schooling, and across historical periods. But preferably, one should develop methods to assess directly the structure and coherence of the knowledge base that generated naive explanations on an individual basis.

There is the alternative, contrasting view that the initial conceptions the student holds are not a theory at all in either the scientific sense or any sense of a theory, but merely an "untidy, unscientific collection of meanings" that are socially acquired and personally constructed (Solomon 1983, p. 5). In order to know that a set of meanings is not coherent, one must determine first what coherence is: Solomon basically meant it designated being consistent and logical, and perhaps also noncontradictory. Because students often seem unbaffled by contradictions in their explanations, Solomon therefore concluded that their world views are incoherent. (As noted earlier, ignoring contradictions need not imply that students' theories are or are not coherent.) DiSessa (1988) likewise holds the view that students' naive conceptions are not theorylike, but instead are disjointed, piecemeal, and fragmented. These pieces of knowledge are derived from phenomenological experiences, such as "overcoming" and "dying away" (to correspond to dissipating forces). Students' explanations are composed of a series of these phenomenological primitives, joined together to explain a particular event.

The reason that both of these theoretical views (a theorylike view and a non-theorylike view) are viable is because the empirical evidence seems to support both. Most of the time, students' explanations are robust, consistent over time, and sometimes consistent across different situations that tap the same concepts. But at other times, students' explanations seem ad hoc, and students seem to be able to accept and produce contradictory explanations, or ignore them. There are two ways to resolve such discrepancies. On the one hand, one could dismiss any discrepancy between the two views by saying that DiSessa's nontheory view, in principle, could be correct, and yet it need not necessarily be viewed as a refutation of McCloskey's and others' claim that naive explanations are theorylike. Instead, one could think of DiSessa's analysis as a reduction of the components of a theory into fundamental primitives. Such reductions need not be viewed as supplanting the claim that the naive beliefs are theorylike (Haugeland 1978). Reductions can be viewed as independent of the issue of whether or not the set of beliefs constitutes a coherent theory. On the other hand, the discrepancy might be resolvable if we adopt a uniform definition of a "theory." For instance, a theory might be a set of coherent primitives, so that the question then becomes: Are the phenomenological primitives coherent in some way?

There is no question that there exists a superficial similarity between contemporary students' beliefs and medieval theories, independent of the issue of whether or not they are theorylike. However, the similarity is not at all surprising if one considers the origins of these naive conceptions. This is the occasion in which historical analogy is useful: for it may suggest what the origins of the misconceptions are, such as whether they are perceptually based on the external world, as in the case of physics concepts, or whether they are based on internalized knowledge, as in the case of some biological concepts. They must have arisen from our and the medieval scientists' daily experiential encounters with the physical world, with "objects in a friction-full world" (Clement 1982; Nersessian 1989). Such experiences began at birth, and Spelke (1990) has some fascinating data showing that even six-month-old infants are sensitive to the fact that solid objects such as a ball cannot pass through another solid object, such as a shelf. Nersessian (1989) goes on to point out, correctly, that such parallels are only weak ones, or exist only at a global level, because there are clearly fundamental differences between naive students and medieval scientists that are trivial to point out: the strongest argument against taking a strict parallel view is that in contrast to the conceptions of naive adults or children, medieval conceptions were "developed in response to problems with Aristotelian physics and metaphysics as well." Furthermore, these "historically naive conceptions are products of centuries of reasoning about problems of motion" (pp. 4–5). Twentieth-century students are not plagued by these metaphysical and epistemological problems. Brewer and Samarapungavan (in press) point out many additional differences between scientists and naive students. Nevertheless, there remains a consistent but weak similarity between the naive reasoning of the twentieth-century students and the reasoning of the medieval scientists. Although such similarity no doubt arose, "in part, from intuitions based on . . . interaction with the natural environment — with how things appear to move when dropped or thrown" (Nersessian 1989), the assumption of this essay would predict that such similarity must have arisen from the underlying assignment of these physical science concepts into the material substance category by both the medieval scientists and the naive students. That is, from their daily experiences, both the naive students and the medieval scientists have formed a substance schema that they use to judge, interpret, predict, and understand physical phenomena. Thus the coherence observed in naive explanations arises from the structure of the underlying substance schema.

I am not proposing that a general case can be made regarding all scientific concepts; that is, that all scientific revolutions are based on an ontological conceptual shift, and that naive conceptions are similar to the medieval ones because they are both based on the same underlying ontological belief. For instance, in biology, Harvey's discovery of the circulatory system, although hailed as a major breakthrough for modern scientific medicine, comparable in significance with the work of Galileo in physics (McKenzie 1960; Pagel 1951), may have characteristics to indicate that this discovery may not have required radical conceptual change of the kind that physical science concepts undergo. To support this point, we can use the four criterion characteristics cited above to see whether students' learning about the circulatory system has the same characteristics as are required for learning physical science concepts. Based on a survey of the scant literature available: (1) there basically appears to be no consistency across naive students' misconceptions about the circulatory system (Arnaudin and Mintzes 1986; Gellert 1962); (2) there was little consistency across different ages; (3) there was little consistency across schooling (in fact, there are usually developmental improvements in what students can understand about the circulatory system; see, for example, Gellert's data [1962]); and most importantly, (4) there seems to be no similarity at all between students' naive conceptions and the medieval misconceptions, as those held by Servetus, Columbus, and others (see progress report by Chi et al. 1989). Thus, I take these characteristics of the evidence to suggest that conceptual change across ontological categories is necessary for certain physical science concepts, whereas in some other types of scientific discipline, drawing on the topic of the circulatory system in particular, conceptual change does not require an ontological shift. In fact, the difficulty students have in learning about the human circulatory system may have nothing to do with the historical barriers that prevented the scientific discoveries from taking place. (The reasons why students may have difficulty learning about systems are spelled out in Chi 1990, and are independent of the issue of radical conceptual change.) Such differences in the nature of the underlying conceptual change among different scientific disciplines (or even different topics within any given discipline) have direct implications for learning and instruction.

The interpretation offered here can make sense out of seemingly discrepant data. Osborne and Wittrock (1983) noted that the proportion of children who held an ancient Greek view of mechanics *increased* from age thirteen to fifteen, whereas the proportion of children who considered a worm and a spider not to be animals *decreased* significantly from age five to eleven. Such discrepant findings can be easily interpreted in the current framework, in which misconceptions about mechanics can increase with age because an ontological shift did not occur with development, whereas misconceptions about animals can decrease because no ontological shift is needed to remove these sorts of misconceptions — they are based on learning that occurs with schooling and development. Examples of this sort are numerous.

In sum, this section addresses the issue that has plagued the research in science education — whether or not naive explanations conform to a theory. Note that the issue of theorylikeness is really quite independent of the issue of theory change. Theorylikeness represents a digression that has been misleading the research agenda, because it does not shed any more light on what constitutes theory change.

2.5. Evidence of Radical Conceptual Change

Now that we have some sense of what radical conceptual change is, is there any empirical evidence of success at eliciting it? There are numerous attempts that failed, especially in the educational literature. I am aware of only two attempts that claim to be somewhat successful, and these two will be discussed to see whether radical conceptual change has been successfully induced, elicited, or instructed. Since radical conceptual change is necessary primarily in physical science concepts, the examples will be taken from that domain (Ranney 1988; Joshua and Dupin 1987).

Ranney's dissertation data (Ranney 1988) attempted to generate evidence to determine the extent to which empirical and analogical feedback can produce coherence-enhancing reorganization among naive beliefs about kinematics and dynamics. Table 1 shows the five tasks Ranney used and the sequence of testing and feedback employed. Task 1 asks students to predict by drawing the trajectories of pendulum bobs that have been released from different points during a swing. Task 2 asks for the trajectories of heavy objects being dropped or thrown. Some of the problems in Tasks 1 and 2 are isomorphic. Task 3 asks students to match each pendulum release problem in Task 1 with a dropping/throwing problem in Task 2. That is, students are asked for a similarity judgment. Task 4 is a near-transfer task, consisting of problems involving a trapeze and a wrecking ball, rather than a pendulum bob. Task 5 is a far-transfer task in which the problems ask for trajectories of projectiles released in the absence of gravity. Table 1 shows a simplified version of Ranney's design: students are given pretests in all the tasks, feedback on Tasks 1 and 3, prediction on Task 2, posttests on Tasks 2, 4, and 5, and delayed posttests on all five tasks. The main result shows that from pretest to delayed posttest, there are significant improvements in all tasks except Task 5, the far-transfer problems.

	Task 1 Pendulum	Task 2 Dropping	Task 3 Similarity	Task 4 Near transfer	Task 5 Far transfer
Pretest	X	X	X	X	X
Feedback	X		X		
Repredict		X			
Posttest		X		X	X
Delayed	X	X	X	X	X

Table 1. Ranney's Tasks and Sequence of Testing and Feedback

This pattern of results is not surprising given that students were given feedback in both the pendulum release trajectories as well as the matching task (of pendulum release and dropping/throwing problems). Feedback in Task 3 on the similarity between problems of Tasks 1 and 2 allows the students to see the similarity in these types of problems without deep understanding. Therefore, they cannot make accurate predictions for the far-transfer problems. Although Ranney (1988) also reported evidence of greater consistency among the responses from pretest to posttest, such as more symmetrical predictions of leftward and rightward swings of the pendulum release problems, such increased consistency does not imply that students have gained greater understanding of the underlying kinematic knowledge. Such symmetrical responses can be gleaned from the symmetrical feedback. In sum, although there is no question that students' responses after feedback did become more consistent and coherent, this does not imply deeper understanding even though deeper understanding usually does lead to consistent and coherent responses. Ranney himself concluded that "deep conceptual changes, i.e., from impetus to inertia, were rare" (p. 6).

Joshua and Dupin (1987) attempted to demonstrate conceptual change in the context of understanding. That is, they attempted to teach understanding of electric current by analogizing it to a moving train. Naive students think of electric current as some kind of flowing fluid (instead of moving charges); thus they cannot understand how this fluid (the current) can be both conserved while it circulates (i.e., the intensity of the current is the same in the circuit), and yet wear out in the battery (i.e., providing the energy). An analogy of a train circulating on a closed loop track is provided. The analogy is made up of cars only (without locomotion) rigidly linked. In a station, workers permanently push on the cars going past in front of them and influence the train speed. Obstacles also exist in the track that influence train speed (corresponding to resistance). The men pushing the train correspond to the battery, the muscular fatigue of men corresponds to the wearing out of the battery, and the car flow-rate corresponds to the current intensity, which should be the same at each point on the track because the cars are rigidly connected.

The claim is that this analogy enabled the students to understand that the current is the same everywhere (just as the car speed is the same everywhere on the track), and the battery can wear out just as the men can get tired from pushing. It is easy to see how students can learn some specific facts from this analogy without generalized understanding. What the students may have understood is the mapping from the local inferences that they can naturally draw about men and muscles, to the electric circuit. For example, they can draw the inference that if men continuously push on the train, then they will get tired, so that by analogy a battery must also wear out after persistent use. Likewise, the students can also draw and map the inference that if cars are rigidly attached, they must travel at the same speed; therefore, if electric charge is comparable to a rigidly attached train, then charge too will have the same intensity (speed of the train) throughout the whole circuit. They thus are asked, in a sense, to take the inferences they can naturally draw (tired men, same speed of train) and apply them directly to the circuit. However, it is not clear what can be gained from such limited and fragmented understanding. No relationship necessarily exists between the concepts of men and train, on the one hand, and the concepts of battery and current, on the other hand. For example, students will not understand that what *caused* current to flow is the potential difference that is provided by the voltage source (the battery), such that when the charge stops flowing, both ends of the electric conductor reach a common potential. There is not a comparable concept in the train-men analogy. Another way to put it is to ask what causes the current to cease flowing. There is a precise, predictable point at which current ceases: when there is no longer a potential difference. Thus, it is the difference in potential between two end points, rather than a single source (the men pushing), that causes current to flow. Thus, although this study claimed to have demonstrated conceptual change, I believe that only very localized mapping of inferences was made.

In sum, neither of the two studies achieved deep conceptual change of the kind discussed in this essay. This is because deep or radical conceptual change is not something that can be achieved easily through one or two sessions of instruction, feedback, or analogical illustrations.

2.6. Fostering Radical Conceptual Change
in the Context of Instruction

Assuming that students' naive intuitions are different in kind from the scientific view (that is, the naive concepts are assigned to a different ontological category than the scientific concepts), what instructional strategy can be offered to revise this intuitive belief? Erickson (1980) summarizes a few standard procedures: (1) provide the students with a lot of experience with the concept (such as heat), (2) discuss and clarify these experiences, (3) create anomalous situations in which the outcome of the situation is counterintuitive to their initial ideas, and (4) restructure their initial ideas through group discussion and teacher intervention in which the inconsistency between accurate interpretation and students' ideas is pointed out. Suggestions 1 and 2 are the aims of traditional classroom instruction, and they presumably have failed to foster understanding of physical science concepts. Suggestion 3 is problematic in numerous ways that I have already cited. But to recapitulate, it has been found repeatedly that counterposing the students' intuitive ideas with the scientific theory does not lead to radical conceptual change: the pre-instructional intuitive ideas may remain, although sometimes hidden or covered up with new words, or the new and old ideas coexist (Driver and Easley 1978). These instructions use a number of different tactics, such as providing counterexamples and conflicting evidence (Driver 1973). Again, the use of conflicting evidence is supposed to provoke a state of "cognitive conflict" or "mental disequilibrium" in the case of instructing children, so that the child may restructure (Posner et al. 1982). (The term "restructure" here is borrowed from Posner et al., and I use it in a generic sense. But in this case, it does refer to radical conceptual change.) However, because cognitive conflict does not yield new understanding, it is taken as evidence in favor of Piaget's notion that learning certain science concepts depends on preexisting logical operations.

The only novel suggestion is the fourth one, namely, that restructuring may be achieved via group and paired discussion. One of the ideas of group or paired problem solving is that it gives students an opportunity to be aware of their preconceptions and encourages them to compare and confront their ideas with the scientific ones. Such confrontations presumably would force them to explain the empirical observations. Thus many people are employing this approach (e.g., Minstrell 1988). But it is not entirely clear what aspects of these activities can produce cognitive changes, nor why such self-realized conflicts are more potent in producing change than those confrontations presented by the teacher, which we already have said are useless. From these group problem solving events, we also cannot know what each of the individual student's initial intuitive ideas are; therefore we cannot possibly know how malleable the individual student's ideas are to change. Although group or pair discussion has certain appeal as an instructional approach (Chi and Bjork, in press), it is not clear how group or pair discussion can promote radical conceptual change per se. Moreover, there is equally convincing negative evidence showing that group discussion in a class is not particularly productive at inducing conceptual change of the radical kind (Joshua and Dupin 1987).

Finally, there are other approaches to instructional interventions beside group discussion. The majority of these use novel technological advances, such as computerized microworlds, in which the concepts can be presented using visual displays (Roschelle 1986), so that they become more concrete (Champagne and Klopfer 1982; Wiser 1987; White and Fredricksen 1986). The overall success of these new approaches has not been systematically evaluated. However, I am skeptical about the tacit assumption underlying these approaches, which is that constraint-based events can be better understood by using concrete substance-based models in the instructional materials.

Any proposed instruction for learning physical science concepts must obey two constraints: First is that human processors understand new information by assimilating it with existing information; and second, that learning physical science concepts requires changing the ontology of the concepts. What kind of instruction can be proposed that fits these two constraints? There are two possible approaches. The first is the revision approach: that is, the attempt to modify a student's initial or naive conception. There are several ways to achieve this. The most obvious and direct way is to force revision by presenting other features and properties of the physics concept by building upon student's naive concepts (through adding, deleting, etc). We had already discussed why this would not make sense, because the naive concepts hang on a different ontological tree than the physics concepts, so that the naive concepts naturally inherit properties from their superordinate categories that would not be correct for the scientific meanings of the concepts. A second revision approach is to present conflicts. Again, we had discussed the discrepant role of conflicts. Conflicts between the student's predictions of a physical phenomenon and a scientific one can lead only to frustration and not to conceptual change (as discussed earlier in the context of the Law and Ki results), because the student cannot possibly *understand* the scientific prediction. Such understanding is the bottleneck to learning the new concept because we are constantly trying to understand this new ontology of concepts by using our substance-based schema. A third and more creative application of the revision approach, proposed by Clement (1982), is to *bridge* the new concepts to the existing concepts. Bridging is a slightly different notion than revising. Bridging means that you find a situation in which the student's naive understanding is consistent with the scientific one, and then build your instruction around this naive but correct understanding. Clement (1982), for example, picked a universally believed idea that is correct (such as that a hand does push up on a rock that the hand is holding) as an anchor from which to bridge an understanding of the idea of normal force, such as that a table would also be pushing on a rock, even though a table is inflexible. The progressive analogies that the student is taught (from the hand, to a spring, to a table) seem to work (that is, students now do believe that a table pushes up on a rock). But it is difficult to see what kind of understanding the students have gained about forces by accepting isolated facts that tables and other hard surfaces apply forces, since they never did understand why a hand pushes up on a rock in the first place. To understand *why* a table or a hand pushes up on a rock is equivalent to understanding the nature of forces.

Besides the revision approach, an alternative approach proposed here is the three-step procedure, corresponding to the processes of radical conceptual change outlined in the list in section 2.2. These processes include merely teaching the new concept and its domain independently (Step 2 of the list in sec. 2.2), much as the way traditional instruction has proceeded, but in addition, to add to such instruction of the specific content domain by emphasizing the ontology of this whole category of concepts (Step 1 of the list in sec. 2.2). So, for instance, such instruction would underscore the ontological attributes of constraint-based events (as opposed to matter) such as being dynamic, time-based, not concrete, often not visible, and so forth, and that many physical science concepts fall into this category. Teaching (or pointing out) such broad-based category knowledge would allow the students to build the bridge that links new domain information about physics concepts with the appropriate ontological tree. This means that the new knowledge can be built or assimilated in the context of the appropriate ontological tree, so that the concepts would not mistakenly inherit properties of the substance category. Thus, the bridge for assimilation is not with existing misconceptions, but with knowledge about the general characteristics of such an ontological category. That is, students could be given instruction first on the nature of this category of constraint-based events, so that additional learning about the concepts within this category can then be assimilated. As Feynman (1963) said, "One has to have the imagination to think of something that has never been seen before, never been heard of before." This instructional approach implicitly implies that students' naive views could be kept intact, since they are quite adequate for predicting everyday phenomena, and an independent knowledge structure should be developed for these new "never been heard of before" concepts.

The notion of having two independent knowledge structures operating — one naive and one formal — seems to have some support. There is evidence to suggest that people who have expert knowledge in physical science domains do seem to maintain two independent knowledge systems about physical science concepts (as if they have two separate microworlds). That is, experts often do fall back and use their naive intuitive notions to make simple predictions about their everyday physical events. This suggests that the naive notions have not been replaced by the scientific theories, but they in fact coexist with the scientific concepts. However, the naive notions are often not accessed, probably because the scientists or experts know that they are inappropriate. However, when scientists are caught off-guard, such as when they are requested to make predictions in an everyday context, or when they are not given time to reflect, they do resort to using their naive notions (Viennot 1979).

There are some informal observations that such an alternative instructional approach is viable. An analogy to the distinction between matter and events may be extensive and intensive quantities — the former is substance-based and the latter is not. Such a distinction between extensive and intensive quantities occurs throughout numerous physics concepts, such as the distinction between heat and temperature, charge and potential, momentum and velocity. A number of instructional approaches in Germany have attempted to introduce the distinction between extensive and intensive quantities, and then have used it to teach the physics concepts. Informal observation by S. Kesidou (personal communication, April 26, 1991) on ten students instructed this way showed that they seemed to have more correct and deeper understanding than students taught in the traditional way.

3. Conceptual Change within an Ontological Category

Conceptual change within an ontological category is by no means mundane. It is often quite difficult to achieve. But the difficulty resides in a number of factors, such as a lack of knowledge, or the complexity of the memory search processes, or the need for extended practice, and so forth, as opposed to a difficulty required by a reassignment across ontological categories. The critical distinction between conceptual change across versus within ontological categories is that concepts in the *within* case can migrate up and down a given tree, but not across the branches, since the branches can represent distinct ontological categories. So, for instance, "plants" and "animals" can merge to form a new superordinate category, "living things," but "plants" cannot become a part of "animals." There are various types of conceptual change within an ontological category. Five will be discussed.

3.1. Revision of Part-Whole Relations

The first type captures a kind of conceptual change in which a concept changes the level it occupies within the tree: Therefore, the concept's part-whole relationship relative to other concepts in the same tree has changed. Let me illustrate with a widely studied topic in developmental psychology. A commonly investigated subject in the developmental literature concerning conceptual change has to do with the concept of "living things." To illustrate the problem, I have depicted a young child's (usually intended to mean preschoolers) conception of "living" and "nonliving" things, based mostly on an interpretation of Carey's results (as shown in her 1985 book in Table 3.1, pp. 80–81). Shown in the top half of Figure 4 is a generic depiction of a young child's concept of "living," which has at least two separate categories, "animals" and "people" (whether "plants" exist as a separate category at this age is not completely clear from the data, so I have depicted that category in broken lines). Actually, a more accurate way to depict this is to say that "animals" and "people" are clearly distinct categories for the child's representation, both probably subsumed under the abstract category "living."

Although it is not clear what kinds of attributes are associated with the "living" node at this age, it is clear that there does exist an implicit contrasting category of "nonliving" things, corresponding to artifacts. Evidence for the existence of the "nonliving" category will be presented below. (However, because of the tenuous existence of the "nonliving" category, I have depicted it in a broken-lined oval also.) Aside from the issue of conceptual change, this literature also is concerned with questions such as whether or not children can make inductions based on category membership, and whether or not children exhibit false semantic mapping between the term *not alive* and the concept *dead*. We will not be dealing with these side issues explicitly.

To justify my depiction of a young child's representation, three sets of converging evidence will now be presented to support it. First, not only do young children exclude people from their category of animals (as when they are asked to pick out all the animals from a pile of pictures that include people), but furthermore, they do not project people

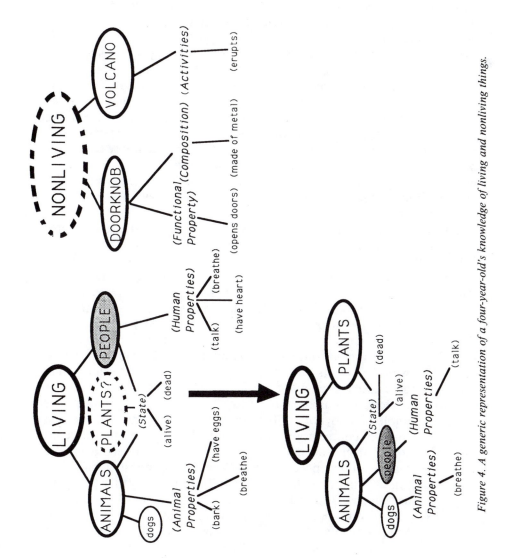

Figure 4. A generic representation of a four-year-old's knowledge of living and nonliving things.

attributes (such as eating and breathing) to other animals (such as hammerhead and stinkoo, which they know are different kinds of animals, when they are asked to pick all the animals out, Carey 1985). Second, although the categories of "living things" and "nonliving things" are never explicitly mentioned, it is obvious that they exist in a young child's representation and may be distinct categories, because properties of living things are *never* attributed to nonliving things such as a volcano or a cloud (Carey 1985; Chi 1988). For example, young children would never agree that "a doorknob breathes" (Chi 1988). Finally, Carey's data are convincing in favoring the interpretation that responses to attribution questions such as "Does a stinkoo breathe (or sleep, get hurt, have a heart, eat)?" are generated by comparing a specific animal in question to its likeness to people, since children know that these probed attributes are attributes of people. Taken together, these pieces of evidence suggest that it is safe to depict "people" as a separate subcategory of "living things," occupying an equivalent level as the subcategory "animals" within the hierarchy.

So far, the top half of Figure 4 simply shows a static and generic representation of a four-year-old's representation of "living" and "nonliving." What changes with development? Presumably, "people" eventually becomes a subset of "animals," and "plants" may gain equal status with "animals" as another kind of "living things." Thus, conceptual change from an inaccurate to an accurate adult representation would involve the migration of the "people" category to a subcategory of "animals," thereby sharing equal status with other instances of "animals," such as "dogs" (shown in the bottom half of Figure 4). This kind of conceptual change, in which concepts either ascend or descend branches of a tree, will be noted as a *change within the same tree* or ontological category, and will be referred to as reorganization.

This example should not confuse two separate claims. On the one hand, one could of course argue that a child who differentiates the concept of "people" from "animals" is fundamentally different from the child who has these concepts coalesced. What I am proposing, basically, is a definition of "fundamentally different." My point is that this kind of conceptual change, although rather dramatic, may not be radical in the sense that no ontological categories were crossed. Thagard (1989) may be making a similar discrimination in a kind of change that he calls *branch jumping*. It makes no difference at what level of the tree the changes are made. The point is that the change is within the same ontological category, occurring in an ascending or descending order. It is not the case, for example, that "people" have become a kind of "plants," which presumably are distinct, "basic" ontological categories.

How might this kind of conceptual change come about? It is clear, from both intuition and Carey's data, that such changes can come about from the acquisition of biological knowledge. One can conceive of necessitating two kinds of acquisitions: First, children need to be told that "people are a kind of animal" so that they can begin to form links between "people" and "animal" categories. Second, they need to learn about many biological features of people as well as other animals, and notice that these features are shared. For instance, if they are instructed in or exposed to the knowledge that giraffes have live babies and suckle their young, and that people do the same, and other animals as well, then over time, the overlapping and correlational nature of these features will be noticed (or stored), and "people" will eventually be one kind of "animals." Thus, storing many new features will ultimately change the nature of the linkages among the concepts, as well as change the nature of their salience (as engendered by frequency of encounters, and so forth), so that a different pattern of linkages will emerge. (See Chi, Hutchinson, and Robin [1989] for more exhaustive explanations, and Chi [1985] for some suggestive evidence of such a developmental progression.)[2] The point is that simple acquisition processes such as encoding or storing factual information, discrimination, generalization, and so forth, can account for reorganization.

3.2. Formation of New Superordinate or Subordinate Categories

An obvious second kind of conceptual change within an ontological category is the formation of new superordinate categories (integrating or coalescing "plants" and "animals" into the "living things" category), or the differentiation of a category into two, such as differentiating "chairs" into "high chairs" and "rocking chairs," in which the categories do not have to be ontologically distinct. Rosch and Mervis (1977), for example, showed that three-year-old children can sort pictures only at the basic level, whereas by age four, they can correctly sort at the superordinate level. My assumption is that the formation of such superordinate categories clearly is learned by simple acquisition processes. This can be supported by the evidence collected using either expert and novice children (five- and six-year-olds who are knowledgeable about dinosaurs), or expert children asked to sort a more familiar and a less familiar set of dinosaurs (Chi and Koeske 1983; Chi, Hutchinson, and Robin 1989). That is, my colleagues and I have shown that children at the same age can sort either at a more basic level or at a more superordinate level, depending on the amount of knowledge they have about dinosaurs. Likewise, children will sort a subset of dinosaurs at a more basic level if they are less familiar with the subset, and sort them at a more superordinate level (on the basis of diet) if they are more familiar with another subset. These results, together, imply that with greater acquisition of knowledge, children learn to coalesce categories on the basis of some more abstract attribute, such as diet.

Again, several possible acquisition processes can explain this kind of conceptual reorganization, such as addition, deletion, generalization, and discrimination. Thus, this kind of formation of superordinate categories and differentiation of subordinate categories seems to be explainable by nonmysterious acquisition processes. Note, however, that differentiation of heat from temperature (Wiser and Carey 1983) requires conceptual change across ontological categories, and cannot be achieved by the processes discussed here.

3.3. Reclassification of Existing Categories

In the previous example of conceptual change, distinct categories (whether at the superordinate or subordinate levels) did not yet exist. And mechanisms of differentiation and integration (or possibly induction) produced the new categories. These new categories are usually produced from the results of acquiring additional features about the concepts. In the case here, categories already existed, and conceptual change in this case requires the formation of new categorical structure without necessarily involving the acquisition of new features. Thus, it is primarily a reorganization of an existing structure. The point is that migration of concepts within a tree can occur with or without the addition of new knowledge. Again, let me illustrate with a specific piece of suggestive evidence from the developmental literature (Chi 1985).

In this study, whose original concern addressed the role of a representation and its interaction with strategy usage, I showed that a five-year-old represented her classmates in an open-classroom type of setting in terms of a spatial layout of the room. (This is not uncommon among young children. See evidence provided by Bjorklund and Bjorklund 1985.) Thus, for concepts that children are very familiar with (classmates), they have their classmates coded hierarchically according to their seating arrangement in the room. In this case, there were four separate sections, with five to six children per section. The children in this case were a mixture of boys and girls, from first and second grades, racially mixed. This representation was assessed by asking the child to retrieve all the names of her classmates. The organization of this particular representation was determined by several converging pieces of evidence. First, retrieval order conformed to a seating arrangement in the sense that the same set of children from each seating section was recalled in a cluster. Second, the cluster or seating section boundaries were further evidenced by long pauses in retrieval. Finally, the child could tell you that this is how she retrieved the names of her classmates.

This evidence shows, first of all, that this particular representation is quite robust. The robustness can be further confirmed by the finding that when the child was asked to retrieve in some other categorical way, such as "all the girls," "all the second graders," or "all the black classmates," the retrieval order obeyed the seating section boundaries. That is, in order to retrieve in a specified categorical way (such as "all the girls"), the child basically searched through each seating section, and named the girls. The reason one could conjecture that a search had occurred was because the retrieval proceeded at a much slower rate, but nevertheless, there was still the same pattern of differential retrieval rates: shorter pauses within a seating section, and longer pauses between sections. Although the study did not continue, there is no question in my mind that with extended practice, the child could eventually reorganize her initial robust representation, from one in which the classmates were organized by their seating section, to one in which they were organized by whatever categorical structure I specified, such as the "girls" versus the "boys" categories. (Actually, with repeated trials of retrieval in a pre-specified way, a new representation may be constructed with the original one — by seating arrangement — remaining intact.) Thus, here is an example of a kind of reorganization that is possible to implement with practice, and it is not the radical kind, in the sense that each individual concept (each classmate) has not changed its meaning, but its status within the tree has changed (rather than belonging to the category of a seating section, it could, with practice, be stored in the category of "boy classmates" and "girl classmates"). Each concept has migrated within the same tree. This example is somewhat analogous to the jury decision work to be mentioned later (Hastie and Pennington 1991), namely that the evidence for the crime has not changed during jury deliberation, but the category to which the crime fits has changed.

The kind of learning mechanism that can account for this kind of migration is similar to the acquisition mechanism postulated earlier: namely, that continued re-formation of new categories according to some selected features can eventually be stored, so that they can be accessed later.

There are probably many other variants of the three types of conceptual change described above. Their similarity lies in the facts that, first, they can all be seen as migration of one sort or another within the same tree; and, second, straightforward acquisition processes can account for their changes, usually based on the similarity of shared features. The outcome of the behavioral manifestation after the conceptual change, however, can be rather dramatic, so that it is often difficult to believe that simple acquisition mechanisms have produced the changes, unless a computational model has been implemented to explicate the learning processes. However, many researchers do recognize this kind of change as more pedantic, so Carey has referred to it as *weak restructuring*, and Keil referred to it as *nonradical*. Further evidence for this kind of reorganization can be seen explicitly in Chi, Hutchinson, and Robin (1989).

3.4. Spreading Associations in Insight Problems

There are many other occasions when conceptual change is required, and it is often difficult to achieve. But this difficulty is of a different sort than the one imposed by radical conceptual change. The difficulty can result, for example, from a complicated search process. A good example is the issue of insight to the solution of puzzle problems. Insight refers to the phenomenon during which a sudden new way of looking at a situation or problem occurs. A classic example is Maier's (1970) two-string problem, in which the problem solver is presented with two strings hanging from a ceiling with a distance between them that is too far to reach one of the strings while holding on to the other. The goal is to tie the two strings together. Lying about is a set of apparently irrelevant objects, such as a hammer, and so forth. The solution is to tie the hammer to one of the strings and swing it like a pendulum weight, so that it can be caught while holding on to the other string.

The Gestalt psychologists saw restructuring primarily as a perceptual problem (although the Gestalt psychologists discuss many other kinds of problems as well; see Ohlsson 1984a for a review). Thus, somehow one is able to "see" an object from a different vantage point, either in what it looks like, or in its function. However, the solution to this type of insight problem clearly fits the definition proposed in this essay for conceptual change, namely that a concept (such as a hammer) has to change its categorization from a kind of tool to a kind of swinging object. How difficult such reconceptualization is depends, according to the view presented here, on how distinct the two categories are psychologically. Ohlsson (1984b), for example, presents a theory that uses the mechanism of memory retrieval as a basis for the reconceptualization. Thus, in order to see a hammer as a pendulum, one may spread activation to (and thus retrieve) related concepts, such as that a hammer is a kind of tool, and tools are kinds of artifacts, and a type of artifact is a clock, and a specific type of clock is a pendulum clock, and a part of a pendulum clock is the pendulum. Note that this activation process is equivalent to what has been referred to in this essay as traversing up and down a tree.

Thus, the idea is that this migration is within a given tree or ontological category. Thus, although the mechanism that accounts for the Gestalt type of reconceptualization is far from trivial, it seems to occur with far greater frequency than conceptual change of the radical kind. In Maier's two-string problem, for instance, typically around 30 percent of students were able to solve it correctly. The percentage of successful solution can increase to perhaps 80–90 percent, depending on the amount of hints given. This rate of success clearly far exceeds the frequency of scientific discovery, and also outnumbers the percentage of the student population who understand physics.

Even though the solution of Maier's two-string problem may be the within-a-tree kind, nevertheless, one can envision what may cause difficulty in such reconceptualization. The path relating hammer to pendulum is quite lengthy in terms of the number of links, and the links are probably not very strongly associated, given that such a path was probably never directly taught, nor activated previously. That is, at every junction where a node is associated with several links, the probability that the correct path is associated during the search is low. Thus it is unlikely that the right path will be found by spreading activation among various chains of weak links that are potentially un-constrained. One could potentially prime the activation by selectively suggesting or activating some of the intermediate links. For instance, suppose we remind the solver that certain clocks have pendulums that swing, a pendulum is usually made of a heavy object, and that swinging changes the location of an object. One would then predict, if such a theory is correct, that systematic reminding of this sort can enhance the probability of achieving "insight" or conceptual change of this kind.

Thus, Ohlsson's model basically conceived of restructuring as a kind of search in which the meaning or representation of a concept is changed so that an existing operator or rule can apply toward reaching a solution. Thus, in the hammer-pendulum case, the solver already has the rule of swinging a pendulum. In order for that rule to apply in the two-string problem, the solver had to reconceptualize the hammer as a pendulum. This would predict that the ease with which an "insight" problem is solved depends upon the degree to which the reconceptualization has to migrate, either across or within a tree. For instance, a simple example of reconceptualization may be to change the concept at hand in terms only of its subordinate-superordinate relation, in order for an operator to apply. An example that Ohlsson gave concerns the situation in which one wants to have peace and quiet, and a canary is making lots of noise. The situation can be represented as follows:

(Goal: PEACE AND QUIET)
(X IS A NOISE SOURCE)
(X IS A CANARY)

The operator that a person has that can solve this situation is SCHOO-AWAY. But the operator SCHOO-AWAY(X) has the following conditions for applicability:

(Goal: PEACE AND QUIET)
(X IS A NOISE SOURCE)
(X IS A BIRD)

Therefore, in order for the SCHOO-AWAY operator to apply, an inference has to be made, on the basis of world knowledge that a canary is a bird. Thus, in this example, the reconceptualization of the situation's concept, canary is a bird, is one of subordinate-superordinate relation. As discussed earlier, this kind of reconceptualization is not radical, since it does not require an ontological change. It is comparable to the kind of reconceptualization needed to comprehend the riddle "A man who lived in a small town married twenty different women of the same town. All are still living and he never divorced a single one of them. Yet, he broke no law. Can you explain?" Again, here, the reconceptualization needed is to spread activation from the node "man" to its subordinate node "clergyman." It seems less difficult than the reconceptualization of a hammer as a pendulum, in terms of the distance (number of links) one has to traverse within one's semantic memory representation. Perhaps due to the distance of search needed to transform a hammer into a pendulum (and, therefore, the low probability that this transformation will succeed), this kind of problem solution is usually credited as "insightful" or "creative." Hence, we may want to restrict the definition of "creativity" to this sort of conceptual change, one that is not necessarily between trees, but merely one that is unlikely to occur, due to the distance that activation has to traverse and spread from the original concept. Hayes (1989), for example, categorizes Galileo's invention of the science of physics along with telling a simple bedtime story as creative acts. Perhaps a distinction of the kind proposed here will separate out the creative acts of different magnitudes. Telling an interesting and unique bedtime story is analogous to the hammer-pendulum kind of creativity, and Galileo's invention would be consistent with the radical restructuring of the across-tree kind.

3.5. Direct Reassignment within Ontological Categories

Occasionally, as mentioned earlier, we may have to reassign a concept from one category to another directly, such as the example of whales being mammals rather than fish. When we are required to make this kind of direct reassignment, we usually think either there is something about the concept that we do not know (in the case of whales, we may not know that they bear live young), or there is something about the category to which it actually belongs (mammals) that we are ignorant of. The next example illustrates this latter case.

Hastie and Pennington (1991) have some preliminary data that may be interpreted as representing conceptual change of this kind. Their work examines jury decision. In particular, I am interested in the case of the jury members changing their opinion during the course of deliberation. In order to understand what causes the jurors' change of opinion, Hastie and Pennington analyzed the contents of the discussion two minutes prior to an individual juror's announcement that he/she has changed his/her opinion about the proper verdict. From such an analysis, what was uncovered is the fact that the discussion preceding changes of opinion did not focus on "story-relevant" events concerning what had happened in the criminal circumstances, but rather, on the law, such as the definitions of the verdict categories and the appropriate procedures the jurors should apply in determining the appropriate classification of the event to the verdict category. Thus, I take this to be an example of conceptual change in which a given concept (in this case a criminal situation) is changed in terms of what category (of verdict) it belongs to. And in this particular example, the trigger of the change is elaborating and defining more clearly what the other category is. This suggests (and is consistent with our proposal about instruction, discussed earlier) that in order to have a conceptual change, adequate knowledge must be available first about the "other" category into which the concept is to be reclassified.

The extent to which this kind of reassignment is difficult depends on a different set of processes than those involved in the other kinds of nonradical conceptual change. In the example presented here, the evidence suggests that the difficulty resides in the processes involved in refining the definition or knowledge about the category to which the concept is being reassigned, or refining the knowledge about a concept (as in the whale-mammal case).

3.6. A Caveat

Nonradical conceptual change is not necessarily straightforward. One cannot simply input an attribute, for example, and hope that it will be encoded and attached at the appropriate place and modify the representation readily. The best example to illustrate this would be Vosniadou and Brewer's study of the encoding of an attribute of a concept that violates a child's existing knowledge of the concept's attributes. In this case, the child is told that the earth is round like a sphere rather than flat like a pancake, as the child initially believed. Although this kind of modification would not be considered radical conceptual change, the data show that children find it difficult nevertheless to accept the new feature, and furthermore they have a tendency to assimilate the information instead, such as thinking that the parts of the earth that people stand on are flat, residing inside the sphere. My only explanation of this resistance to change is that learning a new feature is not as straightforward as what the instruction entails. That is, although it seems easy enough for us simply to tell the children that the earth is round like a sphere, what we do not understand is the fact that any new piece of information presented to the child must be assimilated into the child's *coherent* knowledge structure, which means that the new piece of information has to propagate to other pieces of knowledge, and they all have to make sense. For instance, simply telling the child the earth is round like a sphere does not explain to the child how we can stand erect on a spherical surface. Thus, because a child may not understand the consequences of having a spherical earth — that people stand on its surface rather than its inside, and that the earth's surface is so massive that we do not notice its curvature — these associated pieces of information must also be taught before the child can understand the impact of the factual statement that the earth is round like a sphere. Thus, my point is that the instruction must be coherent as well, and cannot be delivered in a piecemeal fashion. The children's resistance to and difficulty in modifying their concept of the earth are tangential to the factor involved in radical conceptual change, because in this case, no distinct ontological categories have to be crossed in order to change their original conception from a flat earth to a spherically shaped earth. The barrier lies in other problems, such as the correctness and/or coherence of the child's initial mental model (see Chi 1990). In any case, all nonradical conceptual change processes are distinctly different from conceptual change that requires crossing ontological categories.

4. Conclusion

I have proposed that there are basically two kinds of conceptual change: one within an ontological category, and one across ontological categories. Conceptual change within an ontological category, although at times difficult, is not impossible. Insight problem solution, in which the solver can finally see a "hammer" as a "pendulum" in Maier's two-string problem, may be an example of a difficult case of conceptual change within an ontological category. Another example may be a child's ability to see a yellow flower as both a flower (a superset) as well as a yellow flower (a subset). To be able to view a specific item as belonging to two conceptual categories at the same time may be viewed as a kind of conceptual change that has to cross categories, but these categories are not ontologically distinct.

Conceptual change that requires crossing ontological categories is nearly impossible to accomplish, in both physical as well as psychological terms. There are no concrete operations that can transform a physical object, such as a cup, into an ontologically different entity, such as a dream. Likewise, there is no psychological learning mechanism that can modify a concept from one ontological category to another. No mechanism of addition, deletion, generalization, discrimination, specialization, proceduralization, and so forth can change the meaning of a concept from one ontological category to another, because any of these mechanisms applied at a local level can only modify a concept in a local way, which means that it will continue to inherit the properties of its superordinate category. Therefore, the implication is that instruction about a new ontological category must proceed by teaching this new ontological category of concepts independently of the old or existing conceptions. This would be the case for teaching physical science concepts; for example, physical science concepts belong to a distinct ontological category (constraint-based events), whereas naive conceptions of these physical science concepts consider them to be a kind of substance or material kind.

Drawing a distinction between these two kinds of conceptual change has important implications for learning and development. It can clarify a number of contradictory findings. Let me list several instances. Piaget and Inhelder (1974) attributed the formal operational child (or the adult) to have understanding of density, and yet few lay adults understand the concept of weight. One way to resolve this discrepancy is to say that density is an attribute of material kind, and is not an ontologically difficult concept to understand. (Density is mass per unit volume, whereas weight is gravitational force.) Weight, on the other hand, is a concept of the constraint-based kind, and should therefore be difficult even for adults to understand.

A second example is the discrepancy between the notions of whether students' and children's naive conceptions and misconceptions constitute a theory or not. Most of the work arguing for the "theory" view draws evidence from physics. These views, as I have said, may seem theorylike and consistent primarily because they are all based on applying the substance schema to generate the explanations. That is, their consistency derives from using this ontological category of knowledge. On the other hand, Lawson (1988) argues that children's naive views are not theorylike at all. He based his conclusion on interview data with children on various topics of biology. He noted that children most often provided a great deal of "I don't know" responses; their beliefs were not resistant to instruction or to being told; their ideas were not well articulated; and they often were not able to state the reasons for their beliefs. This scenario is in stark contrast with interviews about physical science concepts, as gathered in a large body of literature. Again, the discrepancy can be resolved by postulating that the acquisition of a majority of biological concepts does not require conceptual change that crosses ontological categories. In fact, based on such an assumption, I am currently carrying out work that shows that naive biology misconceptions about the human circulatory systems do not resemble medieval ones. Furthermore, misconceptions about biological concepts can be easily removed by instruction. And finally, biological misconceptions decrease with age and schooling (Chi, Chiu, and de Leeuw 1990). These factors, taken together, suggest that whether or not misconceptions are robust, consistent, and theorylike depends on whether an ontological shift is necessary in their removal.

Finally, by proposing two kinds of conceptual change, it is *not* implied that conceptual change within an ontological category is necessarily a simplistic process. Conceptual change is fairly difficult, and its outcome is fairly dramatic. However, conceptual change within an ontological category does occur with greater frequency than conceptual change across ontological categories, and the former may be more experimentally tractable than the latter as well. Thus, all the data in the science education literature basically constitute empirical evidence showing how difficult it is to achieve radical conceptual change. Furthermore, because it requires extensive learning about the new domain, radical conceptual change cannot be some minute event that one can capture in the laboratory during a few sessions of work. Rather, it can probably come about only after extensive learning about the new domain or ontology of concepts (such as during the process of acquiring expertise). Only after thousands of hours of learning these new conceptions could their mere abundance, coherence, and strength potentially appear to overtake the existing conceptions if one believes in the third type of reassignment process (see 3c of the list in sec. 2.2). Overtaking can be taken to mean merely that the new conceptions are accessed more frequently than the old conceptions, which may still exist and remain intact.

The distinction between the two types of conceptual change cautions against using examples from physical science to illustrate limitations in learning and transfer. For example, J. S. Brown (1989) recently characterized classroom instruction as teaching abstract knowledge that is not particularly transferable, and used physics as an example. Due to its ontological property, I think physics is a unique example that should not be cited as a general case. Also, correctly so, we all tend to think of understanding as seeing (or assimilating) something as an instance of something else that we do understand (Schank 1986; Schon 1987). However, physical science concepts probably cannot be understood that way.

Notes

Preparation of this chapter was supported by the Mellon Foundation, as well as by an institutional grant from the Office of Educational Research and Improvement for the Center for Student Learning. The germ of the idea in this essay was presented at the 1987 meeting of the American Educational Research Association, Washington, D.C., as well as at a 1989 symposium of the Society for Research in Child Development. The opinions expressed do not necessarily reflect the position of the sponsoring agencies, and no official endorsement should be inferred. I would like to thank Susan Carey, Richard Duschl, Ronald Giere, Stellan Ohlsson, Michael Ranney, and Kurt VanLehn for their helpful comments during the preparation of this manuscript. I also benefited greatly from discussions with my students, Nicholas de Leeuw and James Slotta.

1. One could argue that these physical science entities are not constraint-based events. However one wishes to classify them, they are clearly not substances. Feynman (1963) referred to them as "something that has never been seen before, never been heard of before." Therefore, to indicate that they belong to a category other than substance, we will simply consider them to be constraint-based.

2. Notice that this would involve *widening* the attributes associated with the concept of "animals," so that people and other animals will share certain attributes, and stored with each animal and people category will be their individual features, as is possible with a traditional hierarchical semantic network model. Thus, in this sense, we use the term *widening* of the conceptual category to indicate the role of the set of generic features defining the category, and not in the sense of the number of category members, as is intended by Anglin (1977) and Carey (1985).

References

Albert, E. 1978. Development of the concept of heat in children. *Science Education* 62, no. 3:389–99.

Anderson, J. R. 1987. Skill acquisition: Compilation of weak-method problem solutions. *Psychological Review* 94:192–210.

Anglin, J. 1977. *Word, object and conceptual development.* New York: Norton.

Arnaudin, M. W., and J. J. Mintzes. 1986. The cardiovascular system: Children's conceptions and misconceptions. *Science and Children* 23:48.

Bjorklund, D. F., and B. R. Bjorklund. 1985. Organization versus item effects of an elaborated knowledge base on children's memory. *Developmental Psychology* 21:1120–31.

Brewer, W. F., and A. Samarapungavan. In press. Children's theories versus scientific theories: Difference in reasoning or differences in knowledge? In R. R. Hoffman and D. S.

Palermo, eds., *Cognition and the symbolic processes: Applied and ecological perspectives.* Vol. 3. Hillsdale, N.J.: Erlbaum.

Brown, J. S. 1989. Toward a new epidemiology for learning. In C. Frasson and J. Gauthiar, eds., *Intelligent tutoring systems at the crossroads of AI and education*, pp. 266–82. Norwood, N.J.: Ablex.

Carey, S. 1985. *Conceptual change in childhood.* Cambridge, Mass.: MIT Press.

Champagne, A. B., and L. E. Klopfer. 1982. *Laws of motion: Study guide.* Pittsburgh: University of Pittsburgh, Learning Research and Development Center.

Chase, W. G., and H. A. Simon. 1973. The mind's eye in chess. In W. G. Chase, ed., *Visual information processing.* New York: Academic Press.

Chi, M. T. H. 1985. Interactive roles of knowledge and strategies in the development of organized sorting and recall. In S. Chipman, J. Segal, and R. Glaser, eds., *Thinking and learning skills: Current research and open questions.* Vol. 2, pp. 457–85. Hillsdale, N.J.: Erlbaum.

————. 1988. Children's lack of access and knowledge reorganization: An example from the concept of animism. In F. E. Weinert and M. Perlmutter, eds., *Memory development: Universal changes and individual differences*, pp. 169–94. Hillsdale, N.J.: Erlbaum.

————. 1989. Assimilating evidence: The key to revision? (Commentary on P. Thagard's paper entitled Explanatory Coherence). *Behavioral and Brain Sciences* 12:470–71.

————. 1990. The role of initial knowledge in science learning. Technical proposal from the Learning Research and Development Center, Center for Student Learning, University of Pittsburgh, pp. 72–82.

Chi, M. T. H., and R. Bjork. In press. Modelling expertise. In D. Druckman and R. Bjork, eds., *In the mind's eye: Understanding human performance.* Washington, D.C.: National Academy Press.

Chi, M. T. H., M. Chiu, and N. de Leeuw. 1990. *Learning in a non–physical science domain: the human circulatory system.* Office of Educational Research and Improvement Milestone Report. Pittsburgh: Learning Research and Development Center.

Chi, M. T. H., P. J. Feltovich, and R. Glaser. 1981. Categorization and representation of physics problems by experts and novices. *Cognitive Science* 5:121–52.

Chi, M. T. H., J. E. Hutchinson, and A. F. Robin. 1989. How inferences about novel domain-related concepts can be constrained by structured knowledge. *Merrill-Palmer Quarterly* 34:27–62.

Chi, M. T. H., and R. D. Koeske. 1983. Network representation of a child's dinosaur knowledge. *Developmental Psychology* 19:29–39.

Chi, M. T. H., et al. 1989. *Possible sources of misunderstanding about the circulatory system.* Office of Educational Research and Improvement Milestone Report. Pittsburgh: Learning Research and Development Center.

Clement, J. 1982. Students' preconceptions in introductory physics. *American Journal of Physics* 50:66–71.

Darden, L., and R. Rada. 1988. The role of experimentation in theory formation. In D. Helman, ed., *Analogical reasoning: Perspectives of artificial intelligence, cognitive science, and philosophy*, pp. 341–75. Boston: Kluwer Academic Publishers.

Dietterich, T. G., and B. G. Buchanan. 1983. The role of experimentation in theory formation. In *Proceedings of the 2nd international machine learning workshop*, Monticello, Ill.

DiSessa, A. A. 1988. Knowledge in pieces. In G. Forman and P. B. Pufall, eds., *Constructivism in the computer age*, pp. 49–70. Hillsdale, N.J.: Erlbaum.

Driver, R. 1973. The representation of conceptual frameworks in young adolescent science students. Ph.D. diss., University of Illinois.

Driver, R., and J. Easley. 1978. Pupils and paradigms: A review of literature related to concept development in adolescent science students. *Studies in Science Education* 5:61–84.

Erickson, G. L. 1979. Children's conceptions of heat and temperature. *Science Education* 63:221–30.

———. 1980. Children's viewpoints of heat: A second look. *Science Education* 64:323–36.

Falkenhainer, B., and S. Rajamoney. 1988. *The interdependencies of theory formation, revision, and experimentation.* Technical report no. UIUCDCS-R-88-1439. Urbana, Ill.: University of Illinois at Champaign-Urbana.

Faust, D. 1984. *The limits of scientific reasoning.* Minneapolis: University of Minnesota Press.

Feynman, 1963. *The Feynman lectures on physics.* Vol. 2. Reading, Mass.: Addison-Wesley.

Gellert, E. 1962. Children's conceptions of the content and function of the human body. *Genetic Psychology Monographs* 65:293–405.

Gelman, S. 1988. The development of induction within natural kind and artifact categories. *Cognitive Psychology* 20:65–95.

Getzels, J., and M. Csikszentmihalyi. 1976. *The creative vision: A longitudinal study of problem finding in art.* New York: Wiley.

Greenwald, A., et al. 1986. Under what conditions does theory obstruct research progress? *Psychological Review* 93:216–29.

Hastie, R., and N. Pennington. 1991. Cognitive and social processes in decision making. In L. Resnick, J. Levine, and S. Teasley, eds., *Perspectives in socially shared cognition,* pp. 308–27. Washington, D.C.: American Psychological Association.

Haugeland, J. 1978. The nature and plausibility of cognitivism. *Behavioral and Brain Sciences* 2:215–60.

Hayes, J. R. 1989. *The complete problem solver.* Hillsdale, N.J.: Erlbaum.

Hempel, C. 1966. Laws and their role in scientific explanation. *Philosophy of Natural Science,* chap. 5. Englewood Cliffs, N.J.: Prentice Hall.

Joshua, S., and J. J. Dupin. 1987. Taking into account student conceptions in instructional strategy: An example in physics. *Cognition and Instruction* 4:117–35.

Karmiloff-Smith, A., and B. Inhelder. 1975. If you want to get ahead, get a theory. *Cognition* 3:195–212.

Keil, F. 1979. *Semantic and conceptual development: An ontological perspective.* Cambridge, Mass.: Harvard University Press.

———. 1986. The acquisition of natural kind and artifact terms. In W. Demopoulos and A. Marrar, eds., *Language, learning, and concept acquisition,* pp. 133–53. Norwood, N.J.: Ablex.

———. 1989. *Concepts, kinds, and cognitive development.* Cambridge, Mass.: MIT Press.

Klahr, D., and K. Dunbar. 1988. Dual space search during scientific reasoning. *Cognitive Science* 12:1–48.

Kuhn, D. 1989. Children and adults as intuitive scientists. *Psychological Review* 96:674–89.

Kuhn, D., E. Amsel, and M. O'Loughlin. 1988. *The development of scientific thinking skills.* Orlando, Fla.: Academic Press.

Kuhn, T. S. 1970. *The structure of scientific revolutions.* Chicago: University of Chicago Press.

Lakatos, I. 1970. Falsification and the methodology of scientific research programmes. In I. Lakatos and A. Musgrave, eds., *Criticism and the growth of knowledge,* pp. 91–195. Cambridge: Cambridge University Press.

Langley, P. W., et al. 1987. *Scientific discovery: Computational explorations of the creative process.* Cambridge, Mass.: MIT Press.

Laudan, L. 1977. *Progress and its problems.* Berkeley: University of California Press.

Law, N., and W. W. Ki. 1981. A.I. programming environment as a knowledge elicitation and cognitive modelling tool. Paper presented at the Third International Conference on A.I. and Education.

Lawson, A. E. 1988. The acquisition of biological knowledge during childhood: Cognitive conflict or tabula rasa? *Journal of Research in Science Teaching* 25:185–99.

McCloskey, M., A. Caramazza, and B. Green. 1980. Curvilinear motion in the absence of external forces: Naive beliefs about the motion of objects. *Science* 210:1139–41.

McDermott, L. C. 1984. Research on conceptual understanding in mechanics. *Physics Today* 37 (July): 24.

McKenzie, A. E. E. 1960. *The major achievements of science.* Vol. 1. Cambridge: Cambridge University Press.

Maier, N. R. F. 1970. *Problem solving and creativity in individuals and groups.* Belmont, Calif.: Brooks/Cole.

Markman, E. M. 1979. Classes and collections: Conceptual organizations and numerical abilities. *Cognitive Psychology* 11:395–411.

Minstrell, J. 1988. Teaching science for education. In L. B. Resnick and L. E. Klopfer, eds., *Toward the thinking curriculum: ASCD yearbook.* Alexandria, Va.: Association for Supervision and Curriculum Development; Hillsdale, N.J.: Erlbaum.

Nersessian, N. J. 1989. Conceptual change in science and in science education. *Synthese* 80:163–83.

Nersessian, N. J., and L. B. Resnick. 1989. Comparing historical and intuitive explanations of motion: Does "naive physics" have a structure? In *Proceedings of the eleventh annual conference of the cognitive science society,* Ann Arbor, Mich.

Newell, A., and H. A. Simon. 1972. *Human problem solving.* Englewood Cliffs, N.J.: Prentice-Hall.

Ohlsson, S. 1984a. Restructuring revisited. I: Summary and critique of the Gestalt theory of problem solving. *Scandinavian Journal of Psychology* 25:65–78.

———. 1984b. Restructuring revisited. II: An information processing theory of restructuring and insight. *Scandinavian Journal of Psychology* 25:117–29.

Osborne, R. J., and M. C. Wittrock. 1983. Learning science: A generative process. *Science Education* 67:489–508.

Pagel, W. 1951. William Harvey and the purpose of circulation. In *ISIS* 42, pt. 1, April. Reprinted in the Bobbs-Merrill Reprint Series in History of Science, HS-58. Washington, D.C.: Smithsonian Institution.

Pfundt, H., and R. Duit. 1988. *Bibliography: Students' alternative frameworks and science education.* 2d ed. Kiel, Germany: Institute for Science Education.

Piaget, J., and B. Inhelder. 1974. *The child's construction of quantities.* London: Routledge and Kegan Paul.

Posner, G. J., et al. 1982. Accommodation of a scientific conception: Toward a theory of conceptual change. *Science Education* 66:211–27.

Postman, L., and R. F. Jarrett. 1952. An experimental analysis of learning without awareness. *American Journal of Psychology* 65:244–55.

Rajamoney, S. A. 1988. Experimentation-based theory revision. In *Proceedings of the AAAI spring symposium on explanation-based learning.*

Ranney, M. 1987. Restructuring naive conceptions of motion. Ph.D. diss., University of Pittsburgh.

———. 1988. Contradictions and reorganizations among naive conceptions of ballistics. Paper presented at the annual meeting of the Psychonomic Society, Chicago, Ill.

Reiner, M., M. T. H. Chi, and L. Resnick. 1988. Naive materialistic belief: An underlying epistemological commitment. In *Proceedings of the tenth annual conference of the cognitive science society*, pp. 544–51. Hillsdale, N.J.: Erlbaum.

Reiner, M., et al. In preparation. Materialism: An underlying commitment.

Rogan, J. M. 1988. Development of a conceptual framework on heat. *Science Education* 72:103–13.

Rosch, E., and C. Mervis. 1977. Children's sorting: A reinterpretation based on the nature of abstraction in natural categories. In R. C. Smart and M. S. Smart, eds., *Readings in child development and relationships*. New York: Macmillan.

Roschelle, J. 1986. *The envisioning machine: Facilitating students' reconceptualization of motion*. Palo Alto, Calif.: Xerox Palo Alto Research Center.

Schank, R. C. 1986. *Explanation patterns: Understanding mechanically and creatively*. Hillsdale, N.J.: Erlbaum.

Schauble, L., et al. 1991. Causal models and processes of discovery. *Journal of the Learning Sciences* 1, no. 2:201–38.

Schneider, W. 1987. Connectionism: Is it a paradigm shift for psychology? *Behavior Research Methods, Instruments, and Computers* 19:73–83.

Schon, D. 1987. *Educating the reflective practitioner*. San Francisco: Jassey-Bass.

Siegler, R., and E. Jenkins. 1989. *How children discover new strategies*. Hillsdale, N.J.: Erlbaum.

Smith, C., S. Carey, and M. Wiser. 1985. On differentiation: A case study of the development of the concepts of size, weight, and density. *Cognition* 21:177–237.

Solomon, J. 1983. Learning about energy: How pupils think in two domains. *European Journal of Science Education* 5:51–59.

Sommers, F. 1971. Structural ontology. *Philosophia* 1:21–42.

Spelke, E. S. 1990. Origins of visual knowledge. In D. N. Osherson, S. M. Kosslyn, and J. Hollerbach, eds., *Visual cognition and action*. Vol. 2, pp. 99–128. Cambridge, Mass.: MIT Press.

Suppe, F., ed. 1977. *The structure of scientific theories*. 2d ed. Urbana and Chicago: University of Illinois Press.

Thagard, P. 1989. Conceptual change in scientists and children. Panel discussion at the Eleventh Annual Cognitive Science Society Meeting, Ann Arbor, Mich., August.

———. 1990. Explanatory coherence. *Brain and Behavioral Sciences* 12:435–502.

Tiberghien, A. 1980. Models and conditions of learning an example: The learning of some aspects of the concept of heat. In W. F. Archenhold et al., eds., *Cognitive development research in science and mathematics*. Leeds: University of Leeds.

Tweney, R. D. 1985. Faraday's discovery of induction: A cognitive approach. In D. Gooding and F. A. J. L. James, eds., *Faraday rediscovered: Essays on the life and work of Michael Faraday, 1791–1867*. New York: Stockton Press.

VanLehn, K. 1991. Rule acquisition events in the discovery of problem solving strategies. *Cognitive Science* 15, no. 1:1–47.

Viennot, L. 1979. Spontaneous reasoning in elementary dynamics. *European Journal of Science Education* 1:205–21.

Vosniadou, S., and W. F. Brewer. 1989. The concept of the earth's shape: A study of conceptual change in childhood. Unpublished manuscript, Center for the Study of Reading, University of Illinois, Champaign, Ill.

White, A. R. 1975. Conceptual analysis. In C. J. Bontempor and S. J. Odell, eds., *The owl of Minerva*. New York: McGraw Hill.

White, B. Y., and J. R. Fredricksen. 1986. *Progressions of qualitative models as foundation for intelligent environments*. Technical rep. no. 6277. Cambridge, Mass.: Bolt, Beranak, and Newman.

Wiser, M. 1987. The differentiation of heat and temperature: History of science and novice-expert shift. In D. Strauss, ed., *Ontogeny, phylogeny, and historical development*. Norwood, N.J.: Ablex.

Wiser, M., and S. Carey. 1983. When heat and temperature were one. In D. Gentner and A. L. Stevens, eds., *Mental models*, pp. 267–98. Hillsdale, N.J.: Erlbaum.

Information, Observation, and Measurement from the Viewpoint of a Cognitive Philosophy of Science

The cognitive revolution in philosophy of science is well under way, and one point of this volume is to plan an agenda for the future. I believe that part of that agenda should be to rewrite the history of philosophy of science, for no revolution is complete until it has created an appropriate history of its own inevitability and rectitude.

Some kind of observation/theoretical distinction played a large part in the various positivist approaches to philosophy of science, but was swept away by the historicist revolution. Formal studies of foundations of measurement that had never quite connected with other aspects of philosophy of science have survived the revolution, but are still rather disconnected from the mainstream. These are two of the topics I will discuss as part of what I see as unfinished business in tidying up our account of what was happening, right and wrong, in earlier approaches to philosophy of science. I also think that this will lead to some significant and interesting new areas for additional research.

I take as a working assumption that the fundamental premise of the cognitive approach is that humans are natural creatures whose abilities and limitations are subject to empirical study. Furthermore, what distinguishes the study of cognition, as opposed to such other natural human processes and abilities as digestion and reproduction, is a focus on those aspects of human activity that involve information gathering and processing. Some humans — especially scientists — go beyond the mere gathering of information and become active hunters or cultivators. It is important to emphasize also that the processing that we do, especially in nonconscious perceptual mechanisms, is mainly a filtering process that eliminates vast quantities of information to reduce the flow of information to something manageable by our central processing unit.

Thus, I will use as a central concept that of information, arguing that observation and measurement are both processes by which we wrest information about nature from the world and do some processing so that it can then be used in theoretical reasoning. The general, perhaps generic, common, and vague use of the word *information* is ubiquitous, and two more precise and technical specifications of it have generated research programs in epistemology (Dretske 1981) and in semantics (Barwise and Perry 1983). But I will follow instead a once popular but now forsaken path, namely the Shannon/Weaver probabilistic definition of information.

In this kind of context various writers have ignored or dismissed the probabilistic definition, so I need to give at least a brief defense. Gibson (1982) and others argue that the concept applies to restricted kinds of communication situations with artificially constructed messages and an engineered communication channel. Although this is the kind of application Shannon envisioned, the definition of mutual information itself says nothing about what the messages or channels are like, nor indeed that they are in any ordinary sense required to be messages or channels. The definition depends on two partitions of sets of logically independent events A and B, where a partition is a finite division into exhaustive disjoint subsets A_i and B_j. The definition of the mutual information in event A_i of kind A about event B_j of kind B is:

$$-\log_2 \frac{\text{Prob}(B_j/A_i)}{\text{Prob}(B_j)}$$

That is, it is a function of how much more probable B_j is given the occurrence of A_i. Although this looks asymmetric it can be transformed into the symmetric

$$-\log_2 - \frac{\text{Prob}(B_j \ \& \ A_i)}{[\text{Prob}(B_j) \text{x} \text{Prob}(A_i)]}$$

If we wish to assess not the information in one particular event about another particular event but rather the average amount of information that events of the one kind give about events of the other kind, we need to take into account the relative frequency of the different kinds of events. A type of event that conveys a large amount of information may nevertheless occur very infrequently and thus contribute little to the overall systematic information. Thus we use

$$\sum_i \sum_j -\text{Prob}(B_j \ \& \ A_i) \log_2 \frac{\text{Prob}(B_j \ \& \ A_i)}{[\text{Prob}(B_j) \text{x} \text{Prob}(A_i)]}$$

One objection to the usefulness of this definition in philosophy and psychology (Lachman, Lachman, and Butterfield 1979) is that we hardly ever know the probability distributions required in interesting cases. But the general answer to the objection is that we can often discover significant phenomena or provide interesting analyses with only vague general assumptions about the probabilities. In other words, we can often find qualitative differences using the quantitative definition. A more forceful answer, which I hope to provide eventually, is to actually produce such results.

One important point is that the definition applies to the probabilities of the events and the events may be either natural or "artificial." In the latter case, the semantic or intentional properties of the artifact are irrelevant. Let me illustrate. Let us suppose that in Houston the probability of rain on a given day is 1/8, the probability of dark clouds is 1/4, and that it never rains when it is not cloudy. Since the antecedent probability of rain is 1/8, the event of dark clouds being present raises the probability to 1/2, giving 2 bits of information 1/8th of the time; the antecedent probability of nonrain is 7/8 and a noncloudy day raises this probability to 1, giving .2 bits 3/4ths of the time; however, the cloudy but not rainy days in effect give .8 bits of misinformation 1/8th of the time. Thus the average amount of information is about .3 bits.

A partition of events that correlated perfectly with the weather would produce 3 bits 1/8th of the time and .2 bits 7/8ths, giving an average of .55 bits. The amount of information that can in principle be conveyed about a partitioned set will be limited by the number of indices in the partition and the distribution of probabilities; the larger the number of divisions in the partition and the more even the probability distribution, the more potential information exists.

How much information is provided by a weather forecast that asserts a 50 percent chance of rain? The answer is that it depends entirely on the probability of rain given such a forecast. If such a forecast is, in effect, correct (at least on one interpretation of the forecast) and the probability of rain is .5 when such a forecast is given and zero otherwise, then the forecast is as good as the clouds in the previous example. On the other hand, if the forecast is irrelevant so that the probability of rain given the forecast is 1/8, then there is zero information.

To be more cautious and precise, in the last case the forecast conveys no information *about rain*. Of course it very likely does convey information about the forecaster's state of mind, and via whatever meteorological model she or he is invoking, may convey information about the prevailing winds or barometric pressure. This illustrates a second point — the probabilities are to be understood as objective and information about one type of event may be present in another type of event without our knowledge of that fact, and even in spite of our beliefs to the contrary.

It may be helpful to illustrate the very significant difference between information being in an event and our being able to use that information or knowing that it is there. A clock that is stopped has no correlation with the time and so even though it is "right" twice a day (once if it is a 24-hour digital) it carries no information about the time. On the other hand, a clock that is consistently 7 minutes (or 6 hours and 19 minutes) fast correlates perfectly, though unconventionally, with the time and carries complete information about the time. The information is less accessible to us because it is essential to know the nonconventional transformation that translates the states of the clock into a familiar and useful notation. (As adults we probably tend to overlook the convention that has to be learned to make use of clocks at all.) Correcting ancient astronomical observations for atmospheric refraction retrospectively is an example of such transformations.

More extremely, a clock that ran at variable speeds but for which there was a one-one function with the correct time would also carry complete information about the time. In this case a more complex function would have to be known and calculated in order to make the information useful. From this point of view one might consider a theory to be a way of transforming information about current states of the world into information about past or future states.

1. Scales of Measurement and Information

An ongoing controversy in foundations of measurement that has seemed significant to participants but has struck some observers as being 'a semantic dispute' in the pejorative colloquial sense concerns whether scales of measurement other than the ratio and interval scales are truly measurement. I shall consider the alleged scales from the perspective of information.

A nominal scale assigns numbers to objects in the relevant domain meeting the condition of a one-one mapping between the objects and the numbers used but no further conditions. Given the information that two objects, a and b, chosen at random with replacement, are assigned the numbers n and m provides us with the information that they are the same object if n = m and that they are distinct otherwise. How much information this is depends on how many objects are in the domain. The events are identity or nonidentity of chosen objects and of the pairs of numbers. If we assume that the assignment is perfect, i.e., that the numbers are the same exactly when the object(s) are the same, then $P(a = b/m = n) = 1$. If there are only two objects in the domain then $P(a = b) = .5$, and the information about the objects conveyed by the numbers is one bit of information in each case.

On the other hand, if the number of objects is large, say 64, then

$P(a = b) = 1/64$, and while the amount of information conveyed by $m = n$ is large, 6 bits, the frequency of such events is small, 1/64, while the much more frequent event $m = n$ carries only .023 bits of information. Thus the average information in this system is only .12 bits. It is easy to verify that the amount of information varies inversely with the size of the domain.

In an ordinal scale, the numbers reflect not distinctness of objects but their place in a linear ordering R. Using $m(x)$ for the measured value of x on the scale,

$$m(x) > m(y) \leftrightarrow xRy$$

In this case the relevant events are the occurrence or nonoccurrence of Rxy and the relation between $m(x)$ and $m(y)$. If the number of objects is large so that we can ignore the infrequent cases in which $m(x) = m(y)$ and assume again perfect reliability in numerical assignment, then since $P(Rxy) = .5$ there will be one bit of information in each numerical event and in the average in the system.

Moving directly to ratio scales we quickly find things are different. Let us assume that we have a limited range of possible values (e.g., 0,1) for m and also assume that within that range all values are equipossible. Then an assignment of values r and t to objects a and b conveys a great deal of information! Suppose that * is the concatenation operation on the domain and let the $m(a) = .50000\ldots$ and $m(b) = .333333\ldots$ Then we know that $a > b$, but also that if $b' = b$ that $b*b' > a$, and that $b'' = b' = b$ and $a' = a$ entails that $b*b'*b'' = a*a'$ and ... Another way of putting it is that since there are continuum many equipossible values for $m(a)$, an exact specification of $m(a)$ would provide an infinite amount of information.

That, of course, is impossible, but if we consider a determination of $m(a)$ to 3 significant digits, i.e., one that places the true value of $m(a)$ within .001, then there will be about 10 bits of information in that event.

The point of this is to show the differences between the three kinds of scales of alleged measurement. Nominal scales provide at most 1 bit of information per measurement, and the amount drops quickly with larger domains than pairs; ordinal scales provide about 1 bit of information regardless of the size of the domain. Typical ratio scales can easily provide 10 bits of information per measurement and are in principle unlimited (except perhaps by quantum limits) in the amount of information that could be obtained.

I do not care to adjudicate the dispute and settle whether the first two kinds of scale truly represent measurement or not. It seems to me more important to note that whatever kind of operation they are officially designated, they are in principle much more limited in the amount of information that they can provide about the properties and relations of the objects in the domain.

2. Observation, Measurement, and Information

For the purposes of this section I limit the term *observation* to the kind of term the logical positivists thought provided a basis for science. (I do not believe that this use coincides with either scientific or ordinary usage, but that is not in itself an objection to its use.) Typical examples are "blue" or "warm." Probably because they tended to restrict themselves to first-order formulations of theories, the positivists' discussions of the grounding of theories mostly saw the grounding in terms of such sentences that corresponded most naturally to predicate constants. The dominant current view of theories, the structuralist/set-theoretic/model-theoretic/semantic view, mainly treats mathematical theories and sees the empirical grounding mainly in terms of determinate values of quantities in the models. Suppose (never mind why) we wanted to compare the information content of observation and measurement.

Most observation terms (indeed most nouns, at the very least) of natural languages come in contrast sets that are parts of larger semantic fields (see Grandy [1987] for a recent discussion of contrast sets and their larger kin). Thus more commonly "blue" contrasts with other common color terms, "red," "green." ... The number of terms in contrast sets varies considerably, but typical contrast sets most commonly have around 5 to 10 members. And of course blue can occur in more than one contrast set. Thus if we make assumptions of equiprobability of distribution and that reports are highly reliable, we can see that the antecedent probability that a given object is blue is around 1/8 and a typical color report will convey about 3 bits of information.

The numbers are quite variable, as I suggested, but the assumptions could be fairly far off and the result would still obtain that the observation reports are usually considerably more informative than ordinal-scale measurement, though limited in their informativeness in a way that the ratio-scale measurements are not. To the extent that various observers agree in describing varying situations we can be confident that the descriptions convey information about the complex interaction of the situations and observers. On my definition this means that the descriptions are not random, but does not guarantee that they are semantically correct in the usual sense. Nonrandomness or correlation comes in degrees and varies according to contexts, and the attempt to isolate a class of sentences whose nonrandomness is maximal regardless of context is doubly wrongheaded.

3. Observation: From Sensations to Sentences

I am sure it would be controversial whether the cognitive approach to philosophy of science represents a totally novel perspective or whether it should be thought of as a more sophisticated empiricism aided and abetted by the latest psychological innovations. But in any event the empiricist tradition is an important one that needs to be dealt with.

Three versions of the basic empiricist slogan serve to indicate both the continuity and variation in that movement:

"Nothing is in the mind that is not first in the senses."
"All knowledge begins in experience."
"Observation sentences are the basis of meaning and evidence."

They also indicate the progression from psychological phrasing to matters of language and evidence. Knowledge of meanings replaces ideas; evidence replaces knowledge. We believe this is progress because the concepts of "idea" and "knowledge" are vague, but our belief has suppressed premises that hardly bear scrutiny.

Why observation sentences? *Sentences* because talk of the senses and experience is an area of discourse that many philosophers (though by no means all) are willing to cede to psychologists. Or at the very least they are willing to accept a division of labor among philosophers so that those who are concerned primarily with meaning and evidence begin with the sentence that is believed, uttered, assented to, or some equivalent. If one begins with sensations, experience, or perceptions, there is still the job to be done of explaining how people make the transition from those items to linguistic expressions describing them or based upon them. By starting with sentences we leave that problem (and many others) to those who wish to concern themselves therewith.

Observation sentences because those are the sentences closest to the experiences or sensations that we do not want to talk about. Observation sentences are supposed to be those that make the minimum possible demands on linguistic competence and on one's beliefs, past history, and inferential acumen. We do well to note at the beginning that "observation sentence" is a technical, theoretical term — philosophers believe that there is a significant subset of the sentences of a language that play a special role, and they attempt to characterize that subset by an appropriate definition. That belief is based on a particular conception of language, and if that conception is wrong, then the special subset may not exist. Arguments to the contrary may seem convincing prima facie, but they only disguise the assumption:

Observation sentences are the basis of meaning.
English sentences have meaning.
There are observation sentences in English.

The first premise introduces a second technical term, "basis," and a theoretical claim using that term. The premise serves as a persuasive definition, for denying it requires us to deny an implicit theory about the basics of English.

I gave two approximations of definitions of "observation sentence" at the beginning of the last paragraph; here are three others:

3. Observation sentences are those occasion sentences for which all speakers of the language have the same stimulus meaning (Quine 1960).

4. Observation sentences are those that can be taught by ostension (Quine 1975).

5. Observation sentences are those that can be reliably applied after a short period of observation (Hempel 1958).[1]

In all five cases there is danger that the proposed definition will either pick out the empty set of sentences or a class that is clearly wrong. Even my two definitions are subject to the threat of picking out the empty set though they are couched in terms of relations among sentences, such as "closest to experience" and "minimum possible demands on linguistic competence" — for these to pick out a well defined set requires that the requisite relations suitably order the set of English sentences. Note too that the crucial relations in those characterizations are metaphorical — in what sense are some sentence closer to experience? In what sense do sentences make demands on linguistic competence?

Definition 3 contrasts with 5 in that it appears to offer hope of picking out the observation sentences of a language that we do not yet understand, whereas applying 5 requires that we know the meanings or at least truth conditions of sentences in order to assess reliability. Definition 4 is too vague to compare to the others; it could be assimilated to 3 if the criterion for learning is agreement with the community, or with 5 if the criterion is reliable application, or neither if some other test is chosen.

3.1. Communitywide Stimulus Meaning

It is probable that the notion of an observation sentence would have dropped out of both philosophy of science and philosophy of language by now if it had not been for Quine's definition of "observation sentence" in (1960) and his subsequent uses and defenses of the concept. By 1960 and shortly thereafter there were numerous attacks on the notion.

Quine (1969, p. 88) comments on this state of affairs: "It is ironical that philosophers, finding the old epistemology untenable as a whole, should react by repudiating a part which has only now moved into clear focus."

Quine's definition is motivated by the concepts of observation mentioned earlier — that observation sentences are those that are closest to stimulation and that depend least on language learning or on collateral information. (Collateral information is for Quine any information that is not required for understanding the sentence, but that might influence the judgment about the sentence in a given stimulus situation.) One of his steps on the road to his definition reflects the older and inadequate definition of observation sentence: "The sentence is an observation sentence if our verdict (on the sentence) depends only on the sensory stimulation present at the time" (1969, p. 85).

Any responsible verdict on a sentence requires a learning of the language that is hardly a matter of present sensory stimulation alone. Thus strictly construed there are no observation sentences. The plausible revision is to require that the verdict depend only on the present sensory stimulation and the minimum linguistic training required for understanding the sentence. But how are we to adjudicate what is the minimum required for understanding?

> This is the problem of distinguishing between analytic truth, which issues from the mere meaning of words, and synthetic truth, which depends on more than meanings. Now I have long maintained that this distinction is illusory. There is one step toward such a distinction, however, which does make sense.... Perhaps the controversial notion of analyticity can be dispensed with, in our definition of observation sentence, in favor of community wide acceptance. (Quine 1969, p. 86)

Thus Quine arrives at his definition of an observation sentence in terms of a linguistic community: "An observation sentence is one on which all speakers of the language give the same verdict when given the same concurrent stimulation" (Quine 1969, p. 87). For example, the sentence "There is a brown chair about ten feet from you" is an observation sentence in English just if the following condition is met: For any pair of English speakers and any pattern of stimulation, upon presentation of the stimulation either both speakers would assent to the sentence, both would dissent from it, or both would reserve judgment.

Note that there are two parameters in the definitions (as Quine [1960] explicitly recognizes): the community and the length of the stimulation. To have *a* definition of *the* class of observation sentences for a language we must fix both parameters — Quine chooses the widest community but passes over the choice of parameter in silence.

Are there any observation sentences in English? The question can be conclusively answered only by thorough experimentation on sentences, stimulations, and subjects, but I believe that some general considerations strongly suggest that there are not. If we take a sentence like "There is a brown chair about ten feet from you," we will find rather good agreement across a wide range of stimulation patterns, but not complete. A pattern of stimulation produced by my study with very slight illumination will produce a jumble of vague silhouettes that will lead most speakers to suspend judgment, but I might recognize the pattern of silhouettes including the chair that I believe to be brown, and thus I would assent. Other speakers would react similarly for other patterns, thus ruining the uniformity. Discounting patterns that involve only a very low level of radiation (besides looking ad hoc unless we can find a motivation) will not suffice; a clear and bright pattern of stimulation produced by the wall of the room next to my study will equally well produce the idiosyncratic response.

If this general line of argument is correct, a refinement is in order. Let us say that a stimulation pattern is neutral if few or no speakers give a positive or negative judgment given that pattern. Then we can define an observation sentence to be one in which there is communitywide agreement on the nonneutral stimulations. This is another step toward eliminating collateral information. A situation in which a few individuals can render a verdict but most speakers have too little information to respond indicates extra information is present in the environment.

Our refinement still leaves the definition within the austerely (one might say excessively) behavioral Quinian tradition, but we should note that it adds yet another parameter — the size of the collaterally informed group of speakers that we are prepared to discount. For example, "There is a blue flamepoint Assyrian cat about ten feet in front of you" is a sentence that is observational if we discount those who have special knowledge about cats. Most speakers would dissent from the sentence in any situation in which no cat was present and suspend judgment in those in which some cat was at the right distance. Does that make it observational? How many cat specialists are there and are there enough to render the sentence nonobservational? It seems clear that this is a matter of degree and no single choice of cut-off point seems well motivated.

We thus have three parameters — degree of agreement, length of stimulus, size of community of specialists discounted — that are matters of degree, and one for which Quine gives a specific answer, namely how wide a community do we choose. We turn now to the motivation for this choice.

3.2. Why Communitywide?

The reason for opting for the entire community as the standard for an observation sentence is that observation sentences so defined can serve as the termini of disputes. If speakers disagree about the truth of a nonobservational sentence, but they can agree that the disputed sentence A has as a consequence that under circumstances C, observation sentence O_A would be true, whereas the contrary supposition leads to $-O_A$ under those circumstances, then, if they can realize the circumstances, they can test the two hypotheses in a way that will resolve the issue. In this sense the observation sentence provides some empirical content for the nonobservational sentences.

But do they provide *the* empirical content? Can we justify the traditional claim, taken over by Quine, that the observation sentences provide all and the only empirical content that the remainder of the sentences have? Are the observation sentences alone "the repository for scientific evidence" (1969, p. 88), "the cornerstone of semantics" (1969, p. 89)? Are two theories with the same observation-sentence consequence empirically equivalent (1975)?

Let us consider the ways in which a sentence may fail to be an observation sentence. There are many. One way is if there is significant disagreement about a class of stimulation patterns, so long as this disagreement is considerable even if the class of stimulations about which there is disagreement is selective. For example, U.S. English speakers can probably identify robins visually with reasonably good agreement; but I doubt that they can agree very well on identifying robins by their song. Surely many people can, but many cannot. If this is true, the speakers will agree in classifying stimulation patterns that exclude any birdlike sounds, for visual stimulations will produce convergent results. But for stimulation patterns that include birdlike sounds but no visible robins, there will be considerable disagreement. So "There is a robin in the vicinity" fails to be an observation sentence.

In this case a reply is possible — the sentence "There is a robin visible in the vicinity" will be an observation sentence, if there is agreement on this sentence. But it is doubtful if there is such agreement — there is likely a set of situations and stimulation patterns produced by them such that all speakers agree in those cases either in assenting or dissenting. But it is very dubious that there is total agreement — some speakers are better at discerning robins that are partly concealed or swiftly flying by. Shifting to "clearly visible" would shift the problem but not solve it, for speakers are no more likely to agree on when something is *clearly* visible as opposed to *merely* visible.

This line of argument, like that for ruling out or otherwise delimiting neutral stimuli, suggests an extra parameter be added to the definition, one specifying the range of stimulation patterns. A full characterization of observation sentences then will have the form:

O is an observation sentence for community C with respect to stimulation pattern class S of modulus M to degree n if at least n percent of C agree in their patterns of assent, dissent, and suspension of judgment for all stimuli in S of modulus M. Variation in any parameter changes the class of observation sentences. Take a wider community and there will be less agreement and fewer observation sentences; some sentences on which there is agreement with stimulus patterns of length .5 second will not produce agreement with patterns of length .05; broader classes of stimuli of a given modulus will produce less agreement; raising the percentage of agreement required reduces the number of observation sentences.

The four parameters provide dimensions along which we can rank sentences with regard to how observational they are. The four together provide a partial ordering of all the sentences of a language. The moral of all this seems to be that it is unjustifiable to talk of *the* observation sentences providing a cornerstone to meaning or a repository of evidence. Rather, the more observational a sentence is the firmer its semantic grounding is; the more observational a sentence is the more suitable it is for providing an empirical base for confirming or disconfirming other sentences.

Two theorists who disagree about a theoretical matter need not, as we discussed a few pages ago, find an observation sentence in order to resolve the dispute. If they can find a situation and a sentence such that they would agree in their verdicts about the sentence given the stimulus patterns produced in that situation, and given that the sentence has the logical properties enumerated earlier, then they can resolve their dispute. They need not care a whit about other situations in which they would disagree in applying or not applying the sentence, nor about the English-speaking community at large. Of course if they wish to persuade a larger community, then that community is relevant, but the use of the sentence in other situations still is not. There appears to be no reason thus far to identify the empirical content of a theory with the observation sentences, if by the latter we pick out some by fixing cut-off points along the various dimensions of observationality.

3.3. Reliability

Before drawing any general conclusions about observation sentences and empirical content we must survey some other possibilities for defining the concepts. As a first approximation we can say that a speaker is reliable with respect to a sentence S if when the speaker would assent to the sentence it is true and when the speaker would dissent it is false. There are several ways of interpreting "when the speaker would assent/dissent." One is to consider all logically possible circumstances. On this construal I doubt that any speakers are reliable with respect to any sentences that are not simple logical or mathematical truths. There are many logically possible circumstances in which speakers would assent to sentences that were false because of many kinds of deception, hallucination, and so on.

Actual assent/dissent, while it provides the evidence for reliability, is not broad enough. Most speakers have never and will never assent to or dissent from most sentences of English. Thus we are concerned with dispositions to assent in circumstances that are currently physically possible. Can we expect 100 percent reliability even in these cases? Not for any sentences except simple logical and mathematical truths, and perhaps reports of one's own inner states of a simple and obvious kind. And these will not be observation sentences. An observation sentence will be an occasion sentence on which all members of the linguistic community are reliable; logical and mathematical truths are not occasion sentences and the sundry members of the community will not be sufficiently reliable in reporting each other's inner states.[2]

If there are to be any observation sentences it seems that we must construe "reliable" in terms of a high percentage of correct responses, as opposed to complete accuracy. This means that the reliability approach produces the same result as the Quinian approach inasmuch as there will be gradation among sentences, both in reliability and in communitywide agreement. Again we can see that the more reliable sentences provide a firmer foundation for epistemic purposes, but there is no reason to claim that all sentences reliable more than .9732 provide an empirical anchor while those at a level below that do not.[3]

There is a definite relation between the two definitions, for a sentence cannot be reported on reliably by most of the community if there is not significant agreement between the reporters in the community. What reliability does is to add a possibly further consideration — that the reports be reliable as well as being generally agreed upon. Whether this in fact adds anything to intersubjective agreement, whether something that all speakers of a language agree is true could be false, is a matter that I will take up elsewhere.

3.4. Awareness of Observationality

Observation sentences *à la* Quine have certain virtues — if a dispute can be turned into a question about an observation report at a future time in circumstances that can be realized, then the dispute is resolvable. Similarly, if we add reliability to the Quinian conditions we can settle at least the issue of evidence. However, if S is an observation sentence (in either sense) but speakers of the language do not know that it is, then there is no reason to attempt to resolve the dispute by testing S. In other words, unless speakers have some general knowledge about which sentences are observation sentences, in what circumstances, for how wide a subsection of the community, then although they may haphazardly hit upon circumstances and sentences that resolve issues, they will be led more by luck than knowledge. This is a point that has not been much noted, perhaps because in the heyday of positivism and observation sentences the assumption was that the observation sentences came in a special, recognizable vocabulary.

But in the new sociolinguistic formulation, whether a sentence is an observation sentence is not obviously a matter of its syntax or of the terms contained in it. To know that S is a relatively observational sentence under ordinary lighting conditions is to know a great deal about one's language, the world, and one's linguistic community. It requires that S deals with matters typically (in this community) resolved by visual perception and what conditions that requires. And it is knowledge that we do have, though not necessarily in the technical jargon of "observation sentences." Speakers who disagree about an issue have some sense of where to direct the disagreement in order to attempt a resolution. I suspect that numerically most disputes are ended in (alleged) reductions, dictionaries, or reference works, but we can readily think of factual disputes that are eventually resolved by looking at nonlinguistic aspects of the world that lead to what is recognizable as a relatively observational sentence.

Many more disputes would be resolvable that way if it were not for the fact that in many cases, while disputants can agree on an observation sentence and a situation that would resolve the issue, the situation is too distant spatially, is in the past or the distant future.

Awareness of the relative observationality of sentences is valuable not only for the disputatious. Teaching of a natural language depends on at least some awareness of what sentences can be taught first and which are more directly keyed to current situations. Fortunately the process of learning the language, of observing others learning, and simply of conversing with others can all contribute to the same kind of information about observationality. In some of Quine's favorite terminology, observation sentences not only are obvious in truth value in a suitable situation, but everyone in the community knows they are obvious.

If we had not overcome the temptation to define a class of observation sentences, it might be tempting to make general knowledge of the observationality a condition of membership. Since we have overcome that temptation we can simply add to the profile of a sentence an appreciation of how general the community's knowledge is about the observationality of that sentence, or better, the beliefs about its observationality. Not all beliefs in the community, even communitywide ones, about observationality need be correct. Not only may some sentences be highly observational without this being appreciated (perhaps because agreement in use is more widespread than is realized), some sentences may be believed to be more observational than they are. (Shapere's [1982] account of observation in science and philosophy makes similar points in a different form.)

One of the major shifts in the development of scientific method has been increasingly strict standards for inclusion of information about the context in which data are collected. Given enough information about the circumstances of observation others can, contemporaneously or later, make their own informed judgments about the accuracy of the data, and possible sources of error. The use of distant historical data for purposes of current theorizing raises complex issues of reliability. For example, Ribes, Ribes, and Barthalot (1987; 1988) have argued that the size of the sun changed over the last three centuries based on observations of the solar diameter by Picard in the late seventeenth century. Van Helden and O'Dell (1987) have questioned the argument because Ribes, Ribes, and Barthalot are assuming a degree of accuracy in Picard's measurements that is at the theoretical maximum for the instrument used, and they do not take into account the difficulty of making the measurement practically. If the image being measured is moving appreciably across the field of observation, then the reported width of the image is unlikely to be as accurate as it would be for a stationary image. Van Helden and O'Dell argue indirectly against the alleged accuracy of Picard's measurements also by showing that his reports of the observed diameter of Jupiter vary from the known diameter by approximately the amount that Picard's observation of the solar diameter differs from the current value. In other words, since they can show that the information Picard's method provides about the diameter of Jupiter is limited, we can infer that his method provides no better information about the comparable measurements of the sun.

4. Kinds of Theoreticity

The main slogan that killed the project of finding a dichotomy of observational/theoretical terms or sentences was "All observation is theory laden." I believe the slogan is fundamentally true, but it has tended to stifle research toward understanding the element of truth that motivated the original quest, namely that science grounds some claims in others, in favor of undifferentiated holistic epistemologies. There are exceptions, of course; Glymour's bootstrapping is one antiholistic effort, and Hempel and Sneed each introduced relativized distinctions that are intended to capture the insight behind epistemologically foundational approaches.

Sneed, working in the formal context of the structuralist approach to theories, defines a magnitude m to be T-theoretical if all determinations of values of m(x) presuppose theory T. Operating in a less formal mode, Hempel suggests distinguishing between antecedently understood vocabulary and that which is introduced by a new theory T. Both suggestions have some utility, but I suspect they do not discriminate finely enough since they are based on terms rather than various statements using those terms. (I am using a somewhat broader than ordinary sense of statement to include reports m(a) = r of measurements.) For example, a statement of length to the effect that b has length 2.13 meters is not very theoretical, while a statement that distance c is 2×10^{13} meters is probably quite theoretical, while a statement that distance a is 2×10^{-13} meters will also be theoretical but depend on a different theory.

Furthermore, the ways in which a statement can depend on theoretical considerations vary. Consider again the causal chain by which a photon passes from a star through a telescope to a retina, of the complex processes in the nervous system that lead to an observational notation, of how a set of such notations are transformed in a data point, and of how that data point might provide an input to a computation that provides a value for a quite different magnitude. At the very least, then, we should distinguish the following senses of theory dependence:

The statement depends on theory T because it was made using an instrument whose design relies on T.

The statement depends on theory T because the particular neural processing that humans perform has evolved in a way that embodies theory T as an underlying assumption.

The statement depends on theory T of the scale of measurement on which a magnitude is being determined.

The statement depends on theory T because its linguistic expression depends on a theory T implicitly underlying the vocabulary used.

The statement depends on theory T because T was invoked in turning a set of raw data into processed data.

The statement depends on theory T because a computation using T was applied to data to arrive at the statement.

Having lumped all of these as kinds of theory dependence let me hasten to add that they are quite various and importantly different — that is why I have made the distinctions. In the last case the theory is quite explicit and articulated, as in the calculation of planetary distances from periods. In the second case it is likely to be neither explicit nor articulate and in most cases is probably still unknown. I have in mind, for example, the kinds of processing assumptions discussed by Marr (1982), such as that gradient discontinuities tend to indicate boundaries, that objects are generally rigid, that motions are mostly continuous.

It is not always recognized that scales of measurement are theories, even though the semantic or set-theoretic view of theories makes this more plausible. The difficulty may be that the model (for ratio scales) is the already familiar mathematical structure of the real numbers with addition, so that it is not evident that this is functioning as a scientific model of the theorized magnitude. Matters are complicated, of course, by the fact that the same model is used for quite different magnitudes!

Those under the fourth heading are themselves quite diverse since they may involve physical or psychological or statistical assumptions. In astronomy examples would include correction for parallax due to the size of the earth, which was considered from Ptolemy, refraction due to the atmosphere (van Helden 1985), which began to be considered in the late sixteenth century, psychological considerations about systematic human errors in judging when totality of an eclipse begins or ends (Swerdlow and Neugebauer 1984), and individual variations in observations. (See Suppes [1962] for a somewhat more detailed discussion of an example from learning theory.)

In the case of dependence on linguistic vocabulary, the structure of the underlying theory may be relatively unarticulated although some partial knowledge of it is widespread in the linguistic community.

I have omitted from the list the most troublesome kind of theory dependence, when observers see something because it will corroborate their theory or provide a significant new discovery. While this certainly occurs, its long-term effects on the development of science tend to be held in check by the independent observations or measurements of other scientists with other motives, as well as by the fact that to a considerable extent the perceptual system is isolated from beliefs and desires. We often see things that we do not expect and do not want to see; evolution stands us in pretty good stead here.

5. A Program and a Conjecture

If we take the perspective that science is a method of information gathering and processing, then we can ask about the relative contributions of various kinds of developments. How much does the gathering of more information of a familiar kind contribute? How much is contributed by a new theory? How much by new instruments? How much by better data "analysis"? Although I do not think that the application of the concept of information will be unproblematic, it may offer some common measure to compare these kinds of developments in specific cases.

For example, if we accept some standard estimates about the extent to which Brahe improved the accuracy of planetary observations by better instrumentation and attention to the sources of observational variability, he approximately doubled the amount of information. His method of more regular observations contributed another increment that is harder to calculate. The invention of the telescope contributed information about the moons of Jupiter and the phases of Venus that was inaccessible before. Techniques for building larger telescopes contributed more information, especially as advances in applied optics made it possible to compensate for or eliminate sources of error. Kepler's laws, which could like other theories be viewed as a program for transforming information from one form into another, gave us greatly improved information about the planetary distances and relations between planetary distances.

In the later stage during which the quest to measure stellar parallax was being pursued there were again a variety of kinds of contributions. Improved telescopes were essential to the detection of parallax, but improved methods of data analysis were also fundamental in revealing the actual motion within the variations produced by many sources of error; improved understanding of some of the sources of error was also necessary as was the type of theorizing and guessing by Herschel as to which stars were the most promising ones for finding measurable parallax (Williams 1981).

The program I suggest is one of taking improvements in information gathering as a measure of scientific progress and attempting to analyze the various sources of improvements for their relative contributions. I do not expect this to be either easy or straightforward, but it would not promise much insight if it were. I conjecture from the little preliminary work and thought I have devoted to it thus far that one significant result that might well emerge is an increased appreciation for the role of instruments and the finer points of experimentation in the development of science. The role of instruments has been given scarcely any discussion in philosophy of science and the importance of understanding error has been given little more.

We overthrew some time ago the narrow empiricist view of science that saw the atheoretical gathering of facts as the main engine of science, but perhaps we have erred in another direction by emphasizing theoretical advances as the most, almost the only, significant advances in science.

Notes

The approach suggested here has been considerably influenced by reading works by and conversing with John Collier, Ronald Giere, Thomas Kuhn, and Patrick Suppes, and with members of the Cognitive History and Philosophy of Science Group in Houston. None of them would necessarily agree with all of the above. Achinstein (1968) suggested a complex taxonomy of kinds of theoreticity that differs from the above but probably influenced my own.

1. Other, apparently diverse, definitions exist but usually can be shown to reduce to one of these or a variant thereof. For example, Causey (1979, p. 194) defines observation reporting sentences in terms of technological feasibility of realizing a well-defined situation in which the sentence can be used to make a justified observation report. His analysis of "justified" however is in terms of reliability. Actually, Hempel's definition does not use "reliable" explicitly, but this seems to be the intention.

2. The status of inner report sentences seems to have confused some commentators, perhaps because they are usually taken to be sentences with possessives as well. Quine himself errs in saying that the sentence "My cat is on the mat" is not observational because other speakers may not know what *my* cat looks like (Quine and Ullian 1970, p. 24). The sentence is *not* observational, but not for the reason given. Instead it is not observational since the (stimulus) meaning will vary widely from speaker to speaker since speakers will be referring to different cats. Inner state reports in the possessive fail to be observation sentences for the same reason. Inner state reports with proper names, e.g., "W. V. Quine has a toothache," fail because there will be virtually no agreement.

3. See Grandy (1980) where I argue that reliability offers the best analysis of knowledge, but that the analysis cannot reproduce a satisfactory traditional concept of knowledge, and thus we should drop talk of knowledge-recast epistemology in terms of reliability and related notions.

References

Achinstein, P. 1968. *Concepts of Science: A Philosophical Analysis*. Baltimore: Johns Hopkins Press.

Barwise, J., and J. Perry. 1983. *Situations and Attitudes*. Cambridge, Mass.: MIT Press.

Causey, Robert L. 1979. "Theory and Observation." In *Current Research in Philosophy of Science*, ed. P. D. Asquith and H. E. Kyburg, Jr. East Lansing, Mich.: Philosophy of Science Association.

Dretske, F. 1981. *Knowledge and the Flow of Information.* Cambridge, Mass.: MIT Press.

Gibson, J. J. 1982. *Reasons for Realism: Selected Essays of James J. Gibson*, ed. E. Reed and R. Jones. Hillsdale, N.J.: Erlbaum.

Grandy, R. E. 1980. "Ramsey, reliability and knowledge." In *Prospects for Pragmatism*, ed. D. H. Mellor. Cambridge: Cambridge University Press.

———. 1987. "In defense of semantic fields." In *New Directions in Semantics*, ed. E. Lepore. London: Academic Press.

Hempel, C. G. 1958. "The theoretician's dilemma." In *Concepts, Theories, and the Mind-Body Problem*, ed. H. Feigl, M. Scriven, and G. Maxwell. Minnesota Studies in the Philosophy of Science, vol. 2. Minneapolis: University of Minnesota Press.

Lachman, R., J. L. Lachman, and E. C. Butterfield. 1979. *Cognitive Psychology and Information Processing: An Introduction.* Hillsdale, N.J.: Erlbaum.

Marr, D. 1982. *Vision: A Computational Investigation into the Human Representation and Processing of Visual Information.* San Francisco: W. H. Freeman.

Quine, W. V. 1960. *Word and Object.* Cambridge, Mass.: MIT Press.

———. 1969. "Epistemology naturalized." In *Ontological Relativity and Other Essays.* New York: Columbia University Press.

———. 1975. "Empirically equivalent systems of the world." *Erkenntnis* 9:313–28.

Quine, W. V., and J. S. Ullian. 1970. *The Web of Belief.* New York: Random House.

Ribes, E., J. C. Ribes, and R. Barthalot. 1987. "Evidence for a larger sun with a slower rotation during the seventeenth century." *Nature* 326:52–55.

———. 1988. "Size of the sun in the seventeenth century." *Nature* 332:689–70.

Shapere, D. 1982. "The concept of observation in science and philosophy." *Philosophy of Science* 49:485–525.

Sneed, J. 1971. *The Logical Structure of Mathematical Physics.* Dordrecht: D. Reidel.

Suppes, Patrick. 1962. "Models of data." In *Logic, Methodology and Philosophy of Science: Proceedings of the 1960 International Congress*, ed. E. Nagel, P. Suppes, and A. Tarski. Stanford, Calif.: Stanford University Press.

Swerdlow, N. M., and O. Neugebauer. 1984. *Mathematical Astronomy in Copernicus' De Revolutionibus (Part 1).* New York: Springer Verlag.

Van Helden, A. 1985. *Measuring the Universe: Cosmic Dimensions from Aristarchus to Halley.* Chicago: University of Chicago Press.

Van Helden, A., and R. O'Dell. 1987. "How accurate were seventeenth-century measurements of solar diameter?" *Nature* 330:629–31.

Williams, M. E. W. 1981. "Attempts to Measure Annual Stellar Parallax: Hooke to Bessel." D. Phil. diss., University of London.

Foundationalism Naturalized

Foundationalist empiricism — foundationalism, briefly — is the view that knowledge[1] is based solely on the data of sensory cognition: perception, sensation, stimulation, etc.[2] Today, after decades of criticism, traditional versions of this view are virtually without defenders. It is ironic that many of the critics — notably Quine (1951, 1969) — have argued that epistemology should be naturalized and made scientific; for, as this essay will contend, when epistemology is naturalized and made scientific it suggests new interpretations of foundationalism that appear to be viable. My thesis is that a naturalized version of weak foundationalism is the proper framework for an acceptable theory of *conscious*[3] knowledge and that a naturalized version of strong foundationalism is the proper framework for an acceptable theory of knowledge as a whole, *conscious* and *unconscious*.

Some clarification concerning naturalized epistemology and how it bears on my thesis may be useful. Psychobiological theories of knowledge tell us how knowledge is in fact acquired by individuals of various natural species (humans, apes, insects, etc.) and how the acquisition process has evolved in these species. Machine theories of knowledge (artificial intelligence) tell us how knowledge is or can be acquired by artificial systems (machines) and what processes of acquiring it are best for achieving maximum reliability, scope, utility, efficiency, speed, etc.[4] Logico-philosophical theories of knowledge distinguish various types and grades of knowledge (from mere opinion to certainty), describe how such knowledge may be justified or certified, and propose methods for acquiring knowledge of the various types and grades. Theories of this latter sort are not directly concerned with how knowledge is in fact acquired or evaluated by empirical systems, natural or artificial; the distinctions, criteria, and methods they propose are more or less ideal. In rough summary, theories of the above three sorts tell us, respectively, how knowledge is, can be, and should be acquired.

Clearly there can be overlap and interaction between theories of the three types distinguished above. Logico-philosophical theories of knowledge can lead to theories about the way machines acquire knowledge, and these in turn can lead to theories concerning (become models or simulations of) the way natural systems acquire knowledge. Conversely, theories of animal knowledge may lead to theories (become models) of artificial intelligence or even ideal theories that are not instantiated in any actual or possible artificial system. Obviously theories of any of the three types may or may not be advanced in a proper or "scientific" manner.

Epistemology as traditionally practiced by philosophers advances philosophical, ideal theories of knowledge. It typically contains confusing mixtures of empirical, logical, speculative, and unscientific material. A strong tendency of current philosophy of science is to replace traditional epistemology by "scientific epistemology," a discipline that (in a proper, "scientific" manner) advances theories of knowledge of either of the three types distinguished above and recognizes as legitimate any fruitful combination and interaction of the three.

"Naturalized (or naturalistic) epistemology" is a hybrid of traditional and scientific epistemology. As the name suggests, it is an attempt by philosophers to bring their traditional concerns and ideal theories down to empirical earth by applying them to humans and members of other natural species in ways that take account of their natural development. Naturalized epistemology currently tends to assume, on evolutionary grounds, that the methods by which a normal member of a species acquires knowledge required for its survival are among the best available and that a satisfactory theory of knowledge for the species will incorporate these methods as basic and indispensable. Although it may employ the results of logico-philosophical and machine epistemology to suggest theories about the ways natural organisms acquire knowledge, and may distinguish between and examine good (optimal) and bad (suboptimal) ways of acquiring knowledge, the main goal of naturalized epistemology is to understand cognitive processes that do and can occur in natural organisms. In the hands of some practitioners it is coextensive with the psychobiology of cognition, and as a general movement it may prove to be a temporary phase in the attempt to replace traditional epistemology by scientific epistemology. In brief, naturalized theories of knowledge tell us how knowledge is acquired by some species of interest and may (or may not) go on to tell us how (within the limits of psychobiological possibility?) knowledge should be (optimally can be?) acquired by the species.[5]

Traditional foundationalism was created by philosophers before distinctions between types of theories of knowledge such as those above had been articulated. It was always advanced as a theory of human knowledge, but whether as a theory of how knowledge is *in fact* obtained, or can be obtained, or should be obtained was usually unclear. In the present essay, foundationalism is proposed as a theory of scientific human epistemology, more precisely, as a psychobiological theory of how humans do in fact acquire their empirical knowledge. Even more exactly, foundationalism is here proposed as the appropriate form or framework of any adequate psychobiological theory of empirical human knowledge. Which specific foundationalist theory (Gregory's, Marr's, etc.) is most adequate is a question left for psychologists and biologists to answer. Whether foundationalism is the appropriate framework for an adequate theory of how humans *should* (optimally can) acquire knowledge will be briefly addressed at the end.

1. Foundationalist Theories of Conscious Knowledge

Strong foundationalism is the view that a distinct class of data contains the foundation, or basis, of empirical knowledge and that these data are independent and infallible. To say that they are independent (the older term is "self-evident") means that they are not inferred from other data and are neither in need of nor capable of confirmation or disconfirmation by comparison with or inference from other data or by other means. To say that they are infallible means that they are incapable of being nonveridical or false.

A *first version* of strong foundationalism holds that the data of ordinary conscious[6] perception are the basis of empirical knowledge. This version has had few defenders — perhaps only G. E. Moore (1939) and some of his disciples — for good reason. The view is incompatible with the well-known facts that perceptions of physical objects such as tables, chairs, and even one's own hand (Moore's example in the above-cited reference) are sometimes illusory or hallucinatory and that their veridicality may require confirmation or disconfirmation by comparison with other perceptual evidence. For example, my perception of an elliptical saucer on the table, seen at an acute angle in poor light, is disconfirmed by my perception moments later of a round saucer in the same place seen at a right angle in good light.

A *second version* of strong foundationalism is the classical one, which holds that a special type of perception — usually called sensation — is basic. The data provided in sensation are called sensibilia, sensa, or sense-data, and the usual examples are sensations of colored regions, felt heat, felt roughness, sounds, tastes, smells, and so on. For example, my sensation of a round patch of white color in my visual field is held to be infallible, i.e., not capable of being illusory or hallucinatory; and independent, i.e., not inferred from or subject to correction by other data.

Traditional strong foundationalism has been extensively criticized during the past four or five decades. After propounding the theory for many years, Russell (1921) finally abandoned it and concluded that even the most refined perceptions and sensations involve some degree of inference, and that pure, noninferential perception is a theoretical ideal. Hanson (1958), Feyerabend (1965), Kuhn (1962), and in their wake a multitude of others argued that all perceptions and presumably also sensations are interpretive, conceptual, theory-laden, etc. Austin (1962) and others claimed that mistakes in reports of sense-data, reports of how things seem to the observer, are possible. Some writers argued that sense-data do not exist (Barnes 1944–45) or, in a slightly more moderate vein, that sense-data are linguistic artifacts created by a way of speaking (Ayer 1954).

This large body of criticism has been curiously indirect and inconclusive, for several reasons. The point that sense-datum reports are corrigible, if true, does not entail that the reported sensations are fallible, because the sensations themselves may be veridical even if the reports are false. The thesis that sense-data do not exist is incredible, if taken to mean that sensations and their "contents" do not exist. (Of course, the nature of "content" is one of the problems involved.) Finally, although sensations are probably "interpretive" or "inferential" in some sense, it remains to be shown that this feature entails their fallibility.

In sum, the critics have failed to directly address the question at issue: Can one's sensation of a round blue patch in the visual field (for example) be mistaken and be shown to be mistaken? Nonetheless, when the question is directly addressed, its answer seems to be affirmative, at least for conscious sensations. If my sensation is an afterimage produced by the flash of a camera, or a hallucination produced by a psychoactive substance, then the sensation is mistaken. The mistake consists in the fact that the sensation is of (represents) a round blue patch and yet no such patch exists. To the familiar objection that there is no mistake because a "mental" round blue patch does exist, the reply is that the notion of a "mental patch" is a philosophical confusion: "mental patches" are simply hallucinated physical patches and by definition do not exist (the hallucination exists but not the hallucinated object). To the related objection that there is no mistake because a round blue patch exists in "private" or "mental" space, the reply again is that these notions are philosophical confusions: "private" or "mental" space is simply hallucinated public space and consequently does not exist. Once it is accepted that sensations can be mistaken, it becomes obvious that they can be shown to be mistaken. I can show that my sensation of a round blue patch is an afterimage and thus mistaken by observing that it moves with my eyes and is projected on whatever is in my field of vision. I can show that the sensation is a hallucination and thus mistaken by finding that other observers do not see such a patch in the places where I see it. It thus appears that the classical sensationist version of strong foundationalism is indeed unacceptable.

Weak foundationalism is the view that, although none of the data of conscious sensory cognition is completely independent and infallible, the members of a distinct subclass of these data are less dependent and less fallible than all other sensory data and comprise the basis of empirical knowledge. Several versions of weak foundationalism need to be distinguished.

A *first version* holds that the data of conscious sensations — sense-data — are the least dependent and least fallible data available and are therefore the basis of empirical knowledge. This version has had few, if any, adherents, probably because the classical strong foundationalists developed many persuasive arguments purporting to show that sense-data are necessarily independent and necessarily infallible. One version of the latter argument is as follows:

> It is logically possible for me to be mistaken about what I see but not about what I seem to see; for what I seem to see is the same whether I am mistaken or not, and what I seem to see is by definition a visual sense-datum.

The argument is specious: it confuses a mistake about what sensation one has with a mistaken sensation. Even if I cannot be mistaken in my belief that I have a sensation of a round blue patch in my visual field,[7] it is nonetheless possible that my sensation is an afterimage and that no round blue patch exists on the wall or anywhere else, i.e., possible that the sensation is mistaken, nonveridical. Whether sensations are independent or infallible is a factual, empirical question and must be settled, not by a priori arguments, but by unbiased examination of the phenomena. As was noted earlier, such examination quickly reveals that sensations are not infallible. Similarly, whether sensations are less dependent and less fallible than perceptions or other sensory cognitions that might be proposed as a basis is a factual, empirical question. It is, however, a question that has never been properly, i.e., experimentally, addressed, and its answer is not yet known.

A *second version* of weak foundationalism holds that the data of ordinary perceptions of saucers, tables, hands, etc. are the most reliable and independent available and that these are the foundation of empirical knowledge. Proponents of this view sometimes point out that sense-data are the contents of an esoteric type of sensory cognition that requires special training or talent (as in painters) and is therefore interpretive and generally less reliable than ordinary perception. This second version seems to be the view of ordinary common sense and is widely held. It is not the view of scientific common sense, which is willing to employ any type of observation to confirm a theory — ordinary perception, instrumentally assisted observation, sensation of sense-data — as long as it is relevant, careful, replicable, and so on. However, both scientific and ordinary common sense must yield to scientific epistemology regarding the issue. Whether perceptions are the least dependent and least fallible of all sensory cognitions is a factual, empirical question whose answer must be obtained by scientific epistemology. A positive answer would establish the second version of weak foundationalism.

On the other hand, scientific epistemology may establish a *third version*, according to which the data of all types of conscious sensory cognition are in general equally dependent and fallible, though less so than other types of data, and taken together are the basis of empirical knowledge. The following plausible account of the development of human knowledge will motivate this version. At first, humans naturally and un-self-consciously employed the data of their perceptions to guide their actions, discover theories about unperceived objects and processes, and evaluate other perceptual data. Then, the desire to avoid the harmful results of deception induced them to become critical: to select data and draw inferences from these with greater care and to remove from the basis data of perception that had proved unreliable. On the other hand, data from instrumentally aided observation were added to the basis when they proved to be reliable.[8] Still later, some of the data of sensation — sense-data — were added when these were certified by psychology. Finally, data from all three types of sensory cognition were included in the database on the sole criterion of their reliability, which became increasingly easy to evaluate with the growth of scientific epistemology.

The third version is probably correct. However, my main point is that probably one of the three versions will emerge as satisfactory, which means that probably weak foundationalism is the proper framework for an acceptable scientific theory of conscious knowledge.

With strong and weak foundationalism thus defined, it would seem natural to define *antifoundationalism* as a theory of knowledge that is neither strong nor weak foundationalist. The problem with this suggestion is that many antifoundationalists reserve the term "foundationalism" for the view we have called strong foundationalism and refuse to apply the term, even qualified with "weak," to any view on which basic data are dependent or fallible. This terminological practice obscures the differences between strong and weak foundationalism and creates the impression that antifoundationalism has only one, easily refutable competitor — strong foundationalism. It also obscures the differences between weak foundationalism and the undeniably antifoundationalist view that all data are fallible and dependent and thus epistemically on a par. By calling the latter view strong antifoundationalism and conceding that weak foundationalism is just as appropriately called weak antifoundationalism, at least the terminological part of the issue is resolved. Accordingly, *weak antifoundationalism* is here defined as equivalent to weak foundationalism. *Strong antifoundationalism* is defined as the view that all data — the data of perception, sensation, instruments, etc. — are equally dependent and fallible, and that no class of data can be selected as favored or basic. An even stronger antifoundationalism — *very strong antifoundationalism* — would hold that there is no valid epistemic distinction between data and other statements, and that all statements are equally dependent and fallible.[9]

The received theory of knowledge today is antifoundationalist, although it is often defended with such rhetorical exuberance and imprecision that one cannot tell whether the strong or the weak version is being recommended. For example, Churchland (1979, pp. 2–3) says: "That there is a distinction between the theoretical and the nontheoretical...seems to be false.... It appears that all knowledge (even perceptual knowledge) is theoretical; that there is no such thing as nontheoretical understanding." His statement is compatible with either the weak, strong, or very strong version of antifoundationalism.

The strong antifoundationalist doctrine that all data (or statements) are equally unreliable and equally dependent is so incredible that it is difficult to believe anyone seriously subscribes to it. It is well known that perceptual data obtained with sense organs affected by injury, disease, fever, and drugs are less reliable than perceptual data obtained with normal organs; and that data obtained through prophecy and oracular means are less reliable than those obtained through perception and scientific inference. It is also generally agreed that percepts of one's own hand are usually more reliable and less in need of confirmation than theories about quarks, planets, or even the table in the next room, because they involve fewer inferences and hence fewer opportunities for error. Furthermore, psychologists generally agree that sensations of color, heat, roughness, and other fairly simple data are usually more reliable and usually less in need of confirmation than perceptions of hands, tables, and other complex objects, again because they involve fewer inferences. For example, my visual perception of a saucer, seen from an acute angle, as circular in three-dimensional space involves an inference from my (usually unconscious) sensation that the corresponding shape is elliptical in two-dimensional space, and therefore involves one more inference than the sensation.

Let us sum up and proceed to the next topic. Strong antifoundationalism is incredible. Weak foundationalism (i.e., weak antifoundationalism) probably provides the framework for an acceptable naturalized theory of *conscious* knowledge processes. Strong foundationalism is unacceptable when applied only to conscious knowledge processes. However, as we will now proceed to show, strong foundationalism may well provide the framework for an acceptable scientific theory of knowledge processes as a whole, both *conscious* and *unconscious*.

2. A Foundationalist Theory of Conscious and Unconscious Knowledge

Theorists of knowledge have, until recently, virtually ignored unconscious cognition, the unconscious processes of thought and knowledge that underlie conscious processes. Some philosophers have gone so far as to deny that there is any such thing as unconscious cognition. However, contemporary cognitive psychology continually teaches that conscious cognition is only a small part of human cognitive activity. I suggest that strong foundationalism has seemed unacceptable mainly because it was assumed to be a theory of the conscious cognitive activity of humans, and that it may re-emerge as an approximately correct theory of the way humans (and many other animals) unconsciously acquire their perceptual and sensational knowledge. In what follows I sketch a foundationalist theory suggested by recent cognitive, computational theories of perception, a theory on which basic data are provided by a subclass of what I will call presensations, unconscious sensations that precede conscious sensations in the process of perception.

Foundationalist views were classified above on the basis of the epistemic properties they ascribe to basic data: namely, independence and infallibility. They can also be classified on the basis of the types of cognition they assume to provide the basic data. On this classification, *perceptual* foundationalism holds that basic data are provided by perceptions: my visual perception of a white saucer on the brown table, my tactual perception of a hot cup, my auditory perception of a whistling tea kettle, my gustatory and olfactory perception of coffee. This view is most likely incorrect, in view of the consensus among psychologists that perceptions involve inferences from data provided by sensations. For example, my perception from an acute angle of a white saucer on a brown table normally involves an unconscious inference from my normally unconscious sensation of a white elliptical patch inside a brown one.

Sensational foundationalism holds that basic data are provided by sensations: the visual sensation of a white patch inside a brown one, the tactual sensation of heat, the auditory sensation of a whistling sound, the olfactory sensation of a pleasant, acrid smell, and so on. This view is also probably incorrect, given the wide agreement among cognitive psychologists that sensations are the product of a great deal of unconscious, lower-level information processing. In the older terminology, sensations involve unconscious inferences from unconscious data. However, there is little agreement about the nature of these lower-level processes, and the subject is a rapidly expanding area of theory and experiment.

Most psychologists implicitly subscribe to *presensational* foundationalism, which holds that the class of basic data for empirical knowledge is some subclass of those sensory cognitions that precede sensations and perceptions in the causal process of sensory cognition. Perceptions are usually conscious and sensations are sometimes so, but presensations are usually (if not always) unconscious. Consequently, the existence and character of presensations usually cannot be determined by introspection or by any direct method, but must be inferred with the help of a psychological theory of sensory cognition. Such theories can be and have been devised by philosophers.[10] However, on the approach of scientific epistemology, they are empirical theories whose confirmation requires evidence concerning the behavior and cognitive processes of sensory cognizers. Several such theories are presently under construction by cognitive psychologists and specialists in artificial intelligence. I will use as my example the computational theory of vision devised by Marr (1982) and his associates.[11]

In a computational theory of vision, visual cognitions are taken to be images, or imagelike representations, and the process of visual cognition is taken to be a sequence of productions, or computations, of images, images at one level being computed from images at the next lowest level. Perceptual ("three-dimensional") images are computed from sensational ("two-dimensional") images; sensational images are computed from high-level presensational images; these in turn are computed from lower-level images or image elements; and so on, until we come to image elements computed by the retinal receptors — the rods and cones (cones, to use the general term). The receptor image elements and perhaps the images computed directly from them are the first-level visual cognitions, the *ur-sensations* of vision, as we will call them. They are computed by the retinal receptors from optic arrays, that is, from arrays of light energies focused at the surface of the retina by the lens. Optic arrays may be regarded as zero-th level images in vision.

Let us consider an example of perception production according to this theory. My visual image of a white saucer is computed from a visual image of a circular white patch. The latter image is computed from an image of a circle together with images that represent texture elements comprising a white surface. The image of a circle is computed from image elements that represent edges, and these edge elements are computed from elements representing differences in light intensities. Texture elements are similarly computed from receptor-image elements that represent intensity and frequency differences in the optic array. As the example illustrates, there are several levels of presentations below conscious sensations of colored shapes. Whether presensational foundationalism should include only ur-sensations in the class of basic datum cognitions or whether somewhat higher-level presentations should also be included is a question for further investigation. For the purpose at hand it is convenient to assume that only ur-sensations are included.

The central objection to perceptual foundationalism is that perceptions are dependent and fallible and therefore do not constitute an acceptable basis for knowledge. The central objection to sensational foundationalism is that sensations are not infallible, since they can be nonveridical.[12] However, it may be argued that although sensations can be nonveridical, they cannot be *shown* to be nonveridical and are therefore independent. If the argument is sound, then sensations have at least one of the two epistemic characteristics required of foundational data, and sensationalist foundationalism is at least half correct. The argument is as follows:

> Consider my sensation of a round blue patch in the center of my visual field. Because sensations are the most reliable of cognitions, my sensation can be discredited only by comparing it with another, incompatible sensation, for example, the sensation of a round *red* patch in the center of my visual field. Furthermore, the two sensations must occur at the same time; for if one occurs now and the other occurred previously, then I know of the latter by remembering it, and the datum obtained by remembering a sensation is less reliable than the datum obtained by having a present sensation. However, it is impossible for me to have simultaneously a sensation of a round blue patch and a sensation of a round red patch in the center of my visual field. Consequently, my sensation cannot be shown to be nonveridical.

Now of course this argument is specious. Although two incompatible sensations cannot exist simultaneously in a single perceptual field and therefore cannot be directly compared, they can exist at different times and be indirectly compared by means of a cognitive bridge between them. Because we will presently consider an argument that ur-sensations cannot be so compared, it is necessary to examine the comparison process in some detail.

Current theories of memory make it reasonable to suppose that a subject's (S's) sensation of a blue patch in the center of the visual field is stored for a few milliseconds in Very Short Term Memory (VSTM) as a sensation, or possibly as a vivid image slightly more schematic than but otherwise indistinguishable from the sensation. In most instances the sensation or vivid image will be processed and stored for a few seconds in Short Term Memory (STM) as a mental image: a fragmentary, pallid counterpart of the original. In other instances the mental image will be encoded symbolically and stored for longer periods of time in Long Term Memory (LTM) as a proposition[13] associated with the sentence,

1. There was a red circular area in the center of the visual field at time t.

S's (usually unconscious) reasoning processes may subsequently generate the following sequence of propositions.

2. There were no circular areas in the visual field at t-Δt, for Δt small (obtained by scanning visual STM).

3. There were no circular or spherical objects in the immediate environment at t-Δt (inferred from (2) together with the generalization in LTM that where there are no circular areas there are no circular or spherical objects).

4. There were no circular or spherical objects in the immediate environment at t (inferred from (3) together with the generalization in LTM that environmental objects persist for longer than Δt).

5. There were no circular areas in the visual field at t (inferred from (4) together with the generalization in LTM that where there are no circular or spherical objects then there are no circular areas).

6. There was no blue circular area in the center of the visual field at t (inferred from (5) together with the logical principle in LTM that where there are no circular areas there are no blue circular areas).

Now (2) implies (6), and (6) implies not-(1). Thus, a sensation from one time can be discredited by bringing its report, (1), into conflict with the report, (2), of a sensation from another time. Obviously discreditation will require more evidence than the single conflict described above. For this single conflict is equally good evidence that the sensation reported by (2) is nonveridical, or that (2) is an incorrect report, or that one of the inferences drawn from (2) is fallacious. Ideally, deciding which sensations to reject requires a substantial body of sensational data and a sensitive method for employing it. For present purposes we need not consider the various methods that can or have been employed.[14] All such methods involve discrediting sensations by indirectly comparing them with other sensations; consequently, the very possibility of employing such a method shows that sensations are not independent.

When we come to ur-sensations, we find reason to suppose that they cannot be so discredited. Let us take, as our example of an ur-sensation, the excitation of a single color receptor, or cone, in the retina. To discredit this sensation in the manner described earlier, an incompatible ur-sensation must be located and a bridge constructed between it and the sensation to be discredited. It seems certain that ur-sensations are not stored in symbolically encoded form or even in "mental-image" form in LTM or in any other memory. Furthermore, it seems likely that ur-sensations are not stored in STM, or even in VSTM, in any form; for ur-sensations are the inputs for computations of tokens and images at higher levels, and probably are used immediately and then immediately expunged to clear the information-processing channels for the reception and transmission of more data. Only higher-level outputs are stored, some as tokens, some as "mental images," some in symbolic form. Probably, then, there are no memory entries corresponding to ur-sensations. If that is so, then the cognitive bridge between ur-sensations required to discredit one of the two cannot be constructed. An additional difficulty is that the objects of ur-sensations probably should be regarded as concentrations of light energy, either in the optic array or the environment that produces the array, and therefore are momentary entities that do not persist from one moment to the next. If so, human memory will not contain the principle that a cone's sensation of a certain color at one time is incompatible with the cone's sensation of a different color at a slightly different time.

We cannot immediately conclude from the above argument that ur-sensations are indefeasible, because the argument considers only one conceivable method for defeating them. The method was suggested by an assumption that the data of ur-sensations, being the most reliable available, can be discredited only by direct comparison with data of other ur-sensations; but the assumption is questionable. To begin with, ur-sensations have not been shown to be the most reliable cognitions available. Furthermore, even if this had been shown, it would not entail that discreditation requires direct comparison with other ur-sensations. Ur-sensations are the basis for sensations, which are the basis for perceptions, which in turn are the basis for scientific theories. If a scientific theory is well based on perceptions and these are well based on a group of ur-sensations, then the theory might be used to discredit some ur-sensation not in the group.[15]

One type of theory that might be used in this way would describe an instrument for detecting the sensation (cone excitation) and an instrument for detecting the stimulus, or object allegedly sensed (pattern of light energies): if the instruments detect the sensation but not its stimulus, then perhaps the sensation has been shown to be nonveridical and thus discredited. The problem with any such theory is the hypothesis of some psychophysicists that retinal receptors are natural instruments as reliable in detecting their special patterns of light energy as any others, in particular, as reliable as artificial instruments designed to detect these same light patterns.[16] Whether the hypothesis is true is a complicated theoretical question. Obviously there can be no a priori answer: artificial instruments are not by logical necessity more sensitive than natural. Nor, apparently, are they by empirical necessity more sensitive. Which instrument is the more sensitive probably depends on its capacity to respond to its stimulus, not on whether it is constructed by humans or by nature, or whether it is made from transistors or from neurons. It may be that millions of years of evolution have equipped retinas with cones that are the most capable among all empirically possible systems of responding sensitively to their stimuli.

To consider a partially analogous case, the human eye is an extraordinarily sensitive detector of motion from a distance, perhaps more sensitive than any empirically possible artificial detector employing light transmission. There are of course motion detectors that do not employ light transmission, and some of these are more reliable. For example, a system that throws a ball from one pod to another detects the motion of the ball by throwing and receiving it; or, in any case, the experimenter monitoring the system detects the motion by determining that the ball is on one pod at one moment and on the other a moment later. (That it is the same ball must of course be inferred.) This detector may be more reliable than any direct motion detector employing light transmission. But in the case of detectors of patterns of light energy that stimulate retinal cones, there is no superior detector analogous to the ball thrower, because the stimuli for cones cannot be prepared by ordinary manipulation and observation. They must be prepared indirectly with the aid of a sophisticated theory, one that involves inferences to light energies, which, unlike balls, cannot be observed or manipulated by ordinary means. For all we presently know, retinal cones are the most reliable detectors of their stimuli among those that are empirically possible.

Of course there may be purely theoretical methods of discrediting cone sensations, methods that do not employ superior detectors of the stimuli to those sensations. In particular, we might be able to infer from the character of the percepts to which ur-sensations give rise that those sensations are nonveridical. For example, it might be supposed that from the nonveridicality of my percept of an elliptical saucer on the desk (which in reality is a circular saucer seen at an angle in poor light) it can be inferred that the ur-sensations from which the percept was ultimately generated are nonveridical. However, this would be a poor inference. The percept of an elliptical saucer in three-dimensional space is generated from the sensation of an elliptical shape in two-dimensional space, which is generated from presensations of edges, regions, and texture, which are generated from ur-sensations of differences and ratios of light energies. The mistake in the percept was somehow caused by poor light, but it may have occurred in the production of the percept from a sensation, or in the production of the sensation from a presensation, or at some still lower level, without any mistake in the ur-sensations having occurred. Some psychologists would contend that the mistake consisted in the production of a percept of an elliptical object from a sensation of an elliptical shape, owing to the absence of the usual distance cues that indicate a round object. Others would contend that the mistake consisted in the production of the sensation of an elliptical shape without the usual textural elements that indicate the surface of a round object. No psychologist would suggest that the ur-sensations from which the percept was generated were incorrect responses by the retinal cones to the illumination present at the surface of the retina. More or better illumination would have prevented the mistake, but the cones responded correctly to the illumination that was present.

We have seen that it is probably incorrect to infer that ur-sensations are nonveridical from the fact that the percepts they generate are illusory. The probability increases almost to certainty if the inference is from the fact that the percept is hallucinatory. Hallucinations are percepts that have no external cause and are produced by such internal causes as overactive imagination, diseased neural structures, psychoactive substances coursing through the brain, and so on. In cases of percepts such as these, psychologists tend to agree that the retinal receptors are probably not among their causes. Whether hallucinatory visual percepts and images are formed in the cerebral cortex of the brain, or in the optic nerve, or in the interneurons of the retina is still a matter of dispute among experts.[17] But no scientist maintains that they are composed of ur-sensations, retinal cone excitations. It is likely that hallucinations are produced directly at the perceptual level, probably in the cerebral cortex, without being generated from sensations or high-level presensations. It is very unlikely that they are or are generated from ur-sensations.[18]

In sum, ur-sensations may be independent: incapable of discreditation (indefeasible) and incapable of certification. If they are, it does not immediately follow that they are infallible. (This would follow only if they were *logically* incapable of discreditation and certification, rather than *empirically* incapable in the manner conjectured here.) So the question of their infallibility must be separately considered.

To say that sensory cognitions are infallible is to say that they are incapable of being nonveridical.[19] To say what it means for a sensory cognition to be veridical, we must say what entity (state of affairs, process, event, etc.) the cognition represents. If the cognition is accompanied by the represented entity then it is veridical; if it is not accompanied by the represented entity then it is nonveridical. As we have noted, perceptions, sensations, and presensations are regarded in current computational psychology as images, that is, as distributions of image elements that represent features of the environment in image space. They differ from one another with respect to their level in the information-processing hierarchy, and also with respect to the level of the environmental features they represent. Perceptions usually represent the external, molar environment in which observers live and behave; some represent the internal environment of their bodies. Sensations represent special (for example, "two-dimensional") aspects of the observer's internal and external environments. As one descends through the hierarchy of presensations to ur-sensations, cognitions more and more come to represent features of proximal stimulation. For example, visual ur-sensations probably represent features of the array of light energies at the retina of the eye.

Precisely what features of light arrays do they represent? What feature of the array of light energies must exist for a given ur-sensation to be veridical? Where the sensory cognition in question is a conscious percept or sensation it is tempting to say that the cognition represents whatever it resembles, where resemblance is defined as follows. One system resembles another if and only if there exists an isomorphism from the one to the other, that is, if and only if the objects and the relations of the one system have counterparts in the other system. Now it is highly doubtful that ur-sensations resemble their stimuli in the sense of being isomorphic with them. One reason is that ur-sensations do not seem to have the complexity required to define usable isomorphisms, isomorphisms that make it possible to distinguish between ur-sensation S's being isomorphic with feature F_1 of the optic array but not with feature F_2. Furthermore, even if ur-sensations do have the requisite complexity, it is doubtful that they represent features of the optic array with which they happen to be isomorphic. To illustrate these points, most visual ur-sensations can be identified with the results of a receptor's summing, averaging, or performing some other statistical operation on a group of light energies. The result is thus an excitatory response in a system of neurons, which at the molar level is not sufficiently complex to be isomorphic with the energies summed. Furthermore, if we describe the excitation at the micro level in all its complexity, as a spike of several intensities or frequencies, these elements undoubtedly are not isomorphic with the energies summed. Whatever the relation between the result of such an operation and the stimulus, it is probably not isomorphism.

The next most attractive candidate for the representing relation is causation, according to which the feature represented by an ur-sensation is the cause of the ur-sensation. This suggestion begs the question of whether ur-sensations are veridical in the following way. Every event has a cause, according to the (perhaps necessarily true) law of universal causation; consequently, every ur-sensation is accompanied by its cause. From this law and the present suggestion, it immediately follows that every ur-sensation is accompanied by what it represents and is therefore veridical. The objection here is not that ur-sensations are not always veridical: indeed we will shortly speculate that they are. It is, rather, that to establish their veridicality by defining the representing relation in a way that immediately obtains this result is illegitimate. It would be similarly illegitimate to stipulate that the entity represented by a conscious percept is its cause. Here the objection is obvious, since the stipulation, together with the law of universal causation, implies that illusory and hallucinatory percepts — i.e., nonveridical percepts — are veridical. The former have physical processes and the latter have neurophysiological processes as causes, and since all are accompanied by their causes, they are veridical. Thus we arrive at the conclusion that all percepts are veridical, which is obviously false. Because it is not in the same way obviously false that all ur-sensations are veridical, we cannot object to a definition of their representing relation on the ground that it leads immediately to this false conclusion. But we can argue that the definition should not immediately lead either to this conclusion or to the contradictory conclusion that not all ur-sensations are veridical.

The most promising analysis of the representing relation defines the feature represented by an ur-sensation as the feature that explains the generation of conscious sensations and perceptions from the ur-sensation. For example, the generation of the visual sensation of a surface can be explained by positing presensations that represent detectable elements of a surface: edges, vertices, textures, illuminations, etc.; and the generation of these presensations can be explained by positing ur-sensations that represent averages of (or some other operation on) light energies and collections of such averages. Different psychologists have hypothesized different systems of presensations (some use edges and textures to explain sensations of a surface; some mainly use vertices), and they have hypothesized different methods by which sensations and perceptions are computed from presensations. However, causal considerations probably can be employed to decide between the various hypotheses. If there is no (physically) possible feature of the optic array that could cause a hypothesized ur-sensation, then the ur-sensation cannot be part of the process leading to presensations; and if there is no (neurophysiologically) possible ur-sensation that could cause a hypothesized presensation, then the presensation cannot be part of the process.

Sensations produced by environmental stimulation are usually veridical, or at least sufficiently veridical for the purposes of the observer; if they were not, evolutionary pressures probably would have led to their extinction. Consequently, it is highly likely that the causal process by which a sensation of a surface is produced begins with a feature of the optic array that is caused by an element of a surface, on some scheme of analysis of surfaces. Accordingly, the best method for devising hypotheses about the reproduction of sensations of surfaces is to analyze surfaces in such a way that they can be deemed to cause the features of optic arrays that cause the ur-sensations that ultimately result in perceptions of a surface. Within such a hypothesis it is quite natural and no longer question-begging to say that ur-sensations represent the features that cause them and that in consequence ur-sensations are inevitably veridical. Since it is very likely that the finally accepted theory of visual perception will be devised by the method above, it is very likely that the accepted theory will explain all conscious sensations and perceptions, nonveridical as well as veridical, as products of veridical ur-sensations.[20]

In conclusion, it is possible that psychobiological science will discover that ur-sensations cannot be certified or decertified by other data (are independent) and that ur-sensations cannot be nonveridical (are infallible). This discovery would reveal that ur-sensations are analogous to the logically independent and infallible conscious sensational data mistakenly supposed by the classical foundationalists to constitute the basis of empirical human knowledge, and that ur-sensations qualify as the basic data in a new strong foundationalist theory of unconscious and conscious empirical knowledge. To say that ur-sensations would then have been found analogous to the sensations of the traditional foundationalism means of course that they would have been found similar in some respects, different in others. The first difference is that ur-sensations would not be independent and infallible in the traditional senses of these terms. They would be independent as a matter of theoretical empirical fact, not as a matter of logic or definition. The second difference is that ur-sensations could not be consciously employed by the subject as a basis for inference, in the manner envisaged by the classical foundationalists. The first difference should be clear from what has gone before. The next section will deal with the second difference.

3. The Knowing Organism as an Association

Traditional strong foundationalism holds that empirical knowledge is inferred from the basis of conscious sensations; the new version of strong foundationalism considered here would hold that the inferences are unconscious and are made from a basis of unconscious sensations. To throw light on this difference and to clarify the new version, I will employ a societal model of animal information-processing systems. Whether it is merely an analogy or whether it is a correct model of animal cognition will emerge with the further development of cognitive science. Think of an organism as an association of cells, or, more accurately, as an association of information processors built out of cells. Stretching the analogy a bit, think of an organism as an association of clerk-homunculi whose functions are to receive information from clerks in various departments, process it, and send it on to clerks in other departments, with high-level clerks exercising executive functions, and final messages handed to a chief executive who makes decisions for the whole system.

This is a familiar model, and something like it has been employed by Dennett (1983), Minsky (1986), and other writers. Unlike Dennett, but like Minsky, my use of the model does not suggest that the intelligent activity of an organism can be analyzed into the activities of thousands of nonintelligent ("stupid") homunculi, and the mental level thus reduced to a nonmental level. Rather my homunculi have varying degrees of intelligence, or, better, degrees of complexity of cognitive function. Those close to and in the receptors do one or two simple things and do not make decisions. But as one ascends the cognitive hierarchy, the homunculi become more and more sophisticated and complex.

Do they become more and more conscious? The nature of consciousness presents one of the most baffling problems in cognitive science, to which I here offer a provisional solution. In a first and fundamental sense, consciousness is simply the activity of cognition. In this sense the information-processing subsystems of an organism are more or less conscious, just as organisms themselves, from humans to amoebas, are more or less conscious. Lower organisms usually are conscious only of the environment external to their bodies, but higher organisms are conscious also of states of and processes in their bodies. In a second and more specialized sense, the cognitions of my subsystems are not conscious, since "I" am not conscious of them. In this sense the consciousness possessed by a system is the cognition performed by its executive subsystem(s). So, again, consciousness is cognition, but the cognition of a special type of subsystem. Consciousness in the second sense is simply consciousness in the first sense performed by the executive. It is not different in kind from consciousness in the first sense; the difference is merely that it is performed by an executive, a subsystem that identifies itself with the whole system and "thinks, speaks, and acts for it." In lower organisms, which have fewer cognitive layers, the executive is a less developed subsystem, and so the consciousness of the organism is less developed. In some of the lower organisms there is no executive subsystem, and hence no consciousness in the second sense. Self-consciousness, in one of its many senses, is consciousness in the second sense of the state of the system, especially of its relation to other systems.

I have spoken as if there is only one executive in any organism, but probably there are many executive subsystems with different functions and degrees of control. It seems psychologically possible for the executive of a system to die (from cortical injury, for example) and for a less developed subsystem to take over the executive function. It seems likely that in certain subjects several executives of roughly equal power compete with one another for control of the organism (in "multiple personality" for example). It even seems possible that in a normal human several virtually identical executives continually exchange control of the organism, without the organism being aware of it. In that case, a human would be comparable to a nation whose chief executives succeed one another imperceptibly, without any alteration in either domestic or foreign policy; and consciousness in a nation would be comparable to consciousness in a human.

The job of receptor subsystems is to receive a group of very simple messages (such as those of light intensity above a certain threshold at a certain point in the optic array) and combine them in some simple manner (for example, by computing an average of the light intensities from a small region of the optic array). The output message is transmitted to the next subsystem and there used immediately (which may consist in being ignored), and is not stored. Consequently, if something goes wrong with the society's decisions — for example, it decides to move in a direction that leads to collision with an object — the executive cannot examine the low-level messages to determine whether the mistake was made by them. To begin with, the executive does not have access to the operations of the low-level clerks and cannot obtain it by visiting their department. Even if the executive did obtain such access it would not find the messages in any file: they were passed on and no copy was kept. Finally, even if copies of the messages had been kept, they would not have been written in the language the executive uses.

To find out what went wrong, an outside trouble-shooter must be employed who can determine the operation of the subsystems at all levels. This person will of course be a cognitive psychologist. Her or his job is difficult, not merely for the practical reasons that the clerks resist investigation, but for theoretical reasons as well. For the clerks do not speak any language except that of their messages, and it is virtually certain that this language is not like any public language such as English or Mayan. It may not even be a language, in the usual sense of that term. For it will be identical with a system of neural impulses and synaptic junctions, which produces a neural image in the next subsystem and at some sufficiently high level generates an output that can be identified with sentences or similar patterns. It will be more difficult to determine where the mistake occurred in this association than in the executive branch of the U.S. government. If we could have gotten CIA Director William Casey to testify, then perhaps we could have found out who diverted the funds from the Iranian arms sale to the Contras in Nicaragua. But the retina has never been and will never be capable of testifying. Indeed, the notion of a mistake in low levels of the neural system probably will have to be subtly reconstructed in order to apply it to a neural association of information processors. As was suggested earlier, it may turn out that the notion can only be usefully defined for the whole system, and that the notion of a mistake in a particular low-level department is not a theoretical possibility. Again there may be an analogy with associations: it is sometimes impossible to define a mistake on the part of a low-level agent in an association except by reference to the goals and actions of the whole association. For example, a count of all residents with current marriage licenses would be a mistake if the task was to determine the number of households, but not a mistake if the task was to determine the number of legally married residents. (Notice that although the same data — say, that of a general census — could be used for both totals, it need not be; the number of households could be obtained by counting houses with married inhabitants.)

Suppose that the notion of a mistaken ur-sensation has been successfully defined, and that ur-sensations have been discovered to be invariably veridical (infallible). Has empiricist foundationalism thereby been proved true? At first sight it may not seem so. The classical foundationalists wished to discover infallible sensations that the subject can be conscious of and that provide reliable data for conscious inferences by the subject. But we are almost certainly not conscious of our ur-sensations. When I introspect I find nothing like what my ur-sensations would be on any plausible theory of cognition. This fact is not at all surprising on the association-of-information-processors model of visual cognition. For "I" in the sentences beginning "I am (am not) conscious of..." refers to my executive subsystem, and it is highly unlikely that on a plausible cognitivist theory of vision my executive subsystem will be conscious of ur-sensations. Perhaps only a subsystem to which an ur-sensation is directly transmitted can usefully be said to be conscious of the ur-sensation. Or perhaps we can define a notion of indirect consciousness that will permit us to say that subsystems within a given neighborhood of a subsystem are indirectly conscious of its ur-sensations. But it is virtually certain that my executive subsystem is not in any useful sense conscious of my ur-sensations, and that the executive does not employ ur-sensations in its attempts to obtain additional knowledge. It may seem, then, that the infallibility of ur-sensations is of no service to classical philosophical foundationalism.

Not so. In one sense of "I," I am identical with my executive subsystem. But in a second sense, I am identical with the collection of information-processing subsystems of myself. (In still another sense, I am identical with the collection of all my subsystems.) In the second sense of "I" and "my," I make inferences about the world on the basis of my ur-sensations, and if the ur-sensations are infallible, then the inferences based on them may be reliable. The rub is that "I" in the first sense — my executive, who is speaking now — may not have the satisfaction of knowing that "I" in the second sense make reliable inferences. But then again I may. If a psychologist discovers that I use my ur-sensations to obtain reliable knowledge and conveys this information to me, then I in the first sense do know and have the satisfaction of knowing that my knowledge is reliable, though I know it only indirectly and not by introspecting my sensational and cognitive processes. If I myself happen to be the psychologist, my knowledge is still indirect, nonintrospective. Now surely it is an insignificant deprivation that, although I have reliable empirical knowledge, I cannot directly, introspectively know that I do. What does it matter how I know I have such knowledge as long as I have it and somehow know that I do?

There will be a tendency to argue that my way of defeating skepticism is circular and unacceptable. To deal fully with this objection would require a substantial argument. Two observations will indicate the outline. First, my purpose was not to defeat skepticism (which no one believes), but rather to describe the form of a foundationalist theory of sensory cognition that could be confirmed by cognitive science. Second, there is no more circularity in using a cognitive process to ascertain the reliability of cognitive processes than there is in using a person's eyes to determine the reliability of that person's vision, which happens each time my eyes are examined by my ophthalmologist and could happen if I became my own ophthalmologist.

4. Final Remarks

The strong foundationalist theory, or framework, of empirical knowledge presented above seems in retrospect easy to construct. First, assume that empiricism is essentially correct and that all empirical knowledge is obtained through sensation and perception, more generally, through sensory cognition. Second, point out (what every scientist knows) that conscious sensations and perception are products of unconscious neurophysiological processes originating in the organism's sensory receptors. Third, argue that these neurophysiological processes are low-level cognitive processes, processes for acquiring knowledge; and in this vein refer to receptor excitations as sensations, the ur-sensations of the sensory process. Fourth, conjecture that the best-confirmed scientific theory of sensory cognition will contain the following postulates: (1) Ur-sensations in an organism are neither certified nor decertified by sensory processes of the organism or by any other means, and they are in this sense independent. (2) On the definition of representation proposed by the theory, ur-sensations are in fact always accompanied by what they represent, i.e., are always veridical, and are in this sense infallible.

It may seem too easy and too speculative. It may seem too easy in virtue of its assumption that certain neurophysiological processes — those that cause sensation and perception — are cognitive, i.e., essentially similar to conscious perception and thought. There is, however, an argument for the assumption; namely, that psychologists build better theories of perception when they treat the underlying neurophysiological processes as cognitive. When they do not so regard them, there is a virtually unbridgeable gap in the causal, explanatory chain leading to perception and thought; when they do so regard them, there is no such gap. If resistance to regarding the neurophysiological processes underlying perception as cognitive rests on the belief that neurophysiological processes are physical, and perceptions and thoughts nonphysical, then it rests on dualism, here applied to a special case. I hold that monism (antidualism) is properly regarded as an empirical theory for whose truth cognitive science is continually assembling evidence of the following type: in general, theories of cognitive science that assume monism are more adequate (better confirmed by the evidence, predictively more accurate) than theories that assume dualism. I know of no arguments to dissuade those who believe dualism is a true, nonempirical theory.

My strong foundationalist theory, or framework, may also seem too speculative, and therefore again too easy, since speculation is easy. I have endorsed not any particular theory of perception, but only a framework of cognitivism and two postulates for any such theory. Marr's theory was used as an example of a cognitivist theory; but it does not contain the two postulates above (nor does any other contemporary theory of perception, to my knowledge). I have conjectured that the theory that cognitive scientists ultimately accept — Marr's revised or some other — will contain the two postulates, but have provided insufficient evidence that it will. Or so it may be objected.

However, I did offer some evidence for the first half of postulate 1 — that ur-sensations are not certified or decertified by sensory processes. It was that in current cognitive theories of perception ur-sensations are not stored or processed in any way that would permit them to be used in a process of certification or decertification. To put the point in neurophysiological language, in most theories individual receptors are not located in feedback loops that would permit the reinforcement (certification) or inhibition (decertification) of their excitations (ur-sensations). And there may be good epistemological and biological reason for this feature of theories of perception. Perhaps receptors must be located outside such loops if they are to provide the organism with unbiased data about the environment, and perhaps only organisms with access to unbiased data survive.

My only evidence for the second half of postulate 1 — that ur-sensations cannot be certified or decertified by means in addition to sensory means (instruments, theoretical inference) — was that some psychophysicists believe it. And postulate 2 was, admittedly, speculation. The only considerations that prevent it from being sheerly speculative are that some psychologists, Gibson for example, hold that perception is a function of stimulation, and the intuition that successful organisms require accurate data.

Admittedly, my theory that ur-sensations are independent and infallible is highly speculative. However, the most important point is not that the theory is true, but that whether it is true is an empirical question to be settled by cognitive science. Furthermore, at the present stage of development of cognitive science, there is probably more reason to believe the theory will prove true than to believe it will prove false.

I have predicted that the form of an adequate psychobiological theory of empirical human knowledge will be foundationalist in the following sense: it will show that empirical human knowledge is founded on perception and that perceptions are ultimately founded on ur-sensations that are in fact independent and infallible. This claim may seem irrelevant to philosophical epistemology, which is assumed to be normative and thus to prescribe how knowledge should be obtained (must be obtained to achieve maximum reliability, scope, usefulness, etc.) rather than merely to describe how it is obtained. On the contrary, it is relevant in a number of important aspects. To begin with, how humans should acquire knowledge is constrained by how they can acquire it. Apparently humans physically and psychologically cannot acquire empirical knowledge except, ultimately, through perception in its various stages. Perhaps we can imagine a divinity directly transmitting to human memories such propositions as "The sun is presently eclipsed from earth by the earth's satellite that revolves about it every moon," thus avoiding the necessity of perception. But this process is by hypothesis supernatural and therefore does not naturally (psychologically and physically) occur.[21] (Perhaps we cannot imagine it; for the concepts of the proposition are concepts of perceivable objects and processes and it is not clear that these can be acquired except through perception.)

The claim that human empirical knowledge is in fact foundational is also relevant to normative epistemology in a less metaphysical respect. Any adequate theory about the acquisition of human empirical knowledge will imply that such knowledge is acquired through the various stages of perception: initial (presensational), intermediate (sensational), and final (perceptual). Any such theory will have implications concerning ways in which perception can be improved or degraded, and the theory will be partially testable through these implications. Obviously if vision relies on the focusing of light into an eye, then telescopes, glasses, and contact lenses can improve vision; so can substituting clearer or more flexible lenses for the original biological lenses of the eye, or changing the shape of the eye to make the image focus more exactly on the retina. Moving farther inside the visual system and becoming more speculative, visual edge detection might be improved (or degraded) by altering the sensitivity of retinal interneurons to stimulation from receptors, perhaps by affecting the production of neurotransmitters in the brain. Visual edge detection might also be improved (or degraded) by altering the density of retinal receptors, perhaps by affecting their growth with chemicals or altered genes.

The modifications above do not entail that ur-sensations are dependent or nonveridical, and they can be suggested on the basis of either a strong foundationalist theory or some contrary theory. It is difficult to describe modifications that entail that ur-sensations have been or will be dependent or nonveridical and consequently are excluded by a strong foundationalist theory. For example, it is easy to imagine a retinal cone becoming more (or less) sensitive to light through the action of neural stimulants or depressants, which means that it produces an ur-sensation in the presence of low-intensity light more (or less) frequently than previously, which seems to entail that it is more (or less) frequently veridical than previously, thus implying that ur-sensations are sometimes nonveridical. However, the ur-sensation from the less sensitive cone may contribute to sensations that produce a certain percept of a visual edge, and the ur-sensation from the more sensitive cone may contribute to sensations that produce a different percept of a visual edge. If so, then in both cases the ur-sensations may represent their causes — higher intensity light in the former case, lower intensity light in the latter — and thus in both cases be veridical. For, as we saw in a previous section, what an ur-sensation represents is relative to the sensations and perceptions the ur-sensation helps to generate.[22]

The perceptual system of a successful organism constructs representations of its environment on the basis of ur-sensations in a process of acquiring empirical knowledge that enables the organism to function and survive in these environments. The organism does not have and psychophysically cannot have any other source for its empirical knowledge than perception, and it is in this sense dependent on its ur-sensational data. These data can be (but rarely are) impoverished and inconsistent, and most organisms are endowed with perceptual systems that — except in extreme cases — can compensate for such deficiencies in the database and construct useful, veridical percepts nonetheless. But if, as I have speculated, strong foundationalism is correct, individual ur-sensations are inevitably produced independently and veridically; and the perceptual system using them as data cannot be improved or degraded by modifications affecting their veridicality or independence.

I revert to the analogy between an organism and an association for a concluding summary. Strong foundationalism is a theory of how associations do and can acquire empirical knowledge about their environment. It holds that an association must rely on raw data gathered by observers at the interface between the association and its environment, and that such data are never dependent on other knowledge and never mistaken. They are never mistaken because (1) they are always produced by some feature of environment (the observers being too simple-minded to invent them and too thick-skulled to use someone else's invention), and (2) they can only be interpreted — given their use by the association — as representing the features that cause them. The association acquires incorrect knowledge only by using raw data from too few (or too many) observers or by misusing the raw data in producing percepts and empirical propositions. Correlatively, improvements in the association's knowledge acquisition process can be made only by increasing or decreasing its staff of observers or by changing its procedure of processing raw data; they cannot be made by improving individual observers, by improving the quality of individual items of data.

Whether the above theory is true or false is obviously an empirical question. And the same holds for its biological analogue.

Notes

Early versions of this essay were presented at meetings of the American Association for the Advancement of Science (1986), American Philosophical Association (1988), and Society for Philosophy and Psychology (1989).

1. The word "knowledge" is used in this essay to refer to cognitive states or events with the following features: (1) they represent other states and events and are either true or false; (2) even if false, they have etiologies and credentials comparable to true representations. From the perspective of philosophical epistemology, an approximate synonym for "knowledge" would be "warranted belief." From the perspective of naturalized epistemology, an approximate synonym would be "reliable belief," where reliable belief is defined either as belief with a high probability of truth or as belief that would lead to beneficial results if acted upon. (For the distinction between philosophical and naturalized epistemology, see below.) Either proposed synonym would be merely approximate, because to call a cognitive state knowledge is not here meant to imply that anyone believes what it represents. Knowledge is traditionally defined by philosophers as justified, true belief, but the term is used here to refer to something that may be neither justified, nor true, nor a belief. Some other term is needed; but so far no better one has been proposed. Psychologists and workers in artificial intelligence routinely use the term as I do here.

2. Pure foundationalist empiricism holds that all knowledge consists either of data of sensory cognition (experience) or beliefs entirely based on these data. Pure foundationalist rationalism is the view that all knowledge consists either of data of intellectual cognition (intuition) or beliefs based entirely on these data. In Descartes's version of pure foundationalist rationalism, the data are "clear and distinct ideas revealed by the light of reason." In Leibniz's version, the data are logical identities, such as that expressed by "The round table is round." Antifoundationalist empiricism and antifoundationalist rationalism both hold that an epistemic distinction between data and other beliefs is spurious.

Very few philosophers have been pure empiricist or pure rationalist. Most rationalists concede that knowledge of particular empirical events is not based entirely on intellectual cognition. Most empiricists concede that mathematical and logical knowledge is not based entirely on sensory cognition, and limit their claim for a sensory basis to empirical beliefs. Furthermore, although empiricists deny that empirical beliefs are innate, most acknowledge that some of the systems for acquiring such beliefs are innate.

3. "Conscious process" here means "process of which the subject is or could have been conscious, or directly aware, by introspection."

4. Note that a machine need not be made of nuts and bolts, transistors, or other physical components; it can also be made of biological components. Consequently, the primary difference between machine theories and psychobiological theories is that the former deal with artificial and the latter with natural systems. "Machine theories" is therefore a misleading label, but its use is standard in the field of artificial intelligence.

5. What is characterized here is naturalized *individual* epistemology. Social epistemology tells us how systems of individuals do, can, or should acquire knowledge, and when naturalized it focuses on how they do and can acquire it. Compare Kornblith's (1986) characterization. See Goldman (1986) for the first full-scale presentation of a naturalized individual epistemology, with treatments of perception, memory, reasoning, and other cognitive processes, as well as a promise of a naturalized social epistemology later.

6. See note 3.

7. A questionable assumption, as we know anecdotally from cases in which people confuse vivid night or day dreams with sensations, and from the experiments of Perky (1910) with subjects who confused images of imagination with sensations.

8. Philosophical epistemologists and philosophers of science occasionally attempted to improve even the scientific process and recommended radical revisions in the database, sometimes by excluding all but sensational data, sometimes by excluding all but perceptual data. For the most part these attempts were unprofitable: an exclusively sensational base was too narrow, an exclusively perceptual base too broad. However, because scientists paid little attention and the average person none at all, they had little effect, for good or ill. Scientific epistemologists are expected to make more useful contributions.

9. Strictly speaking, coherence theory, or coherentism, is a strong (or very strong) antifoundationalist view. It holds that all data (and in some versions, all statements) are on a par — equally fallible and dependent — and that the largest coherent set of data (or statements) is the set most likely to be true. (Although the coherence relation is usually defined solely in terms of logical consistency, it can be defined in many other ways.) However, nowadays many epistemologists apply the term "coherence theory" to weak foundationalist views as well.

10. For example, by Leibniz ([1702] 1956), who postulated *petites perceptiones*.

11. See Winston (1975), MacArthur (1982), and Ballard and Brown (1982) for background in and additional material concerning computational theories of vision.

12. Here and elsewhere, I use the words "veridical" and "nonveridical" instead of "true" and "false," because the latter are most properly applied to sentences; and percepts, sensations, and other sensory cognitions are not sentences.

13. Philosophers often use the term "proposition" to refer to an abstract, logical entity expressed by certain sentences of some language. In the context of cognitive science, the term refers to a cognitive (mental) entity that is involved in the production of certain sentences of some language, and is roughly synonymous with the term "thought."

14. One relatively insensitive method uses the rule: resolve conflicts between sensations in such a way that the maximum number of sensations is accepted. A somewhat more sensitive rule is: resolve conflicts between sensations in such a way that the maximum number of sensations in a maximum number of sensation types (visual, auditory, etc.) is accepted. Kyburg (1990) proposes a more sensitive version of the latter method, designed for fallible reports of both conscious sensations and perceptions. These are all

methods of a weak foundationalist (i.e., weak antifoundationalist) theory of knowledge, and are not strong antifoundationalist or coherentist in the strict sense. See note 9.

15. It could even be used to discredit an ur-sensation in the group, in much the same way that a group of numerical data points are used to throw out an "anomalous" point far from the average of the group.

16. Stevens (1975) can be interpreted as entertaining this hypothesis in his discussion of absolute thresholds of sensation in such sensory departments as taste, touch, hearing, and vision. "The minute amount of stimulation needed to initiate a sensation makes it appear that nature has pushed close to the theoretical limits in providing us with receptors to detect chemical, mechanical, and electromagnetic events in the environment. Such acute sensitivity no doubt has served a decisive role in evolution and has made it possible for complex, mobile animals to survive in ecological competition" (p. 172). Although absolute thresholds are usually determined on the basis of subject reports of conscious sensations, they can also be determined on the basis of subject behavior affected by unconscious sensations.

17. Those psychologists who hold what Kosslyn (1980, pp. 131ff.) calls the "percept analogy" view maintain that images, and by extension hallucinations, are generated by the same mechanisms that generate percepts; but there is no reason to interpret them as including the excitation of retinal receptors in the perceptual systems to which they refer. They are referring to higher-level perceptual systems, either in the retinal interneurons, or in the optic nerve, or in the central cortex.

18. The exception to this generalization is afterimages. Some neurologists maintain that afterimages are caused by the adaptation or washing out of the light-sensitive substance (rhodopsin) in the retinal cones. If this were true then it might follow that nonveridical ur-sensations, i.e., cone excitations, are involved in the production of afterimages. However, other neurologists maintain that the adaptation that produces afterimages occurs at a higher level. Whether afterimages are produced in the retina or in the cerebral cortex is a question of long standing, still unsettled.

19. See note 12.

20. Fodor's (1983, 1984) thesis of impenetrability is the thesis that perceptions are *causally* independent from the rest of the cognitive system. The thesis is clearly incorrect for conscious sensations and perceptions; for these latter are generated from and hence causally dependent on ur-sensations. But the thesis is probably correct for ur-sensations. Probably the rest of the perceptual system does not contribute to the formation of ur-sensations; they are, as psychologists say, generated bottom-up. It does not follow that ur-sensations are *theoretically* independent from the rest of the cognitive system. As suggested above, what ur-sensations represent is dependent on what cognitions the perceptual system generates when it uses them as a basis. Ur-sensations are thus theoretically dependent (in what they represent) on cognition, dependent in the way a theory is dependent (for its truth) on evidence. Ur-sensations may also be *semantically* or *conceptually* dependent on other concepts in the cognitive system, dependent in the way a concept is dependent on the concepts in its explicit or implicit definition.

21. We can imagine a neurosurgeon implanting in someone's brain a memory cell containing the proposition, "The earth has a satellite called the moon that revolves about it every twenty-eight days," without invoking the supernatural. But to imagine the implantation of a memory cell containing true time-dependent propositions — as in the eclipse example — that the patient will need in order to function and survive after the operation is to imagine a supernatural knowledge acquisition process. For the person or system that prepared the memory cell must have had the foreknowledge of a divinity.

22. Note that, even relative to the same generated sensations and perceptions, a cone

can discharge in the presence of a given stimulus at one time and fail to do so at another and yet produce veridical ur-sensations, the ur-sensations required to produce the sensations and perceptions being supplied during the cone's inactivity by neighboring cones. A sensation or perception is said to be veridical if and only if it is accompanied by what it represents. It is not said to be nonveridical on the ground of not being produced. In one sense a receptor or other sensory system is said to be reliable if and only if the sensations it produces are veridical. In another sense it is said to be reliable if and only if (1) it produces sensations when it is required to, and (2) the sensations it produces are veridical. In the second sense a sensory system can produce veridical sensations and still be unreliable because it fails to satisfy clause (1). Strong foundationalism only claims that receptors are reliable in the first sense.

References

Austin, J. 1962. *Sense and Sensibilia*. Oxford: Clarendon.

Ayer, A. J. 1954. *Philosophical Essays*, chap. 4. London: Macmillan.

Ballard, D. H., and C. M. Brown. 1982. *Computer Vision*. Englewood Cliffs, N.J.: Prentice Hall.

Barnes, W. H. F. 1944–45. "The Myth of Sense-Data." *Proceedings of the Aristotelian Society* 45:89–117.

Block, N., ed. 1981. *Imagery*. Cambridge, Mass.: MIT Press.

Churchland, P. 1979. *Scientific Realism and the Plasticity of Mind*. Cambridge: Cambridge University Press.

Dennett, D. C. 1983. *Brainstorms*. Cambridge, Mass.: Bradford Books.

Feyerabend, P. 1965. "Problems of Empiricism." In R. G. Colodny, ed., *Beyond the Edge of Certainty*, pp. 145–260. Englewood Cliffs, N.J.: Prentice Hall.

Fodor, J. A. 1983. *The Modularity of Mind*. Cambridge, Mass.: MIT Press.

———. 1984. "Observation Reconsidered." *Philosophy of Science* 51:23–43.

Goldman, A. I. 1986. *Epistemology and Cognition*. Cambridge, Mass.: Harvard University Press.

Gregory, R. L. 1970. *The Intelligent Eye*. New York: McGraw-Hill.

Hanson, N. R. 1958. *Patterns of Discovery*. Cambridge: Cambridge University Press.

Kornblith, H. 1980. "Beyond Foundationalism and the Coherence Theory." *Journal of Philosophy* 77:597–612.

Kosslyn, S. S. 1980. *Image and Mind*. Cambridge, Mass.: Harvard University Press.

Kuhn, T. 1962. *The Structure of Scientific Revolutions*. Chicago: University of Chicago Press.

Kyburg, H. E., Jr. 1990. "Theories as Mere Conventions." In C. W. Savage, ed., *Scientific Theories*, pp. 158–74. Minneapolis: University of Minnesota Press.

Leibniz, G. W. [1702] 1956. "Reflections on the Doctrine of a Single Universal Spirit." In L. E. Loemker, ed., *Leibniz: Philosophical Papers and Letters*, vol. 2, p. 905. Chicago: University of Chicago Press.

MacArthur, D. J. 1982. "Computer Vision and Perceptual Psychology." *Psychological Bulletin* 92, no. 2:283–309.

Marr, D. 1982. *Vision*. San Francisco: W. H. Freeman.

Minsky, M. 1986. *The Society of Mind*. New York: Simon and Schuster.

Moore, G. E. 1939. "Proof of an External World." *Proceedings of the British Academy* 25:273–300. Reprinted in G. E. Moore, *Philosophical Papers*, pp. 127–50. London: Allen and Unwin, 1959.

Neisser, U. 1976. *Cognition and Reality*. San Francisco: W. H. Freeman.

Perky, C.W. 1910. "An Experimental Study of Imagination." *American Journal of Psychology* 21:422–52.

Pylyshyn, Z. 1984. *Computation and Cognition.* Cambridge, Mass.: MIT Press.

Quine, W.V.O. 1951. "Two Dogmas of Empiricism." *Philosophical Review* 60:20–43.

———. 1969. "Epistemology Naturalized." In *Ontological Relativity and Other Essays,* pp. 69–90. New York: Columbia University Press.

Rock, I. 1983. *The Logic of Perception.* Cambridge, Mass.: MIT Press.

Russell, B. 1921. *The Analysis of Mind.* London: Allen and Unwin.

Savage, C.W. 1975. "The Continuity Hypothesis of Sensation and Perception." In R. K. Siegel and L.J. West, eds., *Hallucinations: Behavior, Experience, and Theory,* pp. 257–86. New York: Wiley.

———. 1989. "Epistemological Advantages of a Cognitivist Theory of Perception." In M. L. Maxwell and C.W. Savage, eds., *Science, Mind, and Psychology,* pp. 61–84. Lanham, Md.: University Press of America.

Stevens, S.S. 1975. *Psychophysics: Introduction to Its Perceptual, Neural, and Social Prospects.* New York: Wiley.

Winston, P.H., ed. 1975. *The Psychology of Computer Vision.* New York: McGraw-Hill.

PART II

MODELS FROM ARTIFICIAL INTELLIGENCE

Gary Bradshaw

The Airplane and the Logic of Invention

1. Introduction

A major concern in philosophy of science has been to provide norms to scientists that will assist them in their work. Ideally guidance would be provided at every step of the scientific process, but efforts to develop norms are predominantly restricted to the *context of justification*, where a scientist is attempting to prove that a hypothesis is true, and are seldom proposed for the *context of discovery*, where a new hypothesis is developed. This bias originated with the logical positivists, who argued that discovery was an inexplicable process that did not follow any logic or pattern, and therefore could not be subject to guidance (Reichenbach 1966). Many of the postpositivist philosophers, such as Popper (1959), have preserved the distinction between the context of discovery and the context of justification and have argued forcefully for a research program to develop norms only within the context of justification.

Herbert Simon (1966; 1973) rejected the notion that discovery was an inexplicable process. As an alternative, he suggested that scientific discovery is problem solving, differing mainly in scale from the kind of problem solving psychologists have observed in the laboratory. By rejecting the alogical and inexplicable nature of scientific discovery, the possibility arises that we may provide norms within the context of discovery, as well as norms within the context of justification. We will briefly review both psychological models of problem solving and Simon's model of discovery, and consider the potential for discovery norms.

Confronted with a problem, people often engage in mental *search*, where a series of alternatives will be considered, a plan devised, and actions carried out. Consider a person who has a list of purchases to make from various stores around town. Although the task could be accomplished by simply working down the list of items, most people will naturally reorganize the list to minimize travel distances. At the beginning of the search, the problem solver is aware of the current (*initial*) state (the list of items to be purchased, the location of stores around town) and has some understanding of the *goal state* (being back at home after having made all necessary purchases). In addition, the problem solver has several *operators* that can be used to change the current state. The problem solver could drive to each location, or if parking is troublesome, might seek parking spaces in a central location and walk to various stores. Buses, subways, or taxis could also be exploited for transportation at various times.

Minimizing the overall distance traveled is known as the "traveling salesman problem," and is a complex NP complete problem. When planning their trips, few shoppers are compelled to identify the absolute minimum distance, but instead will use simple *heuristics*, or rules of thumb, to improve the efficiency of their route. A simple heuristic for this task is to visit stores close to each other before driving to stores in a different area. This heuristic may not produce an optimal route, but it will be far superior to one where stores are visited randomly.

The problem-solving model of discovery views scientists as engaged in a search for hypotheses, data, and laws to explain the data. The initial state incorporates everything the scientist knows about the phenomena of interest, such as the fact that liquids of different temperatures when mixed together have a resulting temperature between the two initial temperatures. The goal state might be something like a law that describes the final temperature as a function of the initial temperatures, quantities, and types of liquid, and operators include things like making new instruments, heating liquids, conducting experiments, computing means, changing the substances involved, and so on. Heuristics might include using nearly equal quantities of each type of liquid, creating large initial differences in temperature, testing for symmetry of result, and so on.

Although experts often have detailed and specific methods to solve problems, scientists seem to resemble novices, in that they rely primarily on weak but general heuristics in searching for discoveries (Simon, Langley, and Bradshaw 1981). Once a domain has been sufficiently well developed that new discoveries can be generated by rote formula, the activity is considered mere development or application rather than basic scientific inquiry.

Simon (1973) traced out some of the normative implications of his model. One kind of norm can be derived by comparing different forms of search. These will be termed *norms of search*. In searching for a law to fit the data, a scientist might generate a simple formula with a single term, then consider progressively more complex formulas with additional terms and relations, testing each formula against the data. A more powerful set of heuristics was included in the BACON system (Langley et al. 1987), which studied relationships present in the data to guide the construction of potential formulas. Even though BACON's heuristics are weak and general, they can greatly reduce the size of the search space in looking for a scientific law. Scientists who employ heuristic search have a distinct edge over those who work randomly.

Domain norms operate at a more specific level. A scientist might make the assumption that the variables under investigation behave symmetrically with respect to the final outcome (Langley et al. 1987). An example is to assume that masses in collisions are interchangeable. This heuristic can greatly reduce the search needed to discover laws of momentum. Domain norms can be developed either by evaluating the property of the domain, or looking to the past for clues about effective techniques. Unfortunately, domain norms have less force than norms of search, because they require an inductive step in their application. If the characteristics of the domain are misinterpreted or the domain is not similar to those explored in the past, using domain heuristics may mislead investigators and slow the discovery process rather than speeding it up.

The problem-solving model of discovery can be used to develop norms within the context of discovery, but existing norms do not seem to have great practical force. As part of their general psychological makeup, scientists come equipped with heuristic search strategies such as means-ends analysis, hill climbing, and generate-and-test. For this reason, proclamations that search techniques are *the way* to make discoveries would likely be received with some amusement on the part of working scientists. Domain norms are not universal, and so may have problems in their application. That norms of discovery exist supports the contention that discovery has a logic, and therefore is not a random, inexplicable process. Yet until this model can move beyond existing practice in developing norms for discovery, it is unlikely to receive much attention from philosophers who seek to provide guidance to working scientists. Accordingly, the model is likely to appeal to psychologists who seek to understand the discovery process, but will in all probability be neglected by philosophers for lack of substantive normative contributions. Against the backdrop provided by this discussion, I will review the curious history of the invention of the airplane. In the process, I will show that the problem-solving model can make unexpected normative statements under certain circumstances.

2. Why Did the Wright Brothers Invent the Airplane?

The mystery surrounding the invention of the airplane is not a whodunit, for there is little current dispute that Wilbur and Orville Wright invented the airplane. Rather the mystery is "Why the Wright brothers?" From almost any perspective, their effort was a long shot. Wilbur finished high school, but his brother Orville never did. They were bicycle mechanics, with no formal training in engineering, fluid mechanics, physics, or the like. They had no external support for their efforts, funding their work from proceeds generated during the summer months when they made, sold, and repaired bicycles. Many of their contemporaries, such as Dr. Samuel Pierpont Langley of the Smithsonian Institution, enjoyed much more promising circumstances, including better training, a large staff of supporting mechanics and shop workers, extensive funding, and so on. Why did Wilbur and Orville succeed when their contemporaries failed?

Several explanations come quickly to mind. Perhaps the Wright brothers were lucky, stumbling upon the right answer by accident. Perhaps the time was ripe for the invention of the airplane, and the Wright brothers happened to get there just a little faster than their contemporaries. Most biographers and historians have rejected these explanations. Luck can be discounted by noting the extended and detailed nature of their development program: The Wright brothers were careful researchers, keeping detailed notebooks of their research. Even still, they took more than four years to build their first airplane, which was successful from the first tests.

Claims that the time of the airplane was at hand can be discounted by looking at the state of the art both before and after the Wright brothers first flew in 1903. On December 8, 1903, Samuel Langley decided to test his man-carrying Aerodrome. Langley, head of the Smithsonian Institution, was widely recognized as one of the leaders of aviation research. In 1896 he had constructed an unmanned steam-powered airplane that flew for more than half a mile. His Aerodrome had been under construction for several years, at a cost of about $75,000. Langley's pilot, Charles Manley, climbed into the Aerodrome, which was mounted on top of a houseboat, and tried to fly the craft. The plane broke during launching from stresses induced by the catapult and "slid into the water like a handful of mortar." Manley nearly drowned in the wreckage. The War Department, in summarizing this expensive and widely publicized failure, concluded: "We are still far from the ultimate goal, and it would seem as if years of constant work and study by experts, together with the expenditure of thousands of dollars, would still be necessary before we can hope to produce an apparatus of practical utility on these lines."

On December 17, just nine days later, the Wright brothers successfully flew their first powered craft. Gradually word got out about the Wright brothers' accomplishments, but was not widely believed. The Europeans labeled the Wright brothers *bleuffeurs*; their claims to powered flight were just too fantastic to accept. *Scientific American* also rejected the Wright brothers' claims. Indeed, no one was able to match the Wright brothers' accomplishments until 1909, three years after details of their craft were made public through their patents.

Biographers and air historians have offered alternative explanations to account for the success of the Wright brothers. Charles Gibbs-Smith, in his well-known account of the invention of the airplane (1965), distinguishes between *airmen* and *chauffeurs of the air*. Chauffeurs of the air believed that an airplane would resemble a car, and could be "driven" into the sky. Members of this group, including Maxim, the inventor of the machine gun, usually built complete airplanes. Maxim's machine had a wing span of one hundred feet, was powered by a steam engine, and had cast-iron wheels. Airmen often sought to build gliders in advance of attempts to construct powered planes. Members of this tradition recognized that flying was quite different from driving a car, and needed to be understood on its own terms. This distinction is apt, yet it is not sufficient to distinguish the Wright brothers from other unsuccessful inventors. Otto Lilienthal, Percy Pilcher, Octave Chanute, and Sir George Cayley all developed several gliders. None of these attempts was entirely successful, while the Wright brothers invented a record-breaking glider in 1902.

Frank Howard (1987), in his excellent biography of the Wright brothers, describes several factors he felt were important in the Wright brothers' success. Among these are the facts that the Wright brothers were "good with their hands" and that their understanding of bicycle construction helped them design lightweight but solid airframes. Again, these factors were important, but fail to distinguish the Wright brothers from others like Lilienthal and Chanute, who were both engineers and who both created elegant, sturdy gliders.

Tom Crouch (1989), in another excellent biography, pays homage to the family situation of the Wright brothers. Both men were confirmed bachelors partly because their sister Kate took care of the household chores for them. Freed from familial duties, the Wright brothers had extra time to devote to their research. Again time was important, but Lilienthal and Chanute were so successful in their businesses that they effectively retired to work on airplanes. Others such as Langley had extensive research projects with several assistants to help in their work. Thus, manpower alone cannot be the answer.

Wilbur provides a clue to their success in a letter he wrote to his friend Octave Chanute. In 1906, three years after their first flight, the Wright brothers were having difficulty finding a buyer for their airplane, and Chanute was encouraging them to discount their price because their invention might be duplicated by others. Wilbur replied:

> We are convinced that no one will be able to develop a practical flyer within five years. This opinion is based upon cold calculation. It takes into consideration practical and scientific difficulties whose existence is unknown to all but ourselves. Even you, Mr. Chanute, have little idea how difficult the flying problem really is. When we see men laboring year after year on points we overcame in a few weeks, without ever getting far enough along to meet the worse points beyond, we know that their rivalry & competition are not to be feared for many years. (Kelly 1972, p. 181)

This statement is all the more striking since it took the Wright brothers only four years to develop the airplane beginning in 1899, and the field had progressed in the intervening years, so that others had a better starting point. The process of invention used by the Wright brothers must have been far more effective than processes used by their contemporaries. Aware of this advantage, the Wright brothers were so confident that they waited two more years for buyers to meet their terms.

The mystery now turns to the question of how the Wright brothers managed to be so effective in their process of invention. I will briefly review the processes used by a few contemporary inventors, then describe how the Wright brothers approached the problem.

Langley began his research into the development of the airplane in 1887. He first constructed a giant whirling tower with thirty-foot arms, which was used to test the aerodynamic properties of several birds and wing surfaces. Next he turned his attention to small, rubber band-powered models followed by steam-powered models before he tackled the development of a man-carrying airplane.

Langley tested between thirty and forty different rubber band-powered models, though many of the models were extensively modified, so the total number is probably closer to one hundred. Langley varied the number of propellers, wing curvature, number of wings, wing placement, the tips of the wings, and many other factors. Models were tested by being briefly flown, though the distances were usually under twenty meters and the time in flight less than six seconds. From these tests, Langley was able to ascertain some of the factors that contributed to inherent stability and efficiency, although the tandem configuration he finally adopted is not an efficient one. Langley next developed and tested six steam-powered models, followed by a gasoline-powered model, followed by the disastrous tests of the Aerodrome. Once again we find the primary method of testing was to fly the models and observe their time and distance in flight.

Octave Chanute, another prominent American engaged in flight research, also built and tested a number of different models. Chanute frequently commissioned others to create aircraft of their own design, so that much of his research has an unsystematic feel. However, one model, known as the Katydid, was designed by Chanute himself. The Katydid is interesting because it included six pairs of wings that could be placed on the front or rear of the air frame. In a series of tests in 1896, Chanute moved the wings back and forth in a merciless fashion. Once again we find the primary testing method to be attempts at flight, which were measured by their distance and time.

The pattern that emerges from these two investigators is the propensity to construct complete craft that exemplify different designs, then test the craft by measuring distance and time in flight. This pattern can also be observed in most of the other notable figures of the day, such as Otto Lilienthal in Germany and Percy Pilcher in England. Lilienthal created more than eighteen different hang gliders with various numbers, arrangements, and sizes of wings. Pilcher developed four or five different models before his death in a gliding accident.

Studying the Wright brothers, we observe a very different pattern. The Wright brothers built only three gliders before they constructed their successful airplane. All of their craft were of biplane design with a front elevator. The first glider, constructed in 1900, included an effective method of lateral control, something no other investigator included until observing the Wright plane. The 1900 and 1901 craft were almost identical, save that the 1900 craft was smaller than intended due to a lack of suitable wing spars. The Wright brothers never shifted wings around on their plane, explored tandem designs, altered the wing mounting, or even changed the fundamental means of lateral control. Instead they developed a series of increasingly sophisticated craft. Their 1902 glider solved all fundamental problems of manned flight, and flew more than twice the distance achieved by any previous machine. While their counterparts made plodding advances by exploring dozens of different designs, the Wright brothers needed only two attempts to find the right answer. Only luck of the highest order, or a fundamental difference in method, could lead to such a discrepancy.

Fortunately, through the extensive notes and letters preserved from the Wright brothers, we can rule out an explanation based solely on fantastic luck. In contrast to their contemporaries, the Wright brothers never constructed an airplane until they had a good reason to do so. Wilbur began the research by considering the problem of lateral control. Noting that soaring birds often twist their wing tips to restore balance, Wilbur one day twisted a bicycle-tire box, and realized that by warping the wings of an aircraft, he could achieve the same effect. Instead of rushing to the basement to build a new machine, Wilbur constructed a simple biplane kite, and demonstrated that wing-warping had the intended effect.

Using data provided by Lilienthal, the two brothers built their first glider in 1900. They carefully researched flying locations, finally settling on Kitty Hawk, where the steady winds would reduce the amount of time spent dragging their glider back to the starting position. They tested the craft by flying it as a kite. In the process they discovered that their craft had half the drag of previous craft, but the angle of incidence was about twenty degrees, so their wings were not providing the computed lift. Finally, a few free-glide tests were made. The wing-warping controls were tied down during most of these glides: The Wright brothers knew they had a successful solution to lateral control, and were seeking to understand the lift of their machine.

The 1901 craft was larger than the one built in 1900, and by the brothers' computations should have generated sufficient lift to carry a person in moderate winds. However, the glider did not perform as intended. Once again the two brothers spent considerable time flying the glider as a kite, and measuring the lift and drag of their machine. From these measurements, Wilbur discovered that the value of the coefficient of lift for air that was currently in use was too large.

Initially the brothers were discouraged by the performance of the 1901 glider, but soon thought of ways to collect new data to help in the design of future craft. In September of that year, the two brothers created two wind tunnels to measure the lift and drag of wings. Their tests confirmed the finding that the accepted value for the coefficient of lift was incorrect. After testing more than one hundred different air foils, the brothers discovered a shape far more efficient than the Lilienthal design used in their 1901 glider. Using this data, they constructed their breakthrough craft of 1902. The wind tunnel was used again in 1902 to assist propeller development.

The emergent pattern in the Wright brothers' work is to explore solutions to subproblems in flight using directed experiments. A kite was built to explore lateral control. Lift and thrust were solved through the use of wind tunnel experiments. Only when each problem was solved were the Wright brothers willing to invest the time and energy in building a new craft. Another signature characteristic of the Wright brothers' research was the extensive testing they performed on each model. By testing the early gliders as kites, the Wright brothers were able to measure lift and drag, and discovered an important error in aerodynamics overlooked by other investigators.

The activity of Langley, Chanute, and others can be characterized as *design-space search*. To these men, the airplane consisted of a set of structures, such as wings, fuselage, propulsion plant, etc. Developing an airplane meant exploring the set of possible designs. Thus we see Langley considering single-wing and tandem-wing designs, pusher or tractor propellers, curved or straight wing tips, and so on. Chanute explored different wing arrangements on the Katydid, as well as several other basic plane designs.

To guide their search in design space, these investigators focused on two global measurements: time and distance in flight. Time and distance in flight are functions of many factors including characteristics of the design (stability, wing lift, drag) as well as factors outside experimental control (wind velocity, air turbulence). Two different designs can achieve roughly comparable performance for quite different reasons.

Design-space search is inherently inefficient for two reasons: The design space is large, and global measurements provide little guidance in moving through the space. To illustrate the size of the design space, Table 1 shows just a few of the design parameters that were explored around the turn of the century.

Table 1: Some Design Parameters of Early Airplanes	
Number of wings	1–80
Wing position	1–3
Placement	stacked, tandem, staggered
Lateral arrangement	anhedral, flat, dihedral
Camber of wings	1–12, 1–6, etc.
Wing span	6'–104'
Chord	3'–10'
Shape of wings	birdlike, rectangular, batlike, insectlike
Tail placement	forward (canard), rear, mid

Each decision shown in Table 1 is independent, so we can construct at least $80 \times 3 \times 3 \times 3 \times 5 \times 20 \times 5 \times 4 \times 3 = 12,960,000$ different planes. The number of workable designs in this space is small. Indeed, unless some method of lateral control is built into the craft, none of these designs would be capable of true flight. The space of designs is large, and sparsely filled with successful machines.

Even large spaces can be searched effectively if we can use powerful heuristics. The global metrics of time and distance in flight are not diagnostic with respect to design choices. Exactly why did a particular configuration fly two hundred feet in eight seconds? The wings may not have generated much lift, the pilot's position may have caused too much drag, the wings may have been too flexible, and so on. Without more diagnostic information concerning factors that contributed to various aspects of the craft's performance, investigators could only guess about better alternatives. Chanute was a poor guesser: none of the craft he created after 1897 equaled the performance of his Katydid machine or the two-surface glider.

The method used by the Wright brothers can be characterized as *function-space search*. Their major concern was how to achieve certain functions in an airplane: lateral control, sufficient lift, a reduction in drag, etc. When confronted with a problem, the brothers would isolate the problem and search for a solution. In the process, they created new instruments and techniques to study the problem, such as their development of balance scales to measure lift and drag in the wind tunnel. The efficiency of this process can be seen in their rapid and steady progress.

The efficiency of function-space search derives from the comparatively few functions that are needed to build a working airplane, and from the natural independence of aerodynamic factors. Lift could be addressed without regard for lateral control, and vice versa. Without this independence, search in the function space would be difficult. Yet only the Wright brothers intentionally exploited this independence in their efforts, and worked to solve in turn the problems of lateral control, lift, yaw, and thrust.

Function-space search requires a more specific set of tests than global performance measurements. Looking at time and distance in flight is not sufficient to determine the characteristics of the craft, such as its lift or drag. The Wright brothers had to develop a set of instruments and testing procedures that would demonstrate the performance of their craft in accomplishing each function. This was a concern from the beginning, when they wrote to Octave Chanute for advice on procuring an anemometer to measure wind speed during glides and kite experiments. The information obtained via these instruments was diagnostic and very effective in refining the basic design.

On the surface, design-space and function-space searches have much in common. After constructing a wind tunnel and balances, the brothers tested more than one hundred different wing shapes. Fluid dynamic theories were nonexistent, so they had no theoretical guidance about which shapes to test. The brothers used their intuition and previous results to construct new designs. Chanute's experiments with his Katydid machine have the same systematic character. The Katydid was designed so that it would be easy to move wings about, and tests could be made on the effect of wing placement. Yet even in moving a pair of wings from front to back, Chanute was altering many properties of his craft. The obvious effect is to shift the center of pressure, but turbulence between stacked and tandem wings has many subtle effects on lift and stability. These relationships are altered by removing a pair of front wings and adding them to the back. Determining the correct wing placement from time and distance measurements is not a simple task. Even though Chanute was systematic in his work, he was not operating in function space, but in the more complex design space.

3. Conclusions

The notions of design space and function space apply to more problems than simply the development of airplanes, although these concepts are not universally applicable. Some inventions, such as the bicycle, seemed to undergo a process of evolution more akin to hill climbing in design space than development through the function space. One contemporary example of a similar problem is in the invention of speech-recognition systems. No automatic recognition system currently exists that is capable of humanlike skill at interpreting spoken utterances. Many investigators seek to develop such systems by constructing whole systems and testing them with global performance measurements, while a few others seek to identify primitive functions of speech recognition and construct modules that accomplish these functions (Bradshaw and Bell 1990).

A second parallel is the development of complex process models of human thinking. Anderson (1990) notes that cognitive modelers enjoy a wealth of mechanisms to include in their models. Although these mechanisms do not have the same independent character as do components of aircraft, a large design space of cognitive architectures and models can be generated. If global performance characteristics (reaction time, error frequency) are used to compare models against human performance, identifying the "correct" model becomes an almost impossible task, as many models will make quite similar predictions for very different reasons. Under the circumstances, Anderson proposes considering the real-world function of different mechanisms, and how each mechanism would be adapted to best carry out that function.

If the analogy between airplane invention and cognitive modeling is sound, we can apply the same efficiencies of function-space research to inventing appropriate models of human cognition, illustrating the generality of the design-space/function-space distinction and the utility of this notion in creating norms for research and invention.

Note

I thank Marsha Lienert for her assistance in my efforts to understand the invention of the airplane. The Institute of Cognitive Science and the Department of Psychology at the University of Colorado at Boulder provided valuable financial assistance for this project.

References

Anderson, J.R. 1990. *The Adaptive Character of Thought.* Hillsdale, N.J.: Lawrence Erlbaum Associates.

Bradshaw, G.L., and A. Bell. 1990. Toward Robust Feature Detectors for Speech. ICS technical report.

Crouch, T.D. 1989. *The Bishop's Boys: A Life of Wilbur and Orville Wright.* New York: W.W. Norton and Company.

Gibbs-Smith, C.H. 1965. *The Invention of the Aeroplane, 1799–1909.* London: Faber and Faber.

Howard, F. 1987. *Wilbur and Orville: A Biography of the Wright Brothers.* New York: Alfred A. Knopf.

Kelly, F.C. 1972. *Miracle at Kitty Hawk: The Letters of Wilbur and Orville Wright.* New York: Arno Press, Inc.

Langley, P., et al. 1987. *Scientific Discovery: Computational Explorations of the Creative Process.* Cambridge, Mass.: MIT Press.

Popper, K.R. 1959. *The Logic of Scientific Discovery.* London: Hutchinson and Company.

Reichenbach, H. 1966. *The Rise of Scientific Philosophy.* Berkeley: University of California Press.

Simon, H.A. 1966. Scientific discovery and the psychology of problem solving. In R. Colodny, ed., *Mind and Cosmos*, pp. 22–40. Pittsburgh: University of Pittsburgh Press.

———. 1973. Does scientific discovery have a logic? *Philosophy of Science* 40:471–80.

Simon, H.A., P.W. Langley, and G.L. Bradshaw. 1981. Scientific discovery as problem solving. *Synthese* 47:1–27.

Strategies for Anomaly Resolution

1. Introduction

Understanding the growth of scientific knowledge has been one of the major tasks in philosophy of science in the last thirty years. No successful general model of scientific change has been found; attempts were made by, for example, Kuhn (1970), Toulmin (1972), Lakatos (1970), and Laudan (1977). A new approach is to view science as a problem-solving enterprise. The goal is to find both general and domain-specific heuristics (reasoning strategies) for problem solving. Such heuristics produce plausible, but not infallible, results (Nickles 1987; Thagard 1988).

Viewing science as a problem-solving enterprise and scientific reasoning as a special form of problem solving owes much to cognitive science and artificial intelligence (AI) (e.g., Langley et al. 1987). Key issues in AI are representation and reasoning. More specifically, AI studies methods for representing knowledge and methods for manipulating computationally represented knowledge. From the perspective of philosophy of science, the general issues of representation and reasoning become how to represent scientific theories and how to find strategies for reasoning in theory change. Reasoning in theory change is viewed as problem solving, and implementations in AI computer programs provide tools for investigating methods of problem solving. This new approach is called "computational philosophy of science" (Thagard 1988).

Huge amounts of data are now available in on-line databases. The time is now ripe for automating scientific reasoning, first, to form empirical generalizations about patterns in the data (Langley et al. 1987), then, to construct new explanatory theories, and, finally, to improve them over time in the light of anomalies. Thus, the development of computational models for doing science holds promise for automating scientific discovery in areas where large amounts of data overwhelm human cognitive capacities (Schaffner 1986; Morowitz and Smith 1987). The goal is to devise good methods for doing science, whether or not the methods are ones actually used by humans. The goal is not the simulation of human scientists, but the making of discoveries about the natural world, using methods that extend human cognitive capacities. Thus, computational models allow exploration of methods for representing scientific theories, as well as methods for reasoning in theory formation, testing, and improvement. Such exploration holds the promise of making philosophy of science an experimental science. The models can be manipulated to allow "what if" scenarios to be explored; i.e., experiments on scientific knowledge and reasoning can be done.

The reasoning method to be discussed in this essay occurs in the context of resolving anomalies for theories. When a theory makes a prediction that fails, an empirical anomaly for the theory results. There are strategies that can be followed systematically to resolve the anomaly. The next section of the essay will briefly treat anomalies and scientific reasoning. Following that I will discuss prior work by philosophers of science on anomalies. Then systematic strategies for anomaly resolution will be proposed. Reasoning in anomaly resolution may be compared to a diagnostic reasoning and therapy-planning task: find the ailing part of the theory and propose a way to fix it. The analogy between diagnostic reasoning and reasoning in anomaly resolution allows AI work on "model-based diagnosis" (e.g., Davis and Hamscher 1988) to be used to build computational models of theory change. The final sections of the essay describe a pilot AI program to implement an episode of anomaly resolution. A portion of Mendelian genetic theory is represented, methods for localizing a failing component of the theory are demonstrated, and extensions to the current implementation are suggested.

2. Anomalies and Scientific Reasoning

To put the examination of strategies for anomaly resolution into a broader framework of research on scientific reasoning, consider the idea that there are stages in the development of a theory and strategies for making those changes. Figure 1 is a diagram of stages and strategies of theory change. An episode may be considered to begin with a problem to be solved (problem-finding strategies are an interesting issue not addressed here). Then new ideas get built into plausible theories (hypotheses) that are tested and improved over time. Reasoning strategies guide these changes. The types of strategies include: (1) strategies for producing new ideas, (2) strategies for theory assessment, and (3) strategies for anomaly resolution and change of scope. Change of scope can be

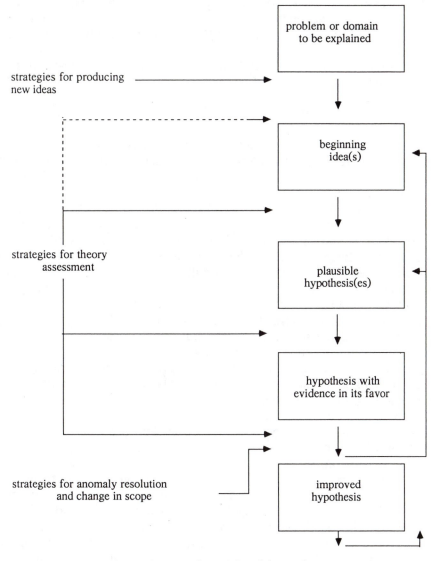

Figure 1. Stages and strategies of theory change.

(1) an expansion of the theory to explain items originally outside its domain (Shapere 1974) of phenomena, (2) expansion to explain new items just discovered, or (3) specialization to exclude anomalies and preserve

adequacy within a narrower domain. Anomaly-resolution strategies are the focus in this essay. Strategies for producing new ideas and strategies for theory assessment will be briefly mentioned. All three types of strategies are the subject of a book that I have just completed (Darden 1991).

To make it easy to recognize the names of specific strategies in this discussion, they will be **bold-faced**.

3. Previous Work on Anomalies

Somewhat surprisingly, methods for anomaly resolution have received comparatively little attention in twentieth-century philosophy of science. Popper (1965), for example, concentrated on falsifying instances as indicators of the inadequacy of a theory, but gave no hints as to how to use the anomaly to localize and correct the problem to produce an improved version of the theory. Instead, he advocated a trial and error process in "conjecturing" hypotheses. Kuhn (1970) discussed the role of puzzle solving in driving the activities in what he called "normal science." Also, he argued, the accumulation of a number of anomalies (puzzles?) sometimes provokes crises; such anomalies, in some mysterious way, lead to the proposal of a new theory or "paradigm." Laudan (1977) proposed an interesting set of categories for classifying ways that anomalies become solved problems, but he provided no strategies for actually generating such solutions. Instead of searching for a method of localizing a problem within a theory (as a first step to generating a solution), Laudan argued for spreading the blame for an anomaly evenly among the parts of the theory (Laudan 1977, p. 43). Laudan's concern was only for how to weight the anomaly in theory assessment, not with how to generate new hypotheses to solve it and produce an improved version of the theory.

My discussion of the role that an anomaly may play in *generating alternative hypotheses* contrasts with the role of anomalies in theory assessment, which was the focus of the philosophers of science mentioned above. One step in anomaly resolution is localization; potential sites of failure within the theoretical components need to be found. The idea of localization of problems within a theory or theoretical model in the light of an anomaly has received some recent attention (e.g., Glymour 1980; Nickles 1981; Darden 1982b; Wimsatt 1987). However, finding one or more plausible locations is only the first step in resolving an anomaly. How the new hypotheses are generated is the next stage of the process of anomaly resolution. Lakatos (1976) suggested heuristics for use in mathematics to improve conjectures in the light of counterinstances; however, when he (1970) discussed scientific reasoning, he spoke only vaguely of domain-specific heuristics associated within particular research programs. Shapere (1974) suggested that simplifications made in the early stages of theory development are likely areas for hypothesis formation, once anomalies arise. Wimsatt (1987) suggested how mechanical and causal models might aid in forming hypotheses for resolving anomalies that arise for them; his analysis extends that of Hesse (1966). She suggested that unexplored areas of an analogy used in the original construction of a theory may function in forming hypotheses to resolve an anomaly at a later stage.

Researchers in AI are developing methods for determining which parts of a complex, explanatory system are involved in the explanation of particular data points. Their techniques for doing "credit assignment" are relevant to the problem of localizing plausible sites for modification (or even all possible sites, given an explicit representation of all knowledge in the system), given a particular anomalous data point (Charniak and McDermott 1985, p. 634). The localization of an anomaly in a theoretical component can be compared to diagnosing a fault in a device or a disease in a patient. Thus, AI methods for diagnostic reasoning can be applied to anomaly resolution (Darden 1990). After one or more components have been identified as potential sites where the theory may be failing, then changing the components is like a redesign task (Karp 1989; 1990). Reasoning in design involves designing something new to fulfill a certain function, in the light of certain constraints. Redesigning theoretical components involves constructing a component that will account for the anomaly, with the constraints of preserving the unproblematic components of the theory and producing a theory that satisfies criteria of theory assessment. Especially important criteria in anomaly resolution are systematicity and lack of ad hocness. It is important that the new theoretical component be systematically connected with the other theoretical components and not be merely an ad hoc addition that serves only to account for the anomaly.

Not surprisingly, AI research has shown that implementing methods for improving faulty modules is more difficult than the first step of localizing the problem. Creatively constructing new hypotheses is, in general, a more difficult task than discovering the need for a new hypothesis. To better understand the reasoning in anomaly resolution, we need to distinguish (1) the problem of localization and (2) the problem of generating a new hypothesis to account for the anomaly. Those issues will be discussed below.

4. Strategies for Anomaly Resolution

I am combining what I call "strategies for anomaly resolution" and "strategies for change of scope." What is inside the domain and what is outside the domain of a theory may be open to debate. Thus, a given item, such as an experimental result, might be considered an anomaly for the theory because the item is inside the domain to be explained by that theory. Alternatively, the same item, it might be argued, is not an anomaly for the theory because that item is outside the scope of the domain of the theory. Perhaps some other theory is expected to account for it. If theory construction begins with a large domain to be explained and then anomalies arise, narrowing the scope of the domain may aid in anomaly resolution. If a proposed hypothesis is very general and an anomaly arises for it, then one strategy is to **specialize the overgeneralization** and exclude the anomalous item from the domain. On the other hand, if theory construction begins with a hypothesis that applies to a very narrow domain, then theory change may occur with expansion of the domain, by including items originally not part of it. Because of this close relationship between the scope of the domain and the identification of an anomaly, strategies for anomaly resolution and change of scope are closely related and are considered together here.

The term "anomaly" usually refers to a problem posed by data within the theory's domain that the theory cannot explain. Often an anomaly is generated when a prediction fails to be confirmed. However, theories may face other kinds of problems besides those posed by empirical anomalies. A theory may be incomplete, even though no empirical anomaly indicates that it is incorrect. (For more discussion of incorrectness versus incompleteness, see Shapere 1974; Leplin 1975.) In addition to problems of incorrectness or incompleteness, a theory may face conceptual problems of various kinds, such as determining the nature of a newly proposed theoretical entity. Discussion of general strategies for resolving all the kinds of problems that a theory may face would make this essay over long. Hence, the focus here will be on empirical anomalies, due to failed predictions, that seem to show that a theory is incorrect.

A general strategy for anomaly resolution entails several stages. List 1 shows four primary stages: (1) **confirm the anomalous data**, (2) **localize the problem**, (3) **resolve the anomaly**, and (4) **assess the resulting theory**. The following subsections will discuss these stages of anomaly resolution.

List 1
Strategies for Anomaly Resolution and Change of Scope

1. Confirm that an anomaly exists
   (a) Reproduce anomalous data
   (b) Reanalyze problem
2. Localize problem
   (a) Outside theory
   (b) Inside theory
3. Resolve the anomaly
   (a) Employ monster-barring
   (b) Alter a component
       (1) Delete
       (2) Generalize
       (3) Specialize
       (4) "Tweak" — alter slightly
   (c) Add a new component using strategies for producing new ideas with added constraints
   (d) For (b) and (c), design an altered or new theoretical component in the light of the following constraints
       (1) The nature of the anomaly — the new theoretical component must account for it
       (2) Maintain a systematic connection with other non-problematic theoretical components, that is, avoid ad hocness
       (3) Other criteria of theory assessment may be introduced as constraints, such as extendibility and fruitfulness
4. If (b) or (c) resulted in a change in the theoretical components, assess the new version of the theory by applying criteria of assessment to the new component(s)
5. If the above steps fail to resolve the anomaly, consider whether the anomaly is sufficiently serious to require abandoning the entire theory, or whether work with the theory can continue despite the unresolved anomaly

4.1. Confirm Anomalous Data or Problem

Before efforts are made to resolve an anomaly, the correctness of the anomalous data needs to be confirmed. If the data are wrong, then no anomaly exists and the subsequent steps in anomaly resolution need not be taken. If experimental error can be blamed for a failed prediction, then the anomaly is resolved without further work or change in the theory.

4.2. Localize the Anomaly

If the anomaly can be localized outside the theory, then the theory will not need to be modified. One way to localize the anomaly elsewhere is to argue that it is outside the scope of the domain of the theory. Thus, the phenomenon is not an anomaly for the theory after all. Another way of localizing the problem without requiring theory change is to show that the instance is an unusual one; while it is inside the domain of the theory, it does not represent a typical phenomenon. This method, which I call "monster-barring," is discussed below.

If the anomaly is to be localized as a failing within the theory, then various methods exist for forming hypotheses about which component(s) is failing. One method for localization is to represent the theory as typical steps in typical processes. Localization is achieved by determining which step failed in the anomalous situation. The anomaly is localized in the failing step. Localization may require additional information not supplied by the experiment that generated the anomaly. Such additional information is needed to determine which steps in the process occurred and which did not. That information may have to be generated by additional experiments to detect which steps were reached and which were not. This kind of localization is analogous to a diagnostic reasoning task. It is like localizing a faulty module in a device by seeing, for example, which module has electricity coming in but puts none out. Queries about inputs and outputs to the steps of the process aid in localization. (For an AI discussion of this kind of diagnostic reasoning about devices, see Sembugamoorthy and Chandrasekaran 1986.)

4.3. Resolve the Anomaly

If an anomaly is considered to be inside the scope of the domain of the theory, then the anomaly may be successfully resolved by a theory in at least three ways: (1) "monster-barring," showing that the anomaly is an exception that does not require theory change; (2) altering an existing component of the theory; and (3) adding one or more new theoretical components. This section discusses each of these in turn.

4.3.1. Employ Monster-barring

I have taken the term "monster-barring" from Lakatos (1976). It is a strategy that he introduced for mathematical reasoning. Monster-barring is a way of preserving a generalization in the face of a purported exception: if the exception can be barred as a monster, i.e., shown not to be a threat to the generalization after all, then it can be barred from necessitating a change in the generalization. Lakatos was concerned with distinguishing between legitimate exception-barring instances and illegitimate barring of instances that really did require a theory change. The strategy I label "monster-barring" is what he called a legitimate instance of "exception-barring." (His "local counterexamples" would be included in what I call "model anomalies," to be discussed below. I use "model" to indicate that the anomaly is not really an exception, but is an exemplary case that will, itself, serve as a model for other such cases. However, because Lakatos was not discussing scientific theories, the term "model" would have been less appropriate for the mathematical, "local" counterexamples that he discussed. Anomalies that serve to falsify an entire theory would correspond, I think, to what Lakatos called "global counterexamples.")

Two kinds of monster anomalies are possible, unique ones and ones that belong to classes. Unique ones can often be difficult to account for. More interesting monsters are ones that occur often, and can, thus, be seen as instances of a malfunction class. Both kinds of monsters, unique ones and malfunction classes, are barred from necessitating a change in the theory. They are explained, or better, explained away. An account is given of what went wrong in the normal process to produce the anomaly.

4.3.2. Alter a Component of the Theory

To have a term to contrast to "monster" anomalies, I introduced the term "model" anomalies, by analogy with "model" organisms (Darden 1990). Model anomalies require a change in the theory. The model anomalies then are shown to be normal, that is, not actually anomalous at all. They serve as models of normal types of processes that are commonly found. Model anomalies are resolved either by changing an existing theoretical component or by adding a new one or both (e.g., one component may be specialized and an additional new component added).

The outline of strategies for anomaly resolution and change of scope (see above) divides strategies for changing a theoretical component into two categories: 3(b) — alter a component; and 3(c) — add a new component. Changing a component already present is usually an easier task than adding an entirely new component. A number of different strategies exist for altering an existing component. They are listed in Step 3(b) of the outline and will now be discussed.

Deleting a component is obviously an easy kind of change to make. If the deleted component explains other items in the domain in addition to the anomalous ones leading to its deletion, then some other component(s) will have to be modified or added to account for those items. If deleting a component leaves a problematic "hole" in the representation of the theory, then one or more other components may have to be added to replace the deletion.

Another method for slightly changing a theoretical component in the light of an anomaly is to **generalize or specialize** the component. Generalization and specialization are methods of modifying a working hypothesis that have been extensively used in studies of induction and concept learning in AI (Mitchell 1982; Dietterich et al. 1982). **Generalization** expands the scope of a hypothesis; **specialization** narrows the scope. (Relations between generalization, abstraction, and simplification are discussed in Darden 1987.)

If bold, general, simplifying assumptions marked the beginning stages of theory construction, then **specialization** and **complication** will be likely strategies to use as anomalies arise. If the theory was constructed originally in a very conservative way, carefully specialized to apply to a narrow domain, then **generalization** will be a way of expanding the scope of its domain, even if no specific anomaly is at issue.

A more systematic strategy for anomaly resolution is, at the outset, to consider explaining the problematic data in both the most general and the most specialized ways consistent with the data. Then alternative hypotheses — varying along a spectrum from general to specific — become candidates for future development. This method of systematically considering a range of hypotheses, from the most general to the most specific, is called the "version space" method of hypothesis formation in AI (Mitchell 1982). The version space is the space of all hypotheses between the most general and the most specific that account for a given set of data. Then refinements are made in the light of new data points or anomalies. Exceptions to generalizations are resolved by adding conditions to make a general hypothesis more specific. A new instance not covered by a specific hypothesis drives the formation of a more general one by dropping conditions from the specific one. I doubt that scientists typically engage in such systematic generation of alternative hypotheses; perhaps they should consider doing so.

"Tweaking" is a term for the strategy of **changing a component slightly** to account for an anomalous or a new instance. Schank (1986, p. 81) used the term similarly when he suggested explaining anomalies by invoking past explanation patterns and changing the patterns slightly to apply to the new situation. However, he was not discussing explanations in science. I am using the term as an eclectic class of strategies for changing a theoretical component. "Tweaking" strategies make slight changes in the theory that do not fit into any of the more specific ways of making slight changes discussed above. An example is slightly changing the parameters in a quantitative model to account for a quantitative anomaly that is just a little off from the predicted value.

4.3.3. Add Something New

The strategies discussed above for altering a component produce slightly new hypotheses. However, those strategies may all prove to be inadequate and a new component of the theory may be needed. In such a case, strategies for producing new ideas will need to be invoked. Such strategies include reasoning by analogy (Hesse 1966; Darden 1982a; Darden and Rada 1988b; Holyoak and Thagard 1989); reasoning by postulating an interfield connection (Darden and Maull 1977; Darden and Rada 1988a); reasoning by postulating a new level of organization (Darden 1978); reasoning by invoking an abstraction (Darden 1987; Darden and Cain 1989); reasoning by conceptual combination (Thagard 1988); and abductive assembly of a new composite hypothesis from simpler hypothesis fragments (Josephson et al. 1987). Detailed discussion of these strategies would take us too far afield here (several are discussed in Darden 1991, chap. 15). If some components of the theory are not candidates for modification, then a consistent and systematic relation with them must be maintained while adding new components. This consistency is a constraint not present in the use of strategies for producing new ideas when an entirely new theory is constructed.

4.4. Assess the Hypotheses to Resolve the Anomaly

Hypotheses proposed as modified theoretical components have to be evaluated using the criteria of theory assessment. Such criteria are given in List 2; they will not be discussed in detail here because most are self-explanatory. The stages diagramed in Figure 1 can be used to represent anomaly resolution. The anomaly is the problem to be solved. The strategies for localization, coupled with strategies for altering a component or the strategies for producing new ideas, provide one or more hypotheses as candidates for the modified theory components. Then the criteria for theory assessment are used in evaluating the hypotheses, with the added constraint that the new components must be compatible with the unmodified components; the criterion of a systematic relation among all the components of the theory ("systematicity") becomes important. Also especially important are the criteria of explanatory adequacy, predictive adequacy, and the lack of ad hocness (see List 2). The new component added to resolve the anomaly should improve the theory in accordance with these criteria. What counts as a legitimate addition to the theory and what is an illegitimate ad hoc change may be a matter of debate, especially when a new component is first proposed to resolve an anomaly. Additional work will be necessary to test the explanatory and predictive adequacy of the newly proposed component and the altered theory of which it is a part.

List 2
Criteria for Theory Assessment

1. internal consistency and absence of tautology

2. systematicity

3. clarity

4. explanatory adequacy

5. predictive adequacy

6. scope and generality

7. lack of ad hocness

8. extendibility and fruitfulness

9. relations with other accepted theories

10. metaphysical and methodological constraints, e.g., simplicity

11. relation to rivals

4.5. Unresolved Anomalies

All of the above strategies for resolving an anomaly may fail. In such a case, scientists working on the theory will have to decide whether the anomaly is sufficiently serious to require abandoning the entire theory or whether it can be shelved as a problem requiring resolution, while work on other parts of the theory continues. Again, no decisive criteria may be present to choose among these alternatives. Even if the entire theory is to be abandoned, the anomaly may well provide a pointer to the components of the theory most at fault and provide hints as to what a new theory should contain in order to avoid having the same anomaly.

5. Representation and Implementation of Anomaly Resolution in Genetics

A subset of these strategies for anomaly resolution has been investigated in an AI system. Revising scientific knowledge is analogous to redesigning tools or other devices. Scientists make use of anomalies in diagnosing faults in a theory and then propose fixes for those faults. The perspective of "theory as device" allows AI representation and reasoning techniques to be applied to the representation of scientific theories and to the simulation of reasoning in anomaly resolution. A pilot AI system has been implemented to represent a portion of Mendelian genetics and simulate the process of resolving one monster anomaly for it.

5.1. Representation of a Scientific Theory

At least some scientific theories can be represented by a series of steps in a mechanistic process. Representing a theory as such a series of steps in a normal process provides a "schematic flow diagram" of the steps in a normal case. Mendelian genetics lends itself to this kind of representation. A series of steps in a normal hereditary process can be used to represent a normal Mendelian segregation process. The steps in a normal process of Mendelian segregation can best be illustrated by an example (see Figure 2). If a pure yellow variety of pea (AA) is crossbred with a pure green variety (aa), in the next generation, all the peas will be hybrid yellow (symbolized by Aa; yellow is called "dominant"). If two hybrid yellows are mated (Aa x Aa), then the next generation produces the ratio of 3 yellow to 1 green (symbolized by AA + 2Aa + aa; AA and Aa both appear yellow). The ratio of 3:1 is produced, given (1) that the Aa of the hybrid separate or segregate during the formation of germ cells (gametes) in a pure, i.e., nonblended, way; (2) that the two types of germ cells (A and a) form in equal numbers; (3) that the fertilization process is random; and (4) that all the types of zygotes are equally viable.

This way of representing the process of Mendelian segregation can easily be put into a computational form using the functional representation (FR) language. FR was created to represent the functioning of devices. The goal in designing the language was to support the problem-solving activities of troubleshooting and of predicting changes in function in a device when there is a change in its components, structure, or constituent behaviors (Sembugamoorthy and Chandrasekaran 1986). The language has been used to represent both concrete devices (bodily systems in medical applications; manufactured devices and manufacturing processes for engineering applications) and abstract "devices" (plans and computer programs) (Sticklen 1987; Chandrasekaran,

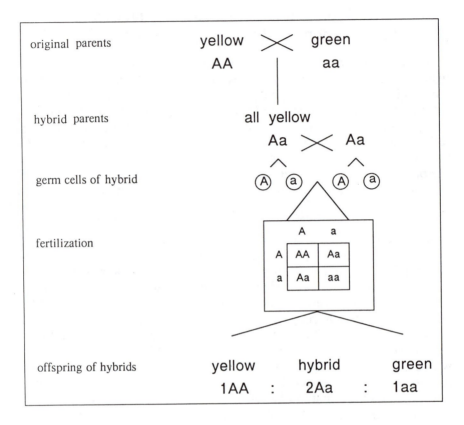

Figure 2. Symbolic representation of Mendelian segregation.

Josephson, and Keuneke 1986; Chandrasekaran et al. 1989; Allemang 1990).

In order to construct an FR, the system (the normal segregation process) must be analyzed into a sequence of states, with steps leading from state to state. The separate steps of the process to be represented are distinguished and labeled. The computational representation of the states supports operations that enable the system to assess (with the aid of a human user) whether the state has been entered. The transitions between states in an FR are typically considered to be causal links, and the nature of the causal responsibility for the transition is specified. In the FR language, the transitions are of various types. The types of transition links and their parameter are (1) *By* [behavior], (2) *Using function of* [a subcomponent], and (3) *AsPer* [background knowledge] (Sembugamoorthy and Chandrasekaran 1986; Keuneke 1989). In other words, a change-of-state transition can be attributed (1) to (*By*) some behavior that occurred, (2) to the function of some subcomponent, or (3) to (*AsPer*) some known process. The "*AsPer*" link, in contrast to "*By*" and "*Using function of*," does not appeal to subbehaviors or subfunctions in the device to provide a causal description of the transition between states; instead, it merely points to a chunk of "knowledge" used for explaining the causal transition being represented.

The process of Mendelian segregation can be analyzed into sequential states and represented in the FR language. An initial FR for Mendelian segregation is shown in Figure 3. The separate steps of the process that lead from parents to offspring are distinguished and labeled. It is a "flat" functional representation of normal genetic segregation (for a single gene locus). The representation is called "flat" because it is at only one hierarchical level, and the only kind of transitions specified are "*AsPer*" links. The term "GametePurity," for example, refers to one chunk of our conjectured knowledge about germ cell (gamete) formation that can be summarized by the rule that genes separate (segregate) completely or "purely" when gametes form; there are no intermediate or blended genes, only pure parental types. The other "*AsPer*" links point off to additional chunks of knowledge according to which the state transitions occur. A way to extend the representation and connect it to a representation of underlying cytological processes would be to change the "*AsPer*" links to "*By*" or "*Using function of*." This ability to package hierarchical relations is a strength of the FR language that has not yet been exploited for the Mendelian segregation representation.

An FR, once constructed, supports simulation. The representation can be "run" by providing an input, the first state, and noting the output, the final state. The final state (or perhaps some intermediate one) can be interpreted as a prediction made by the theory. For example, the FR of normal Mendelian segregation predicts that, given fertile hybrid parents that are mated, offspring will be produced in the ratio of 3 dominant to 1 recessive. This qualitative simulation provides a test of the adequacy of the representation; it also allows simulation of the failure of a step in an anomalous case. Such simulation abilities are useful during anomaly resolution for showing that the hypothesized failing step produces the anomaly.

The analogy of a theory to a device is a useful one. The theory is a device with component parts; the parts have functions that combine to support the functions of the device as a whole. A theory has the functions of explanation and prediction. If the theory encounters an anomaly, then the theory is compared to a faulty device. Some part is

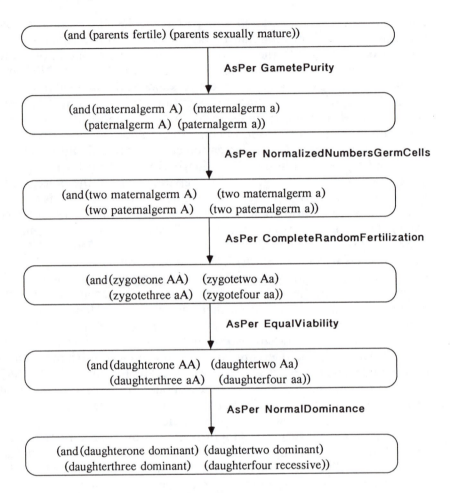

Figure 3. Functional representation of normal Mendelian segregation.

failing and that theoretical component needs to be localized and shown to be capable of producing the fault. The FR language is a particularly useful computational tool for representing a theory and for supporting reasoning in anomaly localization (fault diagnosis).

5.2. Anomaly Resolution: Localization

An instance of the localization of a monster anomaly in Mendelian genetics has been implemented in a simple form. The normal Mendelian segregation process predicts that, when two hybrid (Aa) forms are crossed, a 3:1 ratio will be produced. For example, if hybrid yellow mice (Aa) are crossed (Aa + Aa), then 3 yellow to 1 nonyellow mice are, on the average, predicted to be the ratios for the occurrence of those colors in the litters. The symbolic representation for the outcome is the following: AA + 2Aa + aa. Because those with A are dominant, this equation leads to the prediction of 3 yellow to 1 nonyellow.

When Cuénot did such a cross in 1905, he did not find the expected 3:1 ratios (Cuénot 1905). Instead, he found 2:1 ratios: 2Aa + 1aa, with no AA. Various hypotheses were proposed to account for the anomaly (Darden 1990). The only one to be discussed here is the hypothesis that eventually proved to be the correct one, namely that the AA combination is not a viable embryo, or, in other words, the AA combination is a lethal gene combination. The localization of the failing step in normal segregation, namely the "EqualViability" step in Figure 3, has been demonstrated using the FR of Mendelian segregation.

The FR language supports diagnostic reasoning to localize faults in devices. The representation can automatically be converted into a diagnostic classification hierarchy (Bylander and Mittal 1986; Bylander and Chandrasekaran 1987). This hierarchy is a directed acyclic graph produced by a recursive descent parsing of the functional representation. The transition links (such as *AsPer*) serve as the basis for constructing "knowledge groups" whose values, along with the control messages (normally, an "establish-and-refine" message), govern the fault localization process. According to the establish-and-refine control strategy, only those nodes whose knowledge group satisfies an "establish" threshold, and which have received a "refine" message, will be pursued; other nodes are "rejected," and a saving in search effort is then obtained by not pursuing their subnodes (Moberg and Josephson 1990).

In the present implementation, the knowledge group generates queries that are asked of the human user of the program. Automatic use of the diagnostic system would involve a database that can be suitably queried for data, allowing the knowledge-group node to establish or reject. The generation of useful and appropriate queries is the most significant contribution of the diagnostic hierarchy to the anomaly-resolution process. The user's answers permit the system to localize a potential site of fault in the theory, and thus to provide focus for the computationally costly process of theory redesign (Goel and Chandrasekaran 1989; Moberg and Josephson 1990).

The operation of the diagnostic hierarchy can be illustrated by tracing through the process of localizing the fault that produces the monster anomaly of 2 to 1 ratios. The system takes the FR and makes a classification hierarchy. In order to localize the fault in the hierarchy, the system queries the user about which states have been reached. Those queries are illustrated in List 3. First, the need to engage in diagnostic reasoning is established by asking whether fertile, sexually mature organisms have mated, and whether a 3 dominant to 1 recessive ratio was observed. The human user replies that the initial state (fertile and mature organisms mating) was entered, but that the final state of 3 to 1 ratios of a trait in the offspring population was not observed. This reply establishes the existence of an anomaly. Next, the system refines its diagnostic hypothesis about what part of the theory might plausibly be mistaken. At present, the system traverses the classification hierarchy from the final state toward the initial state, inquiring at each point whether the intermediate state was entered. In this case, localization occurs when the system is informed that four zygotes were produced, but that no AA offspring were produced. Note in Figure 3 that "AsPer EqualViability" was the link before the failed step. This information establishes that the causal behavior referred to as "equal zygote viability" (AsPer EqualViability) is at fault (Moberg and Josephson 1990).

List 3
List of Diagnostic Queries

Queries generated by the diagnostic hierarchy (the system's queries are in *italics*; the human user's answers are in **boldface**; explanatory comments are in [brackets]):

parents fertile and parents sexually mature? **yes**
[the initial state was reached]

daughterone dominant, daughtertwo dominant, daughterthree dominant, daughterfour recessive? **no**
[the final state was not reached; therefore there is an anomaly]

daughterone AA, daughtertwo Aa, daughterthree aA, daughterfour aa? **no**
[the next-to-final state was not reached, so the system proceeds to query about the previous one]

zygote AA, zygote Aa, zygote aA, zygote aa? **yes**
[this state was reached, so the problem is localized in the link "AsPer EqualViability"]

---

n, Lindley, and Joseph A. Cain. 1989. "Selection Type Theories." *Philosophy of
 ience* 56:106–29.

n, Lindley, and Nancy Maull. 1977. "Interfield Theories." *Philosophy of Science*
 :43–64.

n, Lindley, and Roy Rada. 1988a. "Hypothesis Formation Using Part-whole Inter-
 ations." In David Helman, ed., *Analogical Reasoning*. Dordrecht: Reidel, pp. 341–
 .

—. 1988b. "Hypothesis Formation Via Interrelations." In Armand Prieditis, ed.,
 alogica. Los Altos, Calif.: Morgan Kaufmann, pp. 109–27.

, Randall, and Walter C. Hamscher. 1988. "Model-based Reasoning: Troubleshoot-
 .." In H. E. Shrobe, ed., *Exploring Artificial Intelligence*. Los Altos, Calif.: Morgan
 ufmann, pp. 297–346.

rich, Thomas G., et al. 1982. "Learning and Inductive Inference." In Paul R. Cohen
 d E. Feigenbaum, eds., *The Handbook of Artificial Intelligence*. Vol. 3. Los Altos,
 lif.: Morgan Kaufmann, pp. 323–511.

our, Clark. 1980. *Theory and Evidence*. Princeton, N.J.: Princeton University Press.

 Ashok, and B. Chandrasekaran. 1989. "Functional Representation of Designs
 d Redesign Problem Solving." In *Proceedings of the Eleventh International Joint
 nference on Artificial Intelligence*. Detroit, August, pp. 1388–94.

, Mary. 1966. *Models and Analogies in Science*. Notre Dame, Ind.: University of
 tre Dame Press.

ak, Keith J., and Paul Thagard. 1989. "Analogical Mapping by Constraint Satisfac-
 n." *Cognitive Science* 13:295–355.

son, J., et al. 1987. "A Mechanism for Forming Composite Explanatory Hypothe-
 ." *IEEE Transactions on Systems, Man, and Cybernetics* SMC-17:445–54.

 Peter. 1989. "Hypothesis Formation and Qualitative Reasoning in Molecular Bi-
 gy." Ph.D. diss., Stanford University. (Available as a technical report from the
 mputer Science Department: STAN-CS-89-1263.)

—. 1990. "Hypothesis Formation as Design." In J. Shrager and P. Langley, eds., *Com-
 ational Models of Scientific Discovery and Theory Formation*. San Mateo, Calif.:
 organ Kaufmann, pp. 275–317.

ke, Anne. 1989. "Machine Understanding of Devices: Causal Explanation of Diag-
 stic Conclusions." Ph.D. diss., Department of Computer and Information Science,
 e Ohio State University.

am, W. B. 1919. "The Fate of Homozygous Yellow Mice." *Journal of Experimental
 ology* 28:125–35.

, Thomas. 1970. *The Structure of Scientific Revolutions*. 2d ed. Chicago: University
 Chicago Press.

os, Imre. 1970. "Falsification and the Methodology of Scientific Research Pro-
 mmes." In I. Lakatos and Alan Musgrave, eds., *Criticism and the Growth of
 owledge*. Cambridge: Cambridge University Press, pp. 91–195.

—. 1976. *Proofs and Refutations: The Logic of Mathematical Discovery*. Ed. J. Worrall
 d E. Zahar. Cambridge: Cambridge University Press.

ey, Pat, et al. 1987. *Scientific Discovery: Computational Explorations of the Creative
 ocess*. Cambridge, Mass.: MIT Press.

n, Larry. 1977. *Progress and Its Problems*. Berkeley: University of California Press.

, Jarrett. 1975. "The Concept of an Ad Hoc Hypothesis." *Studies in the History and
 ilosophy of Science* 5:309–45.

ell, Tom M. 1982. "Generalization as Search." *Artificial Intelligence* 18:203–26.

This pilot project shows that the causal process specified by part of the theory of the gene can be represented using the FR language. It also demonstrates how diagnostic reasoning about theory faults is supported by the availability of a compiler for a diagnostic classification hierarchy. The diagnostic queries carry out a search for knowledge needed to localize a fault. This search may reveal the need for additional experimental work in order to answer the queries. Historically, when Cuénot found the 2:1 anomaly in 1905, techniques did not exist to discriminate between the states of germ cell formation, zygote formation, and development of zygotes into mature offspring. Later techniques provided experimental access to confirm that small embryos formed (i.e., zygotes formed), but that they did not develop into mature offspring (Castle and Little 1910; Kirkham 1919). The queries from such a diagnostic localization system could direct experimental work to discriminate between states and to determine which state was reached and which failed.

5.3. Extensions to the Current Implementation

A full implementation of all the anomaly-resolution strategies discussed in Section 4 above requires more than just strategies for localization. At the beginning of the process of anomaly resolution, queries are needed to try to resolve the anomaly without localizing it in the theory. In order words, the earlier steps in the list of strategies for anomaly resolution and change of scope (see above) need to be included in an implementation. Also, the difference between monster and model anomalies needs to be exploited by the system. In a monster-anomaly case, the normal process is not questioned, but its application to a particular anomalous instance is questioned. As in the Cuénot case, the normal process of segregation was not changed, but its application to the hybrid yellow mice case was altered. The anomaly was resolved by showing the AA combination was not viable. The FR representation could easily be extended to show that "breaking" the "EqualViability" link for AA forms would produce 2:1 ratios. In other words, the FR simulation capabilities could be used to show that the hypothesized failing state, if it is disabled, does in fact produce the resultant anomaly.

In model-anomaly cases, not only localization is needed but the theory (the representation of the normal process) must be changed. Such anomaly-directed redesign includes both proposing new hypotheses (new theoretical components) and testing to see that the old success of the theory is retained while the anomalous result is removed. The model-anomaly-resolution task can be formulated as an information-processing task:

Given a representation of theory in the form of a series of causal steps that produce a given output and an anomaly (a failed predicted output) for that theory, construct a modified theory that no longer has the given anomaly and that retains previous explanatory successes.

This task is not uniquely specified in that many modified theories may be constructed that retain successes and eliminate the anomaly. Some of the different directions for theory modification are associated with different strategies. A goal for future work is to investigate alternative programs for theory redesign with respect to a sequence of anomalies and selection of strategies. The present implementation efforts have been directed toward finding a good representation for a theory's content and localizing potential failing sites. Such localization for a model-anomaly case makes the redesign task simpler; however, producing the new hypotheses needed for redesign is a difficult task, given current AI techniques (Moberg and Josephson 1990). Such implementation will require considerable further effort and analysis.

The representation of a theory's central causal processes does not yet take advantage of several features of the FR language that would provide considerable power. The pilot project has not yet used a full analysis of components, structure, and behavior of the genetic system, nor the provision for simulation of normal genetic processes. An FR normally involves a hierarchy whose vertical organization provides a way of grouping details of state sequence, subcomponents, and knowledge. These groups then organize process sequences at different levels of detail. These in turn are the parts that can be selected as loci of failure and thus become targets for potential redesign. The strategy of using knowledge from another level of organization to form hypotheses at a given level is one strategy to be investigated. Decisions about hierarchical organization and abstraction will be made to facilitate localization and redesign. These will be investigated in the next phase of this research project.

6. Conclusion

This essay has proposed that computational methods from artificial intelligence are fruitful for representing and reasoning about scientific theories. Sequential strategies for anomaly resolution suggest that quasi-algorithmic reasoning processes can be used systematically in theory refinement. A pilot project to implement anomaly resolution in an AI system has been discussed. This work shows the promise of the analogy between anomaly resolution in science and diagnosis and redesign in

AI. Computational approaches provide a way for p experiments on strategies for theory change.

Note

This work was supported by a General Research Board Awar School of the University of Maryland and by National Science Fc 9003142. The AI implementation was done by Dale Moberg and help from Dean Allemang, at the Laboratory for Artificial Intelligc Ohio State University. My thanks to John Josephson and Dale Mc comments on an earlier draft of this essay.

References

Allemang, Dean. 1990. "Understanding Programs as Devices." Ph.I of Computer and Information Sciences, The Ohio State Universi

Bylander, T., and B. Chandrasekaran. 1987. "Generic Tasks for Kr soning: The 'Right' Level of Abstraction for Knowledge Acquis *Journal of Man-Machine Studies* 28:231–43.

Bylander, T., and S. Mittal. 1986. "CSRL: A Language for Classificatc and Uncertainty Handling." *AI Magazine* 7, no. 3:66–77.

Castle, W. E., and C. C. Little. 1910. "On a Modified Mendelian R Mice." *Science* 32:868–70.

Chandrasekaran, B., J. Josephson, and A. Keuneke. 1986. "Functiona a Basis for Generating Explanations." *Proceedings of the IEEE Cor Man, and Cybernetics.* Atlanta, Ga., pp. 726–31.

Chandrasekaran, B., et al. 1989. "Building Routine Planning Syste Their Behaviour." *International Journal of Man-Machine Studies*

Charniak, Eugene, and Drew McDermott. 1985. *Introduction to Ar* Reading, Mass.: Addison-Wesley.

Cuénot, Lucien. 1905. "Les races pures et leurs combinaisons chez le *de Zoologie Expérimentale et Générale* 4, Serie, T., 111:123–32.

Darden, Lindley. 1978. "Discoveries and the Emergence of New Fiel P. D. Asquith and I. Hacking, eds., *PSA 1978.* Vol. 1. East Lansing, of Science Association, pp. 149–60.

———. 1982a. "Artificial Intelligence and Philosophy of Science: Rea: in Theory Construction." In T. Nickles and P. Asquith, eds., *PSA* Lansing, Mich.: Philosophy of Science Association, pp. 147–65.

———. 1982b. "Aspects of Theory Construction in Biology." In *Procee International Congress for Logic, Methodology and Philosophy of* North Holland Publishing Co., pp. 463–77.

———. 1987. "Viewing the History of Science as Compiled Hindsight. no. 2:33–41.

———. 1990. "Diagnosing and Fixing Faults in Theories." In J. Shrag(eds., *Computational Models of Scientific Discovery and Theory Form(Calif.: Morgan Kaufmann, pp. 319–46.

———. 1991. *Theory Change in Science: Strategies from Mendelian Gei Oxford University Press.*

Moberg, Dale, and John Josephson. 1990. "Appendix A: An Implementation Note." In J. Shrager and P. Langley, eds., *Computational Models of Scientific Discovery and Theory Formation*. San Mateo, Calif.: Morgan Kaufmann, pp. 347–53.

Morowitz, Harold, and Temple Smith. 1987. *Report of the Matrix of Biological Knowledge Workshop, July 13–August 14, 1987*. Sante Fe Institute, 1120 Canyon Road, Sante Fe, NM 87501.

Nickles, Thomas. 1981. "What Is a Problem That We May Solve It?" *Synthese* 47:85–118.

———. 1987. "Methodology, Heuristics, and Rationality." In J. C. Pitt and M. Pera, eds., *Rational Changes in Science*. Dordrecht: Reidel, pp. 103–32.

Popper, Karl. 1965. *The Logic of Scientific Discovery*. New York: Harper Torchbooks.

Schaffner, Kenneth. 1986. "Computerized Implementation of Biomedical Theory Structures: An Artificial Intelligence Approach." In Arthur Fine and Peter Machamer, eds., *PSA 1986*. Vol. 2. East Lansing, Mich.: Philosophy of Science Association, pp. 17–32.

Schank, Roger C. 1986. *Explanation Patterns: Understanding Mechanically and Creatively*. Hillsdale, N.J.: Lawrence Erlbaum.

Sembugamoorthy, V., and B. Chandrasekaran. 1986. "Functional Representation of Devices and Compilation of Diagnostic Problem-solving Systems." In J. Kolodner and C. Reisbeck, eds., *Experience, Memory, and Reasoning*. Hillsdale, N.J.: Lawrence Erlbaum, pp. 47–73.

Shapere, Dudley. 1974. "Scientific Theories and Their Domains." In F. Suppe, ed., *The Structure of Scientific Theories*. Urbana: University of Illinois Press, pp. 518–65.

Sticklen, J. 1987. "MDX2: An Integrated Medical Diagnostic System." Ph.D. diss., Department of Computer and Information Science, The Ohio State University.

Thagard, Paul. 1988. *Computational Philosophy of Science*. Cambridge, Mass.: MIT Press.

Toulmin, Stephen. 1972. *Human Understanding*. Vol. 1. Princeton, N.J.: Princeton University Press.

Wimsatt, William. 1987. "False Models as Means to Truer Theories." In Matthew Nitecki and Antoni Hoffman, eds., *Natural Models in Biology*. New York: Oxford University Press.

Copernicus, Ptolemy, and Explanatory Coherence

We apply in this essay a computational theory of explanatory coherence to an important case in the history of astronomy. The theory has been implemented in a connectionist computer program called ECHO that has been used to model the competition between Copernican and Ptolemaic astronomy. ECHO has also been used to model the acceptance of hypotheses in chemistry, biology, geology, and legal reasoning (Thagard 1989; in press; Thagard and Nowak 1988; 1990). The application reported in this essay is interesting both because of its historical importance and also because it is by far the largest application of ECHO to date, involving upwards of one hundred propositions. This demonstrates that the implementation of the principles of explanatory coherence in the program ECHO can handle even very complex cases of scientific inference. After a brief sketch of the theory of explanatory coherence and the ECHO program, we shall provide a detailed account of the systems of Copernicus and Ptolemy.

1. Explanatory Coherence

Thagard (1989) provides a detailed description of the new theory of explanatory coherence and the program that implements it. Here we provide only a brief excerpt from that essay, which should be consulted for a full discussion of explanatory coherence and specification of the algorithms used in the program.

The theory of explanatory coherence, or TEC, is stated in seven principles that establish relations of explanatory coherence and make possible an assessment of the global coherence of an explanatory system S. S consists of propositions P, Q, and $P_1 \ldots P_n$. Local coherence is a relation between two propositions. The term "incohere" is used to mean more than just that two propositions do not cohere: to incohere is to *resist* holding together. Here are the principles.

Principle 1. Symmetry.
   (a) If P and Q cohere, then Q and P cohere.
   (b) If P and Q incohere, then Q and P incohere.
Principle 2. Explanation.
   If $P_1 \ldots P_m$ explain Q, then:
   (a) For each P_i in $P_1 \ldots P_m$, P_i and Q cohere.
   (b) For each P_i and P_j in $P_1 \ldots P_m$, P_i and P_j cohere.
   (c) In (a) and (b) the degree of coherence is inversely proportional to the number of propositions $P_1 \ldots P_m$.
Principle 3. Analogy.
   (a) If P_1 explains Q_1, P_2 explains Q_2, P_1 is analogous to P_2, and Q_1 is analogous to Q_2, then P_1 and P_2 cohere, and Q_1 and Q_2 cohere.
   (b) If P_1 explains Q_1, P_2 explains Q_2, Q_1 is analogous to Q_2, but P_1 is disanalogous to P_2, then P_1 and P_2 incohere.
Principle 4. Data Priority.
   Propositions that describe the results of observation have a degree of acceptability on their own.
Principle 5. Contradiction.
   If P contradicts Q, then P and Q incohere.
Principle 6. Acceptability.
   (a) The acceptability of a proposition P in a system S depends on its coherence with the propositions in S.
   (b) If many results of relevant experimental observations are unexplained, then the acceptability of a proposition P that explains only a few of them is reduced.
Principle 7. System Coherence.
   The global explanatory coherence of a system S of propositions is a function of the pairwise local coherence of those propositions.

TEC is implemented in ECHO, a computer program written in Common LISP that is a straightforward application of connectionist algorithms to the problem of explanatory coherence. In ECHO, propositions representing hypotheses and results of observation are represented by units. Whenever principles 1–5 state that two propositions cohere, an excitatory link between them is established. If two propositions incohere, an inhibitory link between them is established. In ECHO, these links are symmetric, as principle 1 suggests: the weight from unit 1 to unit 2 is the same as the weight from unit 2 to unit 1. Principle 2(c) says that the larger the number of propositions used in an explanation, the less the degree of coherence between each pair of propositions. ECHO therefore counts the propositions that do the explaining and proportionately lowers the weight of the excitatory links between units representing coherent propositions.

Principle 4, data priority, is implemented by links to each data unit from a special evidence unit that always has activation 1, giving each unit some acceptability on its own. When the network is run, activation spreads from the special unit to the data units, and then to the units representing explanatory hypotheses. The extent of data priority — the presumed acceptability of data propositions — depends on the weight of the link between the special unit and the data units. The higher this weight, the more immune the data units become from deactivation by other units. Units that have inhibitory links between them because they represent contradictory hypotheses have to compete with each other for the activation spreading from the data units: the activation of one of these units will tend to suppress the activation of the other. Excitatory links have positive weights, and inhibitory links have negative weights. ECHO.2 typically uses .04 as the default for the former, and −.06 as the default for the latter; sensitivity analyses reported in section 5 show that the exact values are not important. The activation of units ranges between 1 and −1; we interpret positive activation as acceptance of the proposition represented by the unit, negative activation as rejection, and activation close to 0 as neutrality.

To summarize how ECHO implements the principles of explanatory coherence, we can list key terms from the principles with the corresponding terms from ECHO.

Proposition: unit.

Coherence: excitatory link, with positive weight.

Incoherence: inhibitory link, with negative weight.

Data priority: excitatory link from special unit.

Acceptability: activation.

Recently, TEC and ECHO have been revised by the addition of a new principle concerning competition among hypotheses (Thagard, in press). Hypotheses do not need to contradict each other to be incoherent with each other. We will see in the simulation discussed below that many pairs of Ptolemaic/Copernican hypotheses do not directly contradict each other but are such that scientists would not want to accept both of them. TEC.2 consists of the seven principles of TEC stated above plus:

Principle C. Competition.

If P and Q both explain evidence E, and if P and Q are not explanatorily connected, then P and Q incohere. Here P and Q are explanatorily connected if any of the following conditions holds:

(a) P is part of the explanation of Q.
(b) Q is part of the explanation of P.
(c) P and Q are together part of the explanation of some proposition R.
(d) P and Q are both explained by some higher-level proposition R.

ECHO.2 implements this principle by finding, for each piece of evidence E, all pairs of hypotheses P and Q that together explain E but are not explanatorily related to each other. Then an inhibitory link between P and Q is constructed. In the limiting case, the inhibition is the same as that between units representing contradictory hypotheses. But if P and Q each explain E only with the assistance of numerous other hypotheses, then they incohere to a lesser extent; compare principle 2(c) above. Hence in ECHO.2, degree of inhibition between units representing P and Q is inversely proportional to the number of cohypotheses used by P and Q in their explanations of E, but proportional to the number of pieces of evidence E. Henceforth, ECHO.2 will be referred to simply as "ECHO," and TEC.2 as "TEC."

After input has been used to set up the network, the network is run in cycles that synchronously update all the units. For each unit j, the activation a_j, ranging from -1 to 1, is a continuous function of the activation of all the units linked to it, with each unit's contribution depending on the *weight* w_{ij} of the link from unit i to unit j. The activation of a unit j is updated using the following equation:

$$a_j(t+1) = a_j(t)(1-\theta) + \begin{cases} net_j(max - a_j(t)) & \text{if } net_j > 0 \\ net_j(a_j(t) - min) & \textit{otherwise} \end{cases} \quad (1)$$

Here θ is a decay parameter that decrements each unit at every cycle; *min* is minimum activation (-1); *max* is maximum activation (1); and net_j is the net input to a unit. This is defined by:

$$net_j = \sum_i w_{ij} a_i(t) \quad (2)$$

Repeated updating cycles result in some units becoming activated (getting activation > 0) while others become deactivated (activation < 0).

Let us now turn to a complex historical application of these ideas about explanatory coherence.

2. Ptolemy and Copernicus

In the second century A.D., the Egyptian mathematician Ptolemy developed a theory that dominated astronomy for well over a thousand years. Ptolemy's account agreed with Platonic astronomy and with Aristotelian physical principles in making the earth the center of the universe, and in having the planets, sun, and moon revolve around it. By describing the motion of these bodies in circular orbits, Ptolemy was able to account for a great number of observations of them. Ptolemy's views survived unchallenged except in detail until 1543, when Copernicus's *De Revolutionibus Orbium Coelestium* was published. According to Copernicus, the sun is at the center of the universe, and the earth revolves around it just like the other planets, while the moon revolves around the earth. It took more than a hundred years, but by the end of the 1600s the Copernican view, as expanded and amended by Galileo, Kepler, and Newton, became accepted.

At first look, the competition between the geocentric system of Ptolemy and the heliocentric system of Copernicus does not seem a likely candidate for modeling by coherence relations among propositions. The theories of both Ptolemy and Copernicus are highly mathematical, requiring the specification of many numerical parameters of geometrical constructions. It is possible, however, to use propositions to describe qualitative features of the geometrical constructions and of the evidence, and to identify the explanatory relations between these propositions. We were thus able to construct a relatively comprehensive model of the explanatory structures of both Ptolemy and Copernicus. We shall show that from Copernicus's perspective, his astronomical system gave a more coherent account of the observable features of the heavens.

The original sources for the work of both Ptolemy and Copernicus are remarkably parallel in form. Both astronomers wrote a major, highly technical work consisting mainly of mathematical astronomy, and a smaller work designed to make the author's ideas comprehensible to the less mathematically inclined. One difference is that Ptolemy's *Planetary Hypotheses* came well after the *Almagest* and presented some simplifications of his earlier models, while the *Commentariolus* of Copernicus was an introduction to the heliocentric theory written about thirty years before *De Revolutionibus*, and circulated only in manuscript during Copernicus's lifetime. Our sources for information on these works are Toomer's translation of the *Almagest* (Ptolemy 1984) and Pedersen's (1974) survey of it; Neugebauer (1975) for the content of the *Planetary Hypotheses;* Rosen's translation of *De Revolutionibus* (Copernicus [1543] 1978) and Swerdlow and Neugebauer's (1984) commentary on it; and Swerdlow's (1973) and Rosen's (Copernicus 1939) translations of the *Commentariolus*.

The task of both the *Almagest* and *De Revolutionibus* was to present geometrical models of the motions of the planets, with appropriate parameters that reproduce the appearances in the heavens. Ptolemy's model was the first to do so for all the planets (for Ptolemy the sun and moon were also planets), but he was able to rely on a rich astronomical literature stretching back to the Babylonians as a source of observations. The basic technique of both Ptolemy and Copernicus is to reduce the motion of each planet to a sort of gearwork of circles, of which the main circle is centered more or less at the earth (or the sun, for Copernicus), and the planet travels on the outermost circle. Following the structure of both the *Almagest* and *De Revolutionibus*, the evidence and hypotheses for both sides of the dispute can be divided into four large sections concerning:

1. the superior planets, Mars, Jupiter, and Saturn, which are outside the sun's orbit for Ptolemy and the earth's orbit for Copernicus;

2. the inferior planets, Mercury and Venus, which are inside the earth's orbit for Copernicus, and are distinguished for Ptolemy because they never appear in opposition to the sun;

3. the moon;

4. the earth, sun, and stars.

3. Ptolemy: Evidence and Hypotheses

Ptolemy's *Almagest* is concerned purely with mathematical and observational astronomy, giving observed positions of the planets and constructing models that would move the planets through the observed positions. In contrast, his *Planetary Hypotheses* turns to questions of physical astronomy, and how these motions are actually realized in space. The *Almagest* is divided into thirteen books. The first book is general and mainly nonmathematical, and begins with a discussion of the classification of knowledge and the purposes of astronomy. Next come arguments in favor of a spherical earth and heaven, the immobility of the earth, and a general description of the motion of the planets. The rest of book 1 and all of book 2 are taken up with a discussion of spherical geometry, astronomical measurements, and geography.

For the purposes of ECHO, the Ptolemaic system has to be coded into propositions, some of which are identified as data. In addition, the explanatory relations between the propositions must be specified. The appendixes give the input to ECHO used in the simulation of Ptolemy vs. Copernicus. Evidence propositions have names beginning with "E," and pieces of "negative evidence" that contradict evidence statements begin with "NE." Propositions like P2 with names starting with "P" represent Ptolemy's position, ones like C4 with names starting with "C" represent Copernicus's position, and ones like PC1 with names starting with "PC" are common to both positions. ECHO also requires a series of *explain* statements indicating the explanatory relations. For example,

(explain '(P2 P3) 'E1)

means that P2 and P3 together explain E1. Such input sets up excitatory links between units P2 and P3, P2 and E1, and P3 and E1. When the ECHO network is run, E1 gets activation from a special evidence unit and passes it to P2 and P3, which pass it to each other. Negative evidence propositions starting "NE" contradict evidence statements and therefore become negatively activated, dragging down with them the propositions that explain them.

Our concern in this essay is to describe the explanatory structure of the Ptolemaic and Copernican systems; for introductory exposition of the systems the reader should consult a source such as Kuhn (1959). To elucidate all the explanatory relations we have identified would require a book-length description of the two systems, so here we will only be able to provide enough description of our encoding to enable readers to check its historical accuracy themselves. Our simulation intentionally reflects the ways in which Copernicus thought his system was superior to Ptolemy's; we are comparing the theories from a Copernican viewpoint.

Although some of the details have been omitted from the propositions, we have tried to keep the representations of Ptolemy and Copernicus at the same level of detail. Whether or not a piece of evidence is adequately explained by a set of hypotheses depends upon the numerical parameters used — parameters such as the radii of the epicycles, and the speed of the moon's motion along them. Since both Ptolemy and Copernicus must choose such parameters, however, and since both of them choose only one parameter for each qualitative feature of their models, we may at each opportunity omit consideration of the particular parameters used without concealing from ECHO any evidence necessary to judge the relative complexity of the systems.

Appendix A states the evidence propositions used in our simulation, and Appendix B states the hypotheses, both Ptolemaic and Copernican, that provide competing explanations of the evidence. Most of the evidence about the heavens was presented by Ptolemy and merely recapitulated by Copernicus; in our input to ECHO it is represented by propositions E1–E18. These propositions are simple observations relevant to the shape of the earth and heavens. E1 points out that the day-night cycle includes an apparent motion of the sky. Appendix C, which lists the explain statements input to ECHO, has E1 explained by Ptolemy's hypotheses P2 and P3 concerning the daily rotation of the heavenly spheres. Copernicus, of course, offers a different explanation, using C8, the hypothesis that the earth rotates on its axis once a day. The diligent reader interested in tracing out the input to ECHO's simulation should first note an evidence proposition in Appendix A, then consult Appendix C for the names of hypotheses that explain the evidence, and finally check B for the content of the hypotheses. Note that the ECHO program uses only the explanatory relations of the propositions, not the content, which is provided for information only. Appendix C also gives the input to ECHO concerning contradictory propositions.

E2 is the motion of the sun once that diurnal motion has been taken into account: the sun travels eastward through the zodiac. E4 merely says that the axis of diurnal rotation is not perpendicular to the ecliptic: the orbit of the sky is around a different axis than the orbit of the planets. E7 establishes the importance of eclipse times: since lunar eclipses were correctly understood by Ptolemy as the shadow of the earth on the moon, the fact that a given eclipse might occur shortly after sunset for a western observer but shortly before dawn for an eastern observer meant that dawn was not simultaneous for all observers. This nonsimultaneity is evidence for the spherical shape of the earth, as is E8: if a certain star is visible in the east just after sunset, and you travel some distance to the west, that star may not yet have risen at sunset, so the rising times of stars (as well as for the sun) are different for different observers.

These propositions actually only support the curvature of the earth in an east-west direction; E9 provides similar evidence for north-south curvature. Propositions E11, E12, and E15 establish that the earth appears to be in the center of the heavenly sphere; Copernicus must explain this using C7, the proposition that the orbit of the earth is negligible compared to the size of the heavens. E13 describes the appearance of the sun at the equinoxes, which characterizes the relative motion of the sun and the earth. E16 states that the earth has a pole star, which was a difficult fact for Copernicus to explain due to his imperfect understanding of mechanics. E26 is the result of a minor motion of the earth, which causes it to change pole stars over the course of thousands of years. E18 states the one anomaly of the sun's motion, which we now ascribe to the fact that the earth's orbit is an ellipse with the sun at one focus. Both Ptolemy and Copernicus used an eccentric circle for the sun's and earth's orbits respectively; this is in fact a very good approximation since the ellipses are so close to being circles. "Point of mean motion" refers to speed; another feature of the elliptical orbit is that the planet varies in speed. Ptolemy solved this problem by introducing the abstraction of the "mean sun" that moved exactly 1/365th of its orbit in one day. The points of mean motion were those days on which the actual speed matched the average speed. Since maximum speed is at perigee and minimum at apogee, there are only two such points that we may approximate by the nodes.

Ptolemy and Copernicus shared a number of hypotheses, which are designated in the input with the letters PC. Both agreed that the heavens were spherical (PC1), even though almost any shape was possible since the heavens were taken to be some sort of black solid supporting points of light. PC2 represents the two astronomers' methodological desire to reduce the apparently irregular motions of the planets to circular motions; such motions were appropriate to the heavens because they were eternal and periodic, just like the motions of the planets.

To both Ptolemy and Copernicus, it is apparent that the axis of the stars' rotation and the axis perpendicular to the orbits in the ecliptic is about twenty-three degrees; Ptolemy explains the angle with P7 by saying that it is the angle of the sky's rotation with respect to the ecliptic. P9 is Ptolemy's assertion that changes in position on the surface of the earth are too small for us to see any parallax or foreground-jumping effect in the motions of the planets and stars. P6 is the main proposition of Ptolemaic astronomy, asserting that the earth was at the center of the celestial sphere.

The third book of the *Almagest* discusses the model for the sun. Ptolemy needed to present his model of the sun first, because the motions of the other planets "depend" on it, in a numerical if not a mechanical sense. From the Copernican viewpoint, the earth moves around the sun in a year; to Ptolemy, the annual motion of the sun through the zodiac had to be explained by movements of the sun and stars. Features of the other planets' motions that have a period of exactly one year suggest to us the relevance of the earth's motion, but to Ptolemy, these apparent motions were a puzzling coincidence in which some feature of the planet's motion shared the annual periodicity of the sun's. Ptolemy did not attempt to derive a single geometric configuration for all the planets, but constructed each model separately.

Ptolemy accounts for the sun's motion with a relatively simple eccentric orbit, modeled in our propositions P12 and P14. Since most of the anomalies of planetary orbits are actually caused by the relative motion of the earth and the other planet, it makes sense for the "immobile" sun to have a simple orbit in Ptolemy's system.

Ptolemy's theory of the moon is divided into two books. Book 4 uses a simple lunar model to reproduce those motions of the moon considered by Hipparchus, who also had a geocentric theory. Book 5 discusses a secondary anomaly of the moon's motion, and refines the model to account for this anomaly. Ptolemy has only praise for Hipparchus, and regards it as a great achievement that he was able to improve upon Hipparchus's model. The moon's motion is in fact fairly complicated, due to the gravitational perturbations of the sun, so reducing it to a system of circles was a difficult problem.

The set of propositions E19–E22, E24, and E25 are all accepted features of the moon's orbit, noticed by others before Ptolemy except for E25. E19 describes the daily motion of the moon apart from diurnal rotation. E20 states the first anomaly, which is a type of irregularity in the longitudinal motion of the moon; E25 is the second, smaller anomaly. E22 and E21 say that the line from the center of the moon's orbit (earth) to the perigee or apogee may point at any spot in the zodiac. NE41 and NE42 are conclusions that Copernicus believed followed from Ptolemy's lunar theory, contrary to the observations described in E41 and E42; Copernicus is careful to demonstrate that his system does not suffer these defects.

Both Ptolemy and Copernicus solve the problem raised in E21 and E22 in the same way, assuming that the nodal line of the moon's orbit rotates very slowly about the earth, completing a cycle in about nineteen years. This solution is represented by the common hypothesis PC4. Ptolemy's lunar model is set up in P16–P19 and consists of a gearwork of three circles — a large orbit called the deferent, which is eccentric to the earth; a small, unnamed circle along which the center of the deferent travels closely about the earth; and another small circle called the epicycle, upon which the moon travels and that travels along the deferent. P19S is Ptolemy's solution to one of the periodic disturbances in the moon's motion caused by the sun's gravity. Note that P17 contradicts PC2, the common hypothesis of uniform circular motion. Copernicus criticized Ptolemy for using nonuniform motions, and our model reflects Copernicus's view of Ptolemy's inconsistency. In ECHO, however, internal inconsistency is only one factor affecting the coherence of a theory.

Having developed the models for the sun and moon, Ptolemy devotes book 6 to a discussion of eclipses, presenting the methods needed for their prediction. Books 7 and 8 are concerned with the position of the fixed stars and include a catalogue of the positions of many stars. These books follow the lunar model because the lunar theory is necessary for some measurements of angular distances in the sky; they precede the books on the other planets because the measurements of the positions of the planets often refer in turn to the positions of the stars. These stellar position measurements are represented by, but not directly relevant to, the qualitative descriptions that Ptolemy gives of his evidence; thus only the qualitative propositions are modeled by ECHO.

In books 9, 10, and 11 Ptolemy develops the theory of the longitudinal motion of the planets parallel to the plane of the sun's motion; latitudinal motion perpendicular to this plane is treated separately in book 13. In Ptolemy's time, it was well known that the outer (or "superior") planets, those we now see as outside the earth's orbit, occasionally appear to retrograde, to move backwards. Ptolemy explains retrograde motion by saying that the motion of the planet on its epicycle in the "backward" direction is faster than the epicycle's motion on the deferent or main circle in the forward direction. In contrast, Copernicus explains the retrograde motion as an apparent phenomenon of the earth's passing the planet in their respective trips around the sun. Copernicus's account naturally explains why retrograde motion occurs only when the planet is in opposition (at 180 degrees of angular distance from the sun in the sky), since this is the point at which the earth "passes" the other planet.

Explanation of the inferior planets is more complicated than that of the superior planets, since the planets are harder to observe and exhibit more complicated apparent motions from the point of view of the earth. Ptolemy has to explain why they are never in opposition and seem to follow the sun in its annual orbit. Although Copernicus makes some progress by switching to a heliocentric system, his reliance on Ptolemy's faulty data makes his model needlessly complicated, especially for Mercury. Copernicus, however, can easily explain why the inferior planets are never seen in opposition, since they are inside the earth's orbit. Ptolemy instead has the inner planets curiously "following" the sun around the earth so they all share an orbital period of one year. Book 12 of the *Almagest* is devoted to predicting the positions of the planets at various significant points in their orbits: the retrogradations for the outer planets, and the "greatest elongations" or greatest angular distances from the sun for the inner planets.

In our simulation, evidence about the motions of the inferior planets falls into two groups, with the first group describing less technical features of the planets' motion. E45 states their basic motion apart from diurnal rotation, and E3 their retrograde motion. E40 concerns their positions of maximum latitude. E33 describes the fact that the inferior planets appear to follow the sun. Thus they sometimes are morning stars ("behind" the sun's eastward motion) and sometimes are evening stars (ahead of the sun), but like the sun take a year to traverse the zodiac. E35 establishes the first anomaly of the inferior planets, and E36 is an artifact of Ptolemy's faulty observations of Mercury, which were accepted by Copernicus.

The eight propositions E49–E56 all have to do with the motion in latitude of the inferior planets, one of the most complicated topics in both books. They were stated by Ptolemy and adapted nearly verbatim by Copernicus. The complicated phrasing of the propositions derives from the observational method of waiting until the effect of a certain factor was eliminated and then making an observation to see if there remained any factors not accounted for. Explanation of this evidence by Ptolemy and Copernicus in terms of epicycles and deferents required complex geometrical constructions involving three kinds of angles — deflection, inclination, and slant — that can vary in time.

The propositions P22–P26 state Ptolemy's basic model for the inferior planets, with two circles for Venus and three for Mercury. P26 has the center of Mercury's deferent moving about the sun, in addition to the deferent and epicycle. P24 maintains that the motion along the deferent is uniform not about the center of the deferent but about a newly defined point called the equant. Ptolemy believed he was still postulating uniform motion, since it was uniform at some point. Copernicus held that such motion was *not* uniform; he required the deferent to be a great circle on a (possibly imaginary) sphere with the uniform motion on the circle produced by uniform motion of the sphere. He believed that heavenly spheres made of ether rotated uniformly because it was in their nature to do so, making Ptolemy's equant points physically impossible.

The Ptolemaic propositions P27–P30 deal with the latitude evidence. In short, they assert the existence of deflection, inclination, and slant, their variations, and their particular directions for Mercury and Venus. Deflection, inclination, and slant are terms for three different orientations of a planet's epicycle with respect to its deferent. Each of these angles can appear independently of the others; in the model, separate propositions assert the appearance of each and the ways in which each varies. The level of detail presented here is an example of the way in which an abundance of qualitatively different features as presented in Ptolemy's evidence gives rise to many propositions, each describing a qualitatively different feature of the model.

The motion of the superior planets is easier to explain for Ptolemy since they are "outside" both models of the earth and sun and are easier to observe. The basic features of the motions of the superior planets are described by the following propositions. E46 is the regular motion of the planets apart from the diurnal rotation. E43, like E40, concerns the moon's maximum latitude, except that the maximum latitude does not travel through the zodiac. E30 states the nonuniform motion of the superior planets, which Kepler later explained in terms of elliptical orbits, but which Ptolemy and Copernicus explained using eccentric deferents or epicycles on deferents. E31 states the retrograde motion of the superior planets. E48 and E57–E59 again deal with latitude observations made by Ptolemy.

Propositions P33, P34, and P35 state Ptolemy's two-circle model for the superior planets, which has planets traveling on epicycles that move along eccentric deferents. P33S explains the retrograde motion of the superior planets, implying that backwards motion along the epicycle more than cancels forward motion along the deferent. P33E establishes the periodicity of retrograde motion with respect to the sun. P34 is the main Ptolemaic proposition accounting for the motion of the superior planets using the equant point. Propositions P37–P40 are Ptolemy's theory of latitude of the superior planets, using a variable deflection and inclination, but no slant.

Propositions E60–E73 describe various features of planetary motion used to identify the order of the planets in terms of distance from the sun. Included are the planets' orbital periods, which are seen in opposition, and the conditions under which they so appear. E60, E63, and E64 state that the moon appears in opposition, and has an observable parallax; they suffice to demonstrate to both Ptolemy and Copernicus that the moon is the closest celestial body. For the remaining planets, Ptolemy and Copernicus are forced to rely on more circumstantial evidence about orbital periods (E66, E70, and E71), oppositions (E67M,V, E68M,J,S, and E73M,J,S), and elongations (E69 and E72).

Both Ptolemy and Copernicus believed the common-sense principle PC10 that longer periods required larger orbits. Taken with the observed evidence relating to periods, PC10 gave them confidence in the ordering of the outer planets as Mars, Jupiter, and Saturn, represented by PC5, PC6, and PC7. PC9 represents their common belief in the proximity of the moon. It is difficult to determine how Ptolemy decided upon his ordering of Mercury, Venus, and the sun in between the moon and Mars, although the size of Venus's epicycle might suggest that it should go outside Mercury, and esthetic considerations played a role in Ptolemy's choice of using the sun as a marker between the opposing and nonopposing planets.

We have already mentioned many features of Copernicus's system by comparison with Ptolemy's, but let us now look at it in greater detail.

4. Copernicus: Hypotheses and Explanations

De Revolutionibus is the authoritative source on Copernicus's ideas, superseding the earlier *Commentariolus*. *De Revolutionibus* is self-consciously modeled after the *Almagest*. Book 1 is an introduction to Copernicus's axioms, with an extended argument for the motion of the earth. Copernicus perceives that his ideas are incompatible with Aristotelian physics, and attempts to defend them qualitatively. For example, in response to the criticism that the earth would break up under the stress of daily motion, he points out that under the Ptolemaic system the heavens rotate once daily and are larger than the earth, so they should be much more massive and hence even more subject to stress. The argument for the central hypothesis of the motion of the earth, C9 in our simulation, is physical rather than astronomical, leaving to physics the problem of separating the center of rotation (the sun) from the center of gravity (the earth).

Book 2 of *De Revolutionibus* is devoted to spherical geometry and the basic measurement tasks of astronomy. Book 3 discusses the precession of the equinoxes and the apparent motion of the sun. Copernicus explains the twenty-three-degree angle between the axis of the stars' apparent rotation (caused by the earth's rotation) and the axis of the ecliptic (or the plane in which the planets move) by making twenty-three degrees the angle between the earth's axis of rotation and the axis of the ecliptic in C5 and C8. In C7, Copernicus extends Ptolemy's proposition P9, which states that changes in position on the earth are too small to see the parallax of the stars, to the entire orbit of the earth. C8, asserting the earth's rotation on its axis, contrasts with the assertion in Ptolemy's P2 about the existence of the "primum mobile," the outermost celestial sphere that exhibits only the daily rotation we now ascribe to the earth. C9 is the main Copernican proposition, asserting the annual motion of the earth; it contradicts P6, which assumes immobility. C10 sets up part of Copernicus's model for the earth. C11 reflects his imperfect grasp of mechanics: he believed that if a rotating sphere were orbiting the sun, its axis of rotation would sweep out a cone, or draw a circle in the sky, so an extra motion is required to keep the axis pointed at the pole star. C12 is Copernicus's assertion that the sun is the real center of motion in the universe and is at the center of the heavenly sphere. C12 contradicts Ptolemy's proposition P12, since P12 asserts that the sun moves; C12 also contradicts P6, since they make different claims about what is at the center of the universe. Figure 1 shows some of the explanatory relations involving the major propositions of the Ptolemaic and the Copernican systems, based on the input in Table 1, which is a fragment of the actual input found in Appendix C.

Book 4 of *De Revolutionibus* contains Copernicus's lunar theory, beginning with his critique of Ptolemy's system. Copernicus demonstrates that his model does not share the defects of Ptolemy's and goes on to discuss the prediction of eclipses. Copernicus's model for the moon, described by C13–C15, is a two-epicycle model. PC8 is the shared proposition that the moon's orbit forms a five-degree angle with the ecliptic.

Book 5 constructs the models for motion in longitude of the five planets, working inward from Saturn, and book 6 struggles to develop the latitude theory for the five planets. Copernicus's two-circle models for the inferior planets have the planets moving along deferents whose centers move in small circles around the sun, as described in C19–C21. Propositions C29–C35 concern latitude problems. The deferents of the inferior planets are inclined to the ecliptic at a varying angle; the deferents of Mercury and Venus are continually changing their angles to the ecliptic along two axes at once, with the result that Venus is always north of the ecliptic and Mercury always to the south. The effect is much like that of a plate rocking around quickly on its rim as it settles to the floor. Copernicus prefers not to posit latitude variation through epicyclic motion, instead saving appearances by supposing that the deflections of the inferior planets oscillate, so that Venus is always above the ecliptic and Mercury is always below.

Copernicus is able to dispense with the epicycles used by Ptolemy to account for the retrograde motion of the superior planets. Instead, he simply uses eccentric deferents, as described in C22 and C23. Since C22 asserts that the deferents of the superior planets are centered at the sun, it contradicts P34, in which Ptolemy asserted the existence of equant points about which the motion of the planets was uniform. Copernicus also has a simple latitude theory based on varying deflection and described in C26–C28. It amounts to saying that the deferents of the superior planets are inclined to the ecliptic at a small angle that varies in time with the mean sun. Since Copernicus requires the planets to be in a certain position at oppositions, he in effect keys their motion to that of the earth, a relationship for which he provides no plausible mechanism. Thus both Ptolemy and Copernicus made use of "coincidental" motions; Copernicus cannot be credited with an advantage on this account.

While Copernicus was unable to provide specific distances from the sun to the planets, his system allowed him to provide *relative* distances of the planets from the sun. Thus he was able to derive statements about the ordering of the planets and generalizations such as that the planets with longer orbits were further out (PC10), which were merely claimed

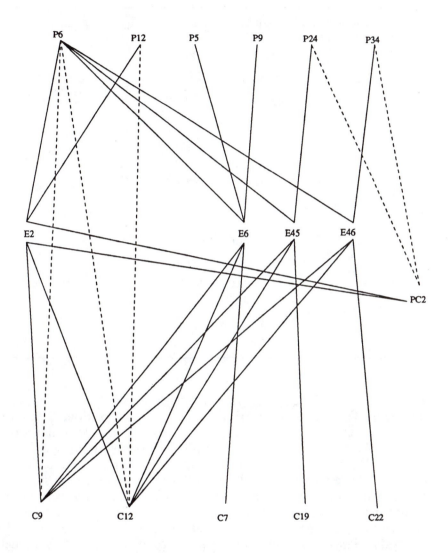

Figure 1. Explanatory relations of selected Ptolemaic and Copernican hypotheses. The solid lines indicate that the hypotheses explain the evidence, while the dotted lines link contradictory hypotheses. From the perspective of ECHO, nodes are units representing the hypotheses, the solid lines are excitatory links, and the dotted lines are inhibitory links. Coherence links between hypotheses that together explain pieces of evidence are not shown; for example, P6 coheres with P12 and PC2. This figure displays only the small part of the total ECHO network produced by the subset of the input to ECHO shown in Table 1.

Table 1. Input to ECHO concerning Propositions Shown in Figure 1

(explain '(P6 P12 PC2) 'E2)

(explain '(C9 PC2 C12) 'E2)

(explain '(P9 P6 P5) 'E6)

(explain '(C7 C9 C12) 'E6)

(explain '(P6 P24) 'E45)

(explain '(C9 C12 C19) 'E45)

(explain '(P6 P34) 'E46)

(explain '(C9 C12 C22) 'E46)

(contradict 'P6 'C12)

(contradict 'P6 'C9)

(contradict 'P12 'C12)

(contradict 'PC2 'P34)

(contradict 'PC2 'P24)

without proof by Ptolemy. Copernicus's orderings follow from his central propositions such as C9. His main ordering proposition, C39, gives the position of the earth with respect to the other planets, which alone explains the fact that only the superior planets appear in opposition, and the elongations of the inferior planets. Copernicus also claims in C41 that of the two inferior planets, Mercury is closer to the sun than Venus, which allows him to explain E69, the fact that Venus gets farther away from the sun than Mercury ever does.

Even though Copernicus thought that the ability to derive the ordering of the planets was one of the major accomplishments of his model, we must acknowledge that at present our ECHO model does not adequately represent this Copernican advantage. This omission is more a result of practical limitations upon the process of supplying ECHO with data than a consequence of any inherent limitations of TEC or ECHO. Since most of the evidence statements used by Ptolemy and Copernicus could be described in qualitative terms (e.g., E49–E56), the determination of explanatory relations was based upon evidence given in this form. Planetary distance observations, however, rest for their confirmation upon specific observations of the positions of the planets that cannot be reduced to qualitative statements. It would have been possible to use such statements in the model, but this would have introduced an asymmetry in the data: qualitative evidence statements would be used everywhere except in the section dealing with distance determinations. To preserve the uniform granularity of the evidence used by the model, we restricted ourselves to qualitative statements. A much more complicated solution would have been to transform all of the qualitative descriptions of evidence into hypotheses explaining the specific planetary observations.

5. Running Echo

The input to ECHO given in the appendixes produces the following units:

Units representing Ptolemaic hypotheses: 39.

Units representing Copernican hypotheses: 29.

Units representing hypotheses belonging to both systems: 12.

Units representing evidence propositions: 61.

Links between the units are created as follows:

Symmetric excitatory links created for explanations: 748.

Symmetric inhibitory links created for contradictions: 9.

Symmetric inhibitory links created for competition: 214.

Excitatory links are created not only between hypothesis units and the evidence units they explain, but also between units representing hypotheses that together explain a piece of evidence. Inhibitory links are created not only between units representing contradictory hypotheses, but also between units representing hypotheses that compete by virtue of the principle of competition. Figure 2 shows the competition-based inhibitory links between the units whose explanatory relations were shown in Figure 1, as well as the excitatory links that ECHO creates between hypotheses that together explain pieces of evidence. Table 2 lists the competitive propositions along with the evidence propositions that they compete to explain. The many inhibitory links produced make the simulation of Copernicus's work much better using ECHO.2 than ECHO.1, whose only inhibitory links were based on contradictions. With ECHO.1, many units representing Ptolemaic hypotheses that Copernicus undoubtedly rejected do not become deactivated.

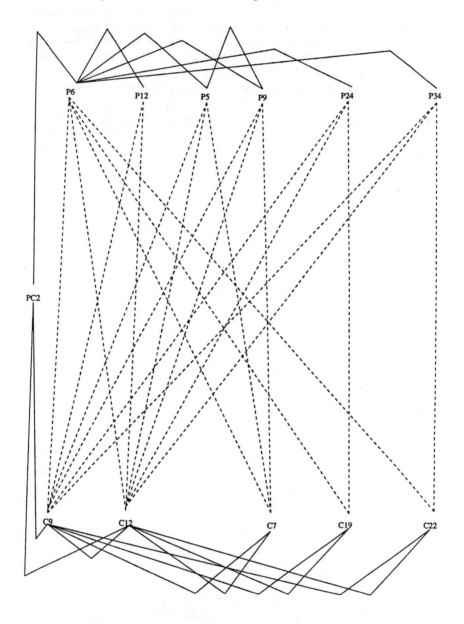

Figure 2. Excitatory and inhibitory links produced by ECHO in addition to those shown in Figure 1. The dotted lines represent inhibitory links formed in accord with the principle of competition, using the explanatory relations shown in Figure 1. (See Table 2 for a list of the competing hypotheses that indicate the evidence propositions explained.) The solid lines are the excitatory links that ECHO creates between hypotheses that together explain pieces of evidence.

hypotheses compete to explain; equations and algorithms are provided elsewhere (Thagard 1989; in press). In addition, symmetric excitatory links are created between each data unit and the special evidence unit.

After ECHO has used the input to create a constraint network, the equations given in section 1 are used repeatedly to adjust the activations of the units on the basis of their excitatory and inhibitory connections with others units. In general, Copernicus scores a decisive victory over Ptolemy, but the interesting question is why. To determine what makes ECHO judge the Copernican system to be superior to the Ptolemaic, given the input provided in the appendixes, we first performed hundreds of runs to determine whether the results were sensitive to particular parameter values. There are four potentially important parameters in ECHO:

> Excitation, the default weight of the link between units representing propositions that cohere, reduced in accord with principle 3(c) if more than one hypothesis is involved in an explanation;

> Inhibition, the default weight of the link between units representing propositions that incohere;

> Decay, the amount that the activation of each unit is decremented on each cycle of updating; and

> Data excitation, the weight of the link from the special evidence unit (whose activation is always 1) to each of the units representing pieces of evidence.

Experiments determined that the last two parameters have little effect on simulations: greater decay values tend to compress asymptotic activation values toward 0, and greater data excitation tends to make the activations of losing units higher, but there are no qualitative differences. In contrast, the relative values of excitation and inhibition are crucial. There are two ways in which the simulation can fail. The less serious occurs if the network does not settle. The more serious failure occurs if at the end of the run the Ptolemaic units are not rejected but have activations greater than 0, or even greater than the activations of the Copernican propositions; call this a "decision failure." Decision failures occur in about 25 percent of the runs with different combinations of parameters, but in every one of these cases the problem was that inhibition was set lower than excitation. So long as the value for the excitation parameter is equal to or higher than the absolute value of the inhibition parameter, the Copernican units dominate their competitors. Similarly, failures to settle in fewer than one hundred cycles of updating are rare so long as excitation exceeds inhibition. For all experiments reported

Table 2. Competition Relations Found by ECHO and Shown in Figure 2

C7 competes with P5 because of (E6 E15).

C7 competes with P6 because of (E6 E11 E12 E15).

C7 competes with P9 because of (E6 E11 E12 E15).

C9 competes with P5 because of (E6 E15 E66).

C9 competes with P6 because of (E2 E3 E6 E10 E11 E12 E13 E14 E15 E45 E46 E66 E68S E68J E68M).

C9 competes with P9 because of (E6 E11 E12 E15).

C9 competes with P12 because of (E2 E5 E10 E13 E14 E18 E25 E27 E60 E66 E68S E68cJ E68M).

C9 competes with P24 because of (E3 E45).

C9 competes with P34 because of (E30 E31 E46 E48 E59 PC5 PC6 PC7 E68S E68J E68M).

C12 competes with P5 because of (E6 E15 E66).

C12 competes with P6 because of (E2 E3 E6 E11 E12 E14 E15 E45 E46 E66).

C12 competes with P9 because of (E6 E11 E12 E15).

C12 competes with P12 because of (E2 E7 E14 E27 E66).

C12 competes with P24 because of (E3 E45).

C12 competes with P34 because of (E30 E31 E46 PC5 PC6 PC7).

C19 competes with P6 because of (E3 E45).

C19 competes with P24 because of (E3 E35 E36 E40 E45 E49 E50 E51 E52 E53 E54 E55 E56).

C22 competes with P6 because of (E46 E68S E68J E68M).

C22 competes with P34 because of (E30 E31 E43 E46 E48 E57 E58 E59 PC5 PC6 PC7 E68S E68J E68M).

The connectivity of a typical unit, C7, is shown in Figure 3, which displays the weights of excitatory and inhibitory links and the asymptotic activations of the units. The weights are not equal to the default values for excitation and inhibition because excitatory links are fractionated for simplicity reasons and inhibitory links based on competition rather than contradiction are fractionated according to the number of other hypotheses involved in the explanations of the evidence that two

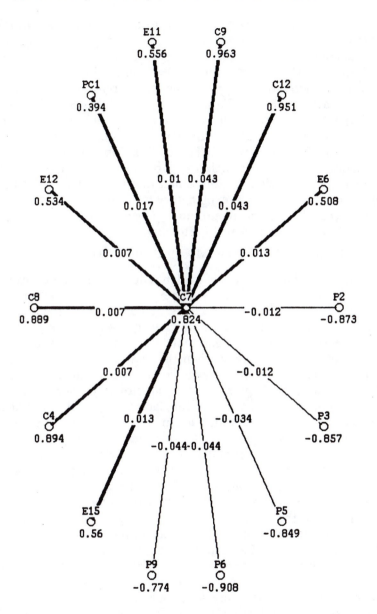

Figure 3. Connectivity of a sample unit, C7. Thick lines indicate excitatory links, while thin lines indicate inhibitory links. Numbers on lines indicate the weights of the links. Numbers under unit names are the truncated asymptotic activations of the units.

below, the parameter values were .05 for decay and data excitation, .04 for excitation, and −.06 for inhibition, but many other combinations of values could have been used as long as excitation was more intense than inhibition. With these values, the twenty-nine Copernican units have a mean activation of .84, while the thirty-nine Ptolemaic units have mean activation of −.77, after the network has settled in sixty cycles of updating. Figure 4 graphs the activations of selected Copernican and Ptolemaic hypotheses over the sixty cycles of updating.

In order to determine why ECHO prefers Copernicus to Ptolemy, we did a series of experiments deleting parts of the input and eliminating the impact of simplicity. We found first that Copernicus beats Ptolemy even when his advantage in explanatory breadth is eliminated by deleting the EXPLAIN statements providing Copernican explanations of ten evidence units not explained by Ptolemy:

E28 E33 E41 E42 E67V E67M E69 E73M E73J E73S

The first two of these pieces of evidence describe aspects of the moon's behavior that were exhibited by Copernicus's model, while Ptolemy's model entailed the negative pieces of evidence NE41 and NE42 (that the moon's size and parallax would change visibly in the course of its orbit). The three E73 statements note the fact that the superior planets only appear to retrograde when they are in opposition to the sun; for Ptolemy these were only strange coincidences, but they were a natural consequence of Copernicus's account of retrograde motion. The E67 statements give the corresponding facts that the inferior planets are never seen in opposition, which again were the result of coincidence for Ptolemy but consequences of Copernicus's theory. Similarly, E69 describes the fact that Venus attains a greater elongation from the sun than Mercury does; for Ptolemy, this is yet another coincidence independent of the preceding one, but it follows from Copernicus's ordering of the planets. E33 expresses the fact that the inferior planets also take one year (just as the sun does) to travel through the zodiac, another coincidence for Ptolemy but a consequence of the ordering of the planets for Copernicus. Finally, E28 expresses the related but independent point that Mercury and Venus are never seen in opposition, which for Copernicus was a consequence of their being inside the earth's orbit. All the other pieces of evidence that Ptolemy does not explain have to do with the motion of the planets and coincidences of their motion involving relative position of the planets and the sun. These coincidences become consequences of Copernicus's assertion that the earth takes a place among the planets in orbiting the sun. With the explanations of these ten evidence statements deleted, the mean activation of Copernican units is slightly

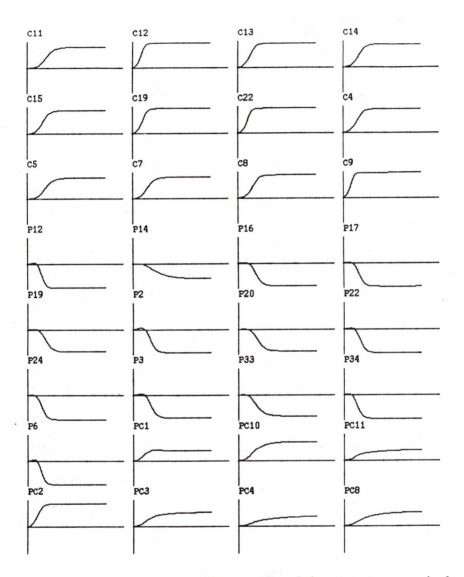

Figure 4. Activation histories of selected units. Each graph shows activation on a scale of 1 to −1, with the horizontal line indicating the starting activation of 0.

lower, but they still completely dominate the Ptolemaic units. Similarly, ECHO's judgment in favor of Copernicus was only slightly affected when we eliminated input statement concerning (1) Ptolemaic explanations of negative evidence, (2) internal contradictions in the Ptolemaic systems, or (3) Copernican explanations of the order of the planets.

Hence the Copernican advantage according to ECHO derives from greater simplicity. When we turned off the simplicity algorithm that implements principle 2(c) by reducing excitatory weights created by explanations with multiple hypotheses, ECHO gave Ptolemaic and Copernican hypotheses virtually the same mean activation. In our encoding, we did not attempt to be more concise for Copernicus than for Ptolemy, but the Copernican system needed fewer hypotheses than the Ptolemaic, twenty-nine versus thirty-nine, mainly because the representation of Copernicus used far fewer propositions to describe the motions of the planets: five as opposed to Ptolemy's nine for the superior planets, and ten as opposed to Ptolemy's seventeen for the inferior planets. Copernicus himself appreciated the simplicity of his theory:

> We therefore assert that the center of the earth, carrying the Moon's path, passes in a great circuit among the other planets in an annual revolution round the Sun; that near the Sun is the center of the universe; and that whereas the Sun is at rest, any apparent motion of the Sun can be better explained by motion of the Earth.... I think it easier to believe this than to confuse the issue by assuming a vast number of Spheres, which those who keep Earth at the center must do. We thus rather follow Nature, who producing nothing vain or superfluous often prefers to endow one cause with many effects. (quoted in Kuhn 1959, p. 179)

It is obvious that incommensurability is not a problem in the case of Copernicus vs. Ptolemy. Since Copernicus often cites Ptolemy as his authority for a qualitative description of the phenomena requiring explanation, it is clear that most of the evidence is common to both systems, and the main issue is simply which provides a better explanation. Given the Copernican triumph in our simulation, we must however ask why it took a hundred years for Copernicus's theory to be generally accepted. In brief, our answer is that our simulation considers only the comparison between the Copernican and Ptolemaic systems of astronomy. The latter system was highly coherent with the well-established principles of Aristotelian physics. Our simulation models Copernicus, who did of course accept his theory, not a typical Ptolemaic astronomer who was well aware of the fit between Ptolemy's geocentric astronomy and the Aristotelian theory of gravity. The demise of the Ptolemaic system was brought about by a host of events that followed Copernicus's theory, which underwent major modifications in the hands of Galileo, Kepler, and Newton: Galileo's discovery of the moons of Jupiter, Kepler's discovery that planetary orbits are elliptical, and especially the development of Newtonian mechanics.

6. Is Echo Necessary?

Some commentators have approved of the theory of explanatory coherence, TEC, but objected to its connectionist implementation in ECHO. Could a simple linear function compute the best explanation just as easily? Hobbs (1989) proposes a "Naive Method": the score of a theory results from subtracting the number of hypotheses in a theory, #H, from the number of pieces of evidence it explains, #E. One theory should be accepted over another if it has higher value for #E − #H.

Hobbs's Naive Method (NM) will in fact work for many cases, since it captures the fundamental aspects of the theory of explanatory coherence — that is, that theories should explain more evidence with fewer assumptions. NM would, like ECHO, prefer Copernicus to Ptolemy, since Copernicus explains more with fewer hypotheses, but NM is obviously not equivalent to ECHO. First, ECHO gives much more detailed output than NM, which evaluates theories — whole sets of hypotheses; ECHO generates an evaluation of individual hypotheses. NM accepts or rejects a set as a whole, while ECHO's output provides an activation value for each hypothesis. ECHO does not need to group hypotheses together into theories in order to evaluate them. It can reject one hypothesis that is part of an accepted theory, accept a hypothesis that is part of a rejected theory, and even reject pieces of evidence (Thagard 1989). ECHO is a more general mechanism of belief revision than NM.

Second, NM does not take into account the effect of being explained (Thagard 1989, sec. 4.2.4). ECHO prefers a hypothesis H1 explained by a higher-level hypothesis H2 to a competing, unexplained hypothesis H3. This violates NM's preference for lower #H. One might deny that a hypothesis becomes more acceptable if it is explained by a higher-level hypothesis. Yet in every domain in which explanatory inference is used, higher-level explanations can add to the acceptability of a hypothesis. In murder trials, for example, questions of motive play a major role since we naturally want an explanation of why the suspect committed the crime as well as evidence that is explained by the hypothesis that the suspect did it. ECHO shows that incorporating this element of explanatory coherence into a computational model does not create any intractable problems. In the Copernicus simulation, we saw that Copernicus could use his basic hypotheses to explain hypotheses about the ordering of the planets.

Third, unification provides an unlimited number of cases where

ECHO exhibits a preference not found in NM (Thagard 1989, sec. 4.2.6). To take one of the simplest, consider a theory T1 consisting of hypotheses H1 and H2, which are employed together to explain evidence E1 and E2. That is, H1 and H2 together explain E1, and together explain E2. The alternative explanations are H3 and H4, but H3 explains E1 alone and H4 explains E2 alone. Suppose that H1 contradicts H3, and H2 contradicts H4. T1 is more unified than the other singleton hypotheses, and ECHO indeed prefers them, despite the fact that the Naive Method calculates #E − #H as 0 in both cases.

Fourth, NM does not take into account the effect of negative evidence. Suppose H1 and H2 are competing explanations of E1, but H1 explains NE2, which contradicts evidence E2. #E − #H is the same, but in ECHO, H2 wins. In the Copernican simulation, the negative evidence concerning the size and parallax of the moon lowers the activation of several Ptolemaic hypotheses. Similarly, NM does not take into account the effects of background knowledge. Suppose one theory explains more with fewer hypotheses than a rival theory, but contains hypotheses that contradict very well-established hypotheses in another domain. For example, Velikovsky (1965) used the hypothesis that Venus passed near the earth nearly five thousand years ago to explain many historical events such as the reported parting of the Red Sea for Moses, but astronomers reject the hypothesis as inconsistent with Newton's laws of motion. Our ECHO analysis of Copernicus's argument reflects his belief that the Ptolemaic case was internally contradictory, with hypotheses P17, P24, and P34 violating PC2, the principle that the motions of heavenly bodies are uniform and circular.

Fifth, NM does not take into account the effects of analogy, which ECHO naturally models by virtue of principle 3 of TEC. Analogy does not, however, appear to play a role in Copernicus's case against Ptolemy. There are thus numerous respects in which ECHO does a more interesting computation than NM.

7. Conclusion

Let us summarize the limitations and the strengths of our application of explanatory coherence theory and ECHO to Ptolemy and Copernicus. The most serious limitation is that the input provided to the program was devised by one of us (Nowak) rather than automatically generated from the texts, and the question of historical bias naturally arises. Ideally, the input to ECHO should come from a natural language processor that analyzes scientists' arguments, or, even better, from another program that generates the hypotheses themselves in response to evidence to be explained; unfortunately, AI programs that could handle language processing or hypothesis generation for cases as complex as the work of Ptolemy and Copernicus are far in the future. Our analysis was based, however, directly on original texts, and every attempt was made to maintain historical accuracy. The result was not a "rational reconstruction" of the sort that prescribes what a scientist should have said, but is tied very closely to the actual hypotheses and evidence cited by the theorists. Our model does not, however, completely recapitulate the arguments of Ptolemy and Copernicus, since, for example, it does not completely represent Copernicus's determination of the order of the planets. Our account does not address many important questions that arise in comparing Ptolemy and Copernicus, for example concerning conceptual change (Thagard 1992, chap. 8).

Nevertheless, we have given a far more detailed description of the structure of and evidence for the Ptolemaic and Copernican theories than has yet been available in the history and philosophy of science. Moreover, despite the historical detail, we have provided algorithms for evaluating, on the basis of several factors, the comparative strengths of complex competing theories. We challenge those who prefer other models of scientific change to produce accounts of similar algorithmic precision and historical detail. The analyses of Ptolemy and Copernicus have illuminated both the historical situation and the theory of explanatory coherence. These cases are much larger than the ones to which ECHO was previously applied, but scaling up produced no problems: on a Sun 4 workstation, each simulation, including both network creation and relaxation, takes less than half a minute. Most important, this experiment provides evidence for the importance of principle C, which establishes competition between hypotheses that explain the same evidence. Without the extra inhibitory links based on competition, in addition to the ones based on overt contradiction, only parts of the Ptolemaic system are rejected. The addition of inhibitory links produced in accord with principle C makes possible a much more historically realistic simulation of Copernicus's rejection of Ptolemy. We thus have the most detailed application to date of what Thagard (1988) dubbed *computational philosophy of science:* the use of artificial intelligence techniques, in this case connectionist ones, to increase philosophical understanding of important episodes in the history of science.

Appendixes: Input to Echo for Simulation of Copernicus vs. Ptolemy

Appendix A: Evidence Propositions

Evidence relating to hypotheses about the heavens:

E1	There is a regular diurnal rotation of the sky along the celestial equator.
E2	The normal motion of the sun is to the east.
E4	The stationary point of the daily rotation of the stars is about 67 degrees from the ecliptic.
E5	The seasons are of slightly unequal length.
E6	The relative distances of the stars are fixed during revolution.
E7	Lunar eclipses happen later further to the east.
E8	The stars rise gradually later for those slightly further west.
E9	Someone traveling northward gradually loses sight of southern stars.
E10	Equinoxes always occur midway between the solstices.
E11	The heavens are always divided in half by the horizon.
E12	Six zodiacal signs are always visible.
E13	At equinoxes, the sun rises due east and sets due west.
E14	Lunar eclipses occur only when the sun and moon are in opposition.
E15	The relative position of the stars is independent of the location of the observer.
E16	There is a fixed pole star for the diurnal rotation throughout the course of a year.
E18	The sun moves from apogee to point of mean motion faster than from point of mean motion to perigee.
E26	The celestial pole moves in a slow circle around the pole of the ecliptic once every 26,000 years.
E66	The sun moves through all the signs of the zodiac in one year.

Evidence relating to hypotheses about the moon:

E19	The moon moves eastward each day in its orbit.
E20	The moon's angular velocity varies from 10 to 14 degrees per day in a one-month cycle (the first anomaly).
E21	The moon's maximum velocity can occur at any point in the ecliptic.
E22	The moon's maximum or mean latitude can occur anywhere on the ecliptic.
E24	The moon does not follow the ecliptic, but varies in latitude to 5 degrees.
E25	The moon exhibits a second anomaly called evection, which is maximum at the quadratures and 0 twice a month.
E27	The moon eclipses the sun.
NE41	The moon varies in apparent size by a factor of 4.

E41   The moon does not vary in apparent size to a large degree.

NE42   The moon's parallax increases greatly at the quadratures.

E42   The moon's parallax does not change.

E60   The moon is seen in opposition to the sun.

E64   The moon has an observable parallax.

Evidence relating to hypotheses about the superior planets:

E46   The normal motion of the superior planets is to the east.

E43   The superior planets' maximum latitudes occur at a fixed longitude.

E30   The superior planets' motion is nonuniform.

E31   The superior planets exhibit retrograde motion.

E48   The superior planets' latitudes are greater at the perigees of epicycles than at apogees.

E57   The superior planets reach their maximum northern latitude in the vicinity of the apogee of the eccentric and their maximum southern latitude in the vicinity of the perigee.

E58   When the center of the epicycle is about 90 degrees in longitude from maximum latitude, and the planet about 90 degrees of true anomaly from the apogee of the epicycle, the planet has no latitude.

E59   When the center of the epicycle is near the apogee and perigee of the eccentric, and thus near the northern and southern limits of latitude, the planet appears farther from the ecliptic at opposition in the perigee of the epicycle than near conjunction near the apogee.

E68M   Mars often appears in opposition.

E68J   Jupiter often appears in opposition.

E68S   Saturn often appears in opposition.

E73M   Mars appears in opposition only during retrograde motion.

E73J   Jupiter appears in opposition only during retrograde motion.

E73S   Saturn appears in opposition only during retrograde motion.

Evidence relating to hypotheses about the inferior planets:

E3   The inferior planets occasionally retrograde to the west.

E45   The normal motion of the inferior planets is to the east.

E40   The inferior planets' maximum latitudes occur at a fixed longitude.

E28   Mercury and Venus never appear in opposition.

E33   The inferior planets have mean motion in longitude equal to that of the sun.

E35   The inferior planets' motion in longitude is nonuniform.

E36   Mercury experiences two perigees and one apogee.

E49   When the centers of the epicycles of the inferior planets are $+/-90$ degrees from the apsidal line as seen from earth (true eccentric anomaly), the maximum latitudes above or below the ecliptic occur near the inferior and superior conjunctions, the larger latitude near the inferior conjunction.

E50   When the center of Venus's epicycle is at $+90$ degrees from the apsidal line, the latitude of Venus at inferior conjunction is south and at superior conjunction north, and at -90 degrees the reverse.

E51   When the center of Mercury's epicycle is at $+90$ degrees from the apsidal line, the latitude of Mercury at inferior conjunction is north and at superior conjunction south, and at -90 degrees the reverse.

E52   When the centers of the epicycles of the inferior planets are on the apsidal lines, they reach equal maximum latitudes on either side of the ecliptic at opposite greatest elongations.

E53   At apogee, the evening elongation of Venus is north and the morning elongation is south, and at perigee, the reverse is true.

E54   At apogee, the evening elongation of Mercury is south and the morning elongation is north, and at perigee, the reverse is true.

E55   When the center of Venus's epicycle is in the apsidal line and the planet is near inferior or superior conjunction, Venus always has a small northern latitude.

E56   When the center of Mercury's epicycle is in the apsidal line and the planet is near inferior or superior conjunction, Mercury always has a small southern latitude.

E67M   Mercury is never seen in opposition.

E67V   Venus is never seen in opposition.

E69   Venus's greatest elongation is larger than Mercury's.

Appendix B: Hypotheses in the Ptolemy-Copernicus Simulation

Hypotheses about the heavens:

PC1   The heavens are shaped like a sphere.

P2   There is a ninth heavenly sphere, the primum mobile, experiencing only diurnal rotation.

P3   The eight innermost spheres are all carried along by the primum mobile in its daily rotation.

PC2   Motions of heavenly bodies are uniform, eternal, and circular or compound-circular about the center of the universe.

P5   The stars are fixed in position on the eighth sphere, which rotates once every 26,000 years.

C4   The sphere of stars is immobile.

Hypotheses about the earth:

P6   The earth is always at the center of the heavenly sphere.

P7   The axis of the celestial diurnal rotation is inclined about 23 degrees to the ecliptic.

P9   The earth has the ratio of a point to the heavens.

PC3   The earth is spherical.

C5   The earth's axis of rotation is inclined about 23 degrees to the perpendicular to the ecliptic.

C7   The earth's orbit has the ratio of a point to the heavens.

C8   The earth rotates on its axis once a day.

C9   The earth revolves around the sun once a year, uniformly in a circle.

C10   The orbit of the earth is eccentric to the center of the sun.

C11   The earth's axis rotates with a conical motion slightly less than once a year.

Hypotheses about the sun:

P12   The sun moves eastward along a circle about the earth in one year.

P14   The sun's orbit is eccentric.

C12   The sun is immobile at the center of the universe.

PC12   Solar eclipses are caused by the moon passing between the earth and the sun.

Hypotheses about the moon:

PC4   The nodal line of the lunar orbit rotates around the center of the deferent 3 minutes of arc per day.

P16   The moon moves along an epicycle.

P17   The moon's epicycle moves eastward along a deferent about the earth.

P18   The moon's deferent is eccentric.

P19   The center of the moon's eccentric deferent revolves around the earth.

P20   The center of the moon's eccentric deferent revolves around the earth in synchronization with the sun.

PC11   Lunar eclipses are caused by the earth's shadow falling on the moon.

C13   The moon moves eastward on a deferent about the earth's center once a month.

C14   The moon's deferent is the path of its first epicycle.

C15   The first epicycle carries the moon along on a second epicycle.

PC8   The moon's deferent is inclined to the ecliptic by about 5 degrees.

Hypotheses about the inferior planets (Mercury and Venus):

P22   The inferior planets each move along an epicycle.

P23   The inferior planets travel in epicycles whose centers have the same longitude as the "mean sun."

P24   The epicycles of the inferior planets travel eastward about the earth along eccentric deferents about the earth, uniformly as seen from the center of another eccentric circle called the equant.

P25   The centers of the epicycles of the inferior planets move uniformly with respect to an equant point along the apsidal line.

P26   The center of Mercury's deferent moves in a small circle about a point on the apsidal line.

P27   The deferents of the inferior planets are inclined (deflection) to the ecliptic along a fixed nodal line.

P28   The angle of deflection of the deferents of the inferior planets varies along the nodal line.

P28E   The angle of deflection of the deferents of the inferior planets varies in phase with the motion of the epicycle center on the deferent, being 0 at the nodes and maximum at the apses.

P28V   The deflection of the deferent of Venus is always above the ecliptic; Mercury is always below.

P29   The epicycles of the inferior planets have an angle of inclination (angle of first diameter) with respect to the deferent.

P29I   The epicycles of the inferior planets have a varying angle of inclination.

P29E   The varying angle of inclination of the inferior planets is maximum at the nodes, and 0 at the apses.

P29V   The maximum of the inclination of the inferior planets is positive for Venus and negative for Mercury.

P30   The second diameter of the epicycles of the inferior planets makes an angle (slant) with their deferents.

P30I   The slant of the inferior planets varies.

P30E   The slant of the inferior planets varies in phase with the motion of the epicycle center, being 0 at the nodes and maximum at the apses.

P30V   The maximum angle of the second diameter of the epicycles of Venus is positive for Venus and negative for Mercury.

C19   The inferior planets have one major longitudinal motion, eastward along a deferent about the sun.

C20   The centers of the deferents of the inferior planets move in small circles about the sun.

C21   The small circles carrying the orbits of the inferior planets are eccentric to the sun.

C29   The deferents of the inferior planets are deflected to the ecliptic.

C31   The nodal line of the deflection of the inferior planets' deferents rotates about the sun.

C32   The deflections of the inferior planets' deferents vary in such a way that they are 0 whenever the planets cross their nodal lines.

C32V   The deflection of Venus is always to the north and the deflection of Mercury is always to the south.

C33   The inferior planets are inclined along their apsidal lines.

C34   The inclination of the inferior planets along their apsidal lines varies.

C35   The inclination of the inferior planets along their apsidal lines varies from a fixed minimum at the apses to a fixed maximum at the nodes.

Hypotheses about the superior planets (Mars, Jupiter, and Saturn)

P33   The superior planets move along epicycles.

P33E   The superior planets move along their epicycles at rates keyed to the sun.

P33S   The superior planets move along their epicycles faster than the epicycles move along the deferents.

P34   The epicycles of superior planets travel eastward along circular deferents about the earth, uniformly as seen from the center of another eccentric circle called the equant.

P35   The deferents of the superior planets are eccentric.

P37   The deferents of the superior planets have a fixed inclination to the ecliptic, with the apogee being inclined to the north.

P38   The epicycles of the superior planets are inclined to their deferents.

P39   The inclination of the epicycles of the superior planets to their deferents varies.

P40   The inclination of the epicycles of the superior planets varies so that it is maximum at the apses and minimum at the nodes, where they are parallel to the ecliptic.

C22   The superior planets have one major longitudinal motion, eastward on deferents about the sun.

C23   The deferents of the superior planets are eccentric.

C26   The deferents of the superior planets are inclined to the ecliptic.

C27   The inclination of the superior planets' deferents varies within a range of positive inclination.

C28   The inclination of the superior planets' deferents varies from maximum to minimum inclination between each true opposition and conjunction of the planet with the mean sun.

Hypotheses about the order of the planets:

PC5   Saturn is the outermost planet.

PC6   Jupiter is the next planet inward after Saturn.

PC7   Mars is the next planet inward after Jupiter.

PC9   The moon is the closest planet to the earth.

C39   Earth is inside the orbits of Mars, Jupiter, and Saturn, and outside the orbits of Mercury and Venus.

C40   Venus is the next planet out from the sun after Mercury.

C41   Mercury is the first planet counting out from the sun.

PC10   Longer planetary periods require larger orbits.

Appendix C: Explanations and Contradictions
in the Ptolemy-Copernicus Simulation

Explanations:
(explain '(P2 P3) 'E1)
(explain '(C4 C8) 'E1)
(explain '(P6 P12 PC2) 'E2)
(explain '(C9 PC2 C12) 'E2)
(explain '(P6 P22 P24 PC2) 'E3)
(explain '(C9 PC2 C12 C19) 'E3)
(explain '(P2 P3 P7) 'E4)
(explain '(C4 C5 C8) 'E4)
(explain '(P14 P12) 'E5)
(explain '(C9 C10) 'E5)
(explain '(P9 P6 P5) 'E6)
(explain '(C7 C9 C12) 'E6)
(explain '(PC3 P12 P17 PC11) 'E7)
(explain '(PC3 C12 C13 PC11) 'E7)
(explain '(PC1 P2 P3 P5 PC3) 'E8)
(explain '(PC1 PC3 C4 C8) 'E8)
(explain '(PC1 PC3 C4) 'E9)
(explain '(PC1 PC3 P5) 'E9)
(explain '(PC1 P2 P6 P7 P12) 'E10)
(explain '(C9 C5 C8) 'E10)
(explain '(PC1 P6 P9) 'E11)
(explain '(PC1 C7 C9 C12) 'E11)
(explain '(PC1 P2 P3 P6 P9) 'E12)
(explain '(PC1 C4 C7 C8 C9 C12) 'E12)
(explain '(P2 P3 P6 P7 P12) 'E13)
(explain '(C5 C8 C9 C11) 'E13)
(explain '(P6 P12 P17 PC11) 'E14)
(explain '(C9 C12 C13 PC11) 'E14)
(explain '(PC1 P5 P6 P9) 'E15)
(explain '(C7 C9 C12) 'E15)
(explain '(P2 P3 P5) 'E16)
(explain '(C4 C8 C11) 'E16)
(explain '(P12 PC2 P14) 'E18)
(explain '(C9 PC2 C10) 'E18)
(explain '(P2 P3 P5 P17) 'E19)
(explain '(C4 C13) 'E19)
(explain '(P16 P17 P18 P19 P20 P12) 'E20)
(explain '(C13 C14 C15) 'E20)
(explain '(P16 P17 P18 P19 P20 P12) 'E21)
(explain '(C13 C14 C15 PC8) 'E21)
(explain '(PC4 P16 P17 P19 PC8) 'E22)
(explain '(C13 C14 C15 PC4 PC8) 'E22)
(explain '(PC4 P16 P17 PC8 P19) 'E24)
(explain '(PC8 C13 PC4) 'E24)
(explain '(P16 P17 P18 P19 P20 P12) 'E25)
(explain '(C13 C14 C15 C9) 'E25)
(explain '(P2 P5) 'E26)
(explain '(C4 C8 C11) 'E26)
(explain '(P12 P17 PC9 PC12) 'E27)
(explain '(C9 C12 C13 PC12) 'E27)
(explain '(C9 C12 C19 C39) 'E28)
(explain '(P33 P33E PC2 P33S P34 P35 P36)
  'E30)
(explain '(C12 C22 C23 C9 PC2) 'E30)
(explain '(P33 P33E P33S PC2 P34) 'E31)
(explain '(C9 C12 C22 PC2 C39) 'E31)
(explain '(C9 PC2 C12 C19 C39) 'E33)

(explain '(P22 P23 P24 P25 PC2) 'E35)
(explain '(C19 C20 C21 PC2) 'E35)
(explain '(P22 P23 P24 P25 P26 PC2) 'E36)
(explain '(C19 C20 C21 PC2) 'E36)
(explain '(P24 PC2 P27 P28 P28E) 'E40)
(explain '(C19 C29 C31 C32 PC2) 'E40)
(explain '(P16 P17 P18 P19 P20) 'NE41)
(explain '(C13 C14 C15 PC2) 'E41)
(explain '(P16 P17 P18 P19 P20) 'NE42)
(explain '(C13 C14 C15 PC2) 'E42)
(explain '(P33 P34 P37 PC2) 'E43)
(explain '(C22 C26 PC2) 'E43)
(explain '(P6 P24) 'E45)
(explain '(C9 C12 C19) 'E45)
(explain '(P6 P34) 'E46)
(explain '(C9 C12 C22) 'E46)
(explain '(P33 P34 P37 P38 P39 P40) 'E48)
(explain '(C9 C22 C26 C27 C28) 'E48)
(explain '(P22 P24 P29 P29I P29E) 'E49)
(explain '(C19 C33 C34 C35) 'E49)
(explain '(P22 P24 P29 P29I P29E P29V) 'E50)
(explain '(C19 C33 C34 C35) 'E50)
(explain '(P22 P24 P29 P29I P29E P29V) 'E51)
(explain '(C19 C33 C34 C35) 'E51)
(explain '(P22 P24 P30 P30I P30E) 'E52)
(explain '(C19 C33 C34 C35) 'E52)
(explain '(P22 P24 P30 P30I P30E P30V) 'E53)
(explain '(C19 C33 C34 C35) 'E53)
(explain '(P22 P24 P30 P30I P30E P30V) 'E54)
(explain '(C19 C33 C34 C35) 'E54)
(explain '(P22 P24 P27 P28 P28E P28V) 'E55)
(explain '(C19 C29 C31 C32 C32V) 'E55)
(explain '(P22 P24 P27 P28 P28E P28V) 'E56)
(explain '(C19 C29 C31 C32 C32V) 'E56)
(explain '(P33 P34 P36 P35 P37) 'E57)
(explain '(C22 C23 C26 C27) 'E57)
(explain '(P34 P38 P36 P39 P40) 'E58)
(explain '(C22 C26 C27 C28) 'E58)
(explain '(P34 P36 P37 P38 P39 P40) 'E59)
(explain '(C22 C26 C27 C28 C9) 'E59)
(explain '(C13 C9) 'E60)
(explain '(P12 P16 P17) 'E60)
(explain '(C22 C9 C12) 'PC5)
(explain '(PC10 P34) 'PC5)
(explain '(C12 C22 C9) 'PC6)
(explain '(PC10 PC5 P34) 'PC6)
(explain '(C12 C22 C9) 'PC7)
(explain '(PC10 PC5 PC6 P34) 'PC7)
(explain '(C13 C19 C22) 'PC9)
(explain '(C9 C19 C22) 'PC10)
(explain '(PC9) 'E64)
(explain '(PC10 C19 C9 C12) 'C40)
(explain '(PC10 C19 C9 C12) 'C41)
(explain '(C4 C9 C12) 'E66)
(explain '(P2 P3 P5 P6 P12) 'E66)
(explain '(PC5 PC6 PC7 PC10 C40 C41) 'C39)
(explain '(C40 C41 C9 C12 C19) 'E67M)
(explain '(C40 C9 C12 C19) 'E67V)

(explain '(C22 C39 C9) 'E68S)
(explain '(P6 P12 P34) 'E68S)
(explain '(C22 C39 C9) 'E68J)
(explain '(P6 P12 P34) 'E68J)
(explain '(C22 C39 C9) 'E68M)
(explain '(P6 P12 P34) 'E68M)
(explain '(C9 C19 C40 C41) 'E69)
(explain '(C22 C12 C39 C9) 'E73S)
(explain '(C22 C12 C39 C9) 'E73J)
(explain '(C22 C12 C39 C9) 'E73M)

Contradictions:
(contradict 'E41 'NE41)
(contradict 'E42 'NE42)
(contradict 'P6 'C12)
(contradict 'P6 'C9)
(contradict 'P3 'C4)
(contradict 'P12 'C12)
(contradict 'PC2 'P17)
(contradict 'PC2 'P34)
(contradict 'PC2 'P24)

References

Copernicus, N. 1939. *Commentariolus.* In E. Rosen, ed. and trans., *Three Copernican Treatises.* New York: Columbia University Press, pp. 55–90.

———. [1543] 1978. *On the Revolutions.* Trans. E. Rosen. Baltimore: Johns Hopkins University Press.

Hobbs, J. 1989. Are Explanatory Coherence and a Connectionist Model Necessary? *Behavioral and Brain Sciences* 12:476–77.

Kuhn, T. S. 1959. *The Copernican Revolution.* New York: Random House.

Neugebauer, O. 1975. *A History of Ancient Mathematical Astronomy.* 3 vols. New York: Springer-Verlag.

Pedersen, O. 1974. *A Survey of the Almagest.* Denmark: Odense University Press.

Ptolemy. 1984. *Ptolemy's Almagest.* Trans. G. Toomer. London: Duckworth.

Swerdlow, N. M. 1973. The Derivation and First Draft of Copernicus's Planetary Theory: A Translation of the Commentariolus with Commentary. *Proceedings of the American Philosophical Society* 117, no. 6:423–512.

Swerdlow, N. M., and O. Neugebauer. 1984. *Mathematical Astronomy in Copernicus's De Revolutionibus.* 2 vols. New York: Springer-Verlag.

Thagard, P. 1988. *Computational Philosophy of Science.* Cambridge, Mass.: MIT Press.

———. 1989. Explanatory Coherence. *Behavioral and Brain Sciences* 12:435–67.

———. 1992. *Conceptual Revolutions.* Princeton, N.J.: Princeton University Press.

———. In press. The Dinosaur Debate: Explanatory Coherence and the Problem of Competing Hypotheses. In R. Cummins and J. Pollock, eds., *Philosophy and AI: The Developing Interface.* Cambridge, Mass.: MIT Press.

Thagard, P., and G. Nowak. 1988. The Explanatory Coherence of Continental Drift. In A. Fine and J. Leplin, eds., *PSA 1988.* Vol. 1. East Lansing, Mich.: Philosophy of Science Association, pp. 118–26.

———. 1990. The Conceptual Structure of the Geological Revolution. In J. Shrager and P. Langley, eds., *Computational Models of Discovery and Theory Formation.* San Mateo, Calif.: Morgan Kaufmann, pp. 727–72.

Velikovsky, I. 1965. *Worlds in Collision.* New York: Dell.

Understanding Scientific Controversies from a Computational Perspective: The Case of Latent Learning

Computational approaches to scientific reasoning have traditionally focused on attempts to simulate scientific discoveries, such as Ohm's Law (Langley et al. 1987), oxygen (Thagard 1989a), and the wave theory of sound (Thagard and Holyoak 1985), on the computer. Based on these findings, Slezak (1989) has argued that such simulations of scientific discovery refute the basic assumption of the strong programme within the sociology of scientific knowledge that scientific discoveries occur within a social context. While the computational approach to science may not refute the strong programme, it does serve to elucidate the role of cognitive processes in science. The computational approach assumes that it is possible to simulate the cognitive processes by which scientific discoveries are made. Yet, because these simulations of scientific discovery assume the inevitability of the final discovery, they tell only half of the story. Any complete cognitive theory of science should explain how cognitive processes are related both to scientific discoveries and to scientific controversies. As I have previously argued (Freedman and Smith 1985a; Gholson, Freedman, and Houts 1989), the same cognitive processes that lead to scientific advances also lead to systematic biases and errors in scientific judgment. By accounting for scientific controversies, we can increase explanatory power of the cognitive science of science.

Scientific controversies often develop over the findings of experiments that match two theories against each other. Philosophers, such as Popper (1972), have argued that crucial experiments should be used to decide between competing theories if the theories make opposite predictions of the outcome of the crucial experiment. Unfortunately, as Lakatos (1970) has noted, the history of science reveals that crucial experiments seldom decide between competing theories. According to Quine (1961), the reason that crucial experiments do not resolve the differences between theories is that hypotheses are not tested in isolation. Instead, a hypothesis is tested as part of a theory that has numerous auxiliary hypotheses associated with it. Therefore, disconfirmation of a particular hypothesis may not lead to rejection of the theory. Nevertheless, it is generally agreed that evidence can be used to evaluate rival theories.

The inability of crucial experiments to resolve differences between competing theories has led sociological accounts of scientific controversies to emphasize that evidence does not determine the outcome of debates; rather, the debate decides how to interpret the evidence (Collins 1983). Mulkay, Potter, and Yearly (1983) have argued that scientific discourse plays a central role in the resolution of scientific debates. The meaning of evidence is assumed to be contingent on the interpretive context, which varies for competing theoretical communities. Similarly, Pinch (1985) has argued that crucial experiments do not settle scientific controversies because there is no agreement over basic statements. Agreement over basic statements occurs through a process of social negotiation. Thus, for sociologists, it is the scientific community and not the evidence that settles scientific controversies.

While the cognitive science of science does not deny the social context of scientific controversies, the cognitive approach emphasizes that social action requires scientists with certain cognitive capacities (De Mey 1981; Giere 1987). Consequently, the theories and methods of cognitive science can be used to explain the characteristics of scientific controversies. Application of the computational approach to scientific controversies is particularly germane to cognitive explanations of science because the sociology of science has viewed scientific controversies as prime evidence for the social construction of scientific knowledge. A cognitive approach to scientific controversies assumes that scientists' mental processes play a crucial role in the development and resolution of scientific controversies. Specifically, scientists who advocate competing theories form different mental representations of the available evidence. Thagard (forthcoming) has recently attempted to explain the controversy over the extinction of the dinosaurs by suggesting that scientists use a separate set of hypotheses and evidence to form distinct mental representations. By providing separate sets of hypotheses and evidence for the competing dinosaur-extinction theories, Thagard (forthcoming) has demonstrated that both theories could provide a coherent explanation of the extinction of the dinosaurs. The present conception of scientific controversies differs from Thagard's approach in that a single set of hypotheses and evidence was employed. It is postulated that scientists form mental representations that include the hypotheses and evidence of the rival theory. In other words, all of the participants in a scientific debate have a corresponding set of hypotheses and evidence in their mental representations. This assumption seems reasonable because scientists normally are aware of the hypotheses and evidence of the competing theory. However, the activations and connections among these hypotheses and evidence differ in scientists advocating opposing theories. The present approach emphasizes that scientific controversies reflect the differential application of reasoning processes. Chiefly, the acceptance and rejection of competing theories depend on a number of identifiable cognitive constraints that scientists use to construct and modify their mental representations. Cognitive constraints — such as the relative weight that scientists assign to a particular hypothesis or piece of evidence, skepticism, and tolerance for competing hypotheses — will determine changes in scientists' mental representations when they are faced with a particular piece of evidence. When scientists modify these constraints, diverse mental representations of the same evidence can be constructed. Therefore, the degree to which scientists can modify these constraints will determine the course of a particular controversy.

In the present essay, the types of cognitive mechanisms that may have led many behaviorists to maintain their theories despite disconfirmatory evidence obtained by the latent-learning studies will be examined by applying Paul Thagard's (1989a) program, ECHO (Explanatory Coherence by Harmany Optimization). The latent-learning controversy pitted the behaviorists, specifically Clark Hull and Kenneth Spence, against cognitivists, Tolman and his colleagues, concerning the role of reinforcement in learning. This historical episode is especially suitable because participants generated radically different conclusions from the results of the experiments. Before providing a description of ECHO, a brief summary of the latent-learning controversy is necessary.

1. Latent-learning Study

Tolman (1932) developed the first major cognitive theory of learning. He held that behavior is purposive, that is, goal-oriented. Through exploration of their environment, organisms developed expectations (i.e., mental representations) of the spatial layout of the environment, referred to as cognitive maps. For Tolman, reinforcement was, therefore, not necessary. However, he acknowledged that reinforcement was very effective for learning because it provided powerful goals for the organism. Additionally, reinforcement can be employed to elicit previously learned responses. For most of the major behaviorists (e.g., Hull, Spence, and Skinner), reinforcement was a necessary condition for learning. Hull and other behaviorists assumed that reinforcement produced incremental changes in the strength of conditioned responses. Hull proposed that learning is a function of the number of reinforcements and that reinforcement involves reduction of a biological drive.

Tolman and Honzig (1930) designed the latent-learning task to investigate the necessity of reinforcement in conditioning. In the original study, separate groups of hungry rats were run each day, for twenty-one days, in a complex T-maze. The complex T-maze had a series of fourteen choice points, and each choice point had a correct decision that got the rat closer to the goal box and an alternate choice that led to a dead end. In order to prevent the rats from seeing down the end of each alley, screens covered the alley. Tolman and Honzig measured the number of errors, that is, the number of dead ends, that the rats made on each trial. Groups of rats received one of three experimental conditions: (1) rats that were always reinforced with food when they reached the goal box; (2) rats that were never reinforced (they were removed from the maze when they reached the goal box); and (3) rats that were reinforced only during the last half of the experiment, that is, beginning on day eleven (they were not reinforced during the first ten days). The third experimental condition was pivotal. On the one hand, Tolman predicted that a dramatic drop in the number of errors after the reinforcement was introduced on the eleventh day would be expected because he assumed that the rats had acquired information about the spatial layout of the maze. On the other hand, behaviorist, such as Hull and Spence, maintained that there would only be a slight drop in errors because learning occurred only through reinforcement. The results indicated that errors dropped dramatically in the group that was reinforced starting on the eleventh day (see Figure 1). This finding suggested that the rats had learned about the spatial layout of the maze during the first half of the experiment and that reinforcement was not, therefore, necessary.

Although this study was considered to be a "crucial experiment," most behaviorists maintained the necessity of reinforcement despite the disconfirmatory evidence that the latent-learning study provided. Furthermore, this study led to forty-eight additional experiments that attempted to determine the existence of latent learning. Of the nine direct replications of Tolman and Honzig's (1930) experiment, seven were interpreted as positive and two as negative (Hilgard and Bower 1956). One explanation of Tolman's findings was that some conditioning may have occurred during the nonreinforced trials because the rats were run in the maze hungry. Support for this hypothesis was obtained by the fact that there was a slight, although insignificant, drop in the errors in the rats that were never reinforced. In order to control for this possibility, nine experiments were conducted in which the rats were satiated to both hunger and thirst during the nonreinforced trials and then were tested during the reinforced trials under the appropriate drive. This type of latent-learning design led to seven positive findings and two negative

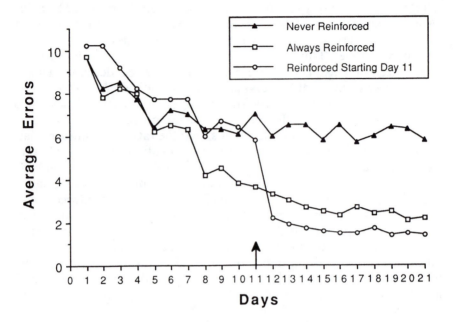

Figure 1. Evidence for latent learning (adapted from Tolman and Honzig 1930; reproduced by permission of the University of California Press).

findings (Thistlethwaite 1951). In order to determine whether rats' responses have to be associated with an appropriate biological drive, a third type of latent-learning task examined learning under a strong irrelevant drive (e.g., thirst) present during the nonreinforced trials, and then the rats were switched to a relevant drive (e.g., hunger) and reinforcer (e.g., food) on the crucial reinforced trials. On the one hand, if conditioning involves reducing some appropriate drive, then no learning would occur when rats are run under an irrelevant drive. Additionally, the rats' responses would be associated with the irrelevant drive present during the nonreinforced trials, thereby interfering with performance when the relevant reinforcement is introduced. If, on the other hand, drives are irrelevant and acquiring expectations about one's environment is the critical factor, then running rats with an irrelevant drive during the nonreinforced trials should have no effect on the performance when the relevant reinforcement is introduced. Of the eighteen studies conducted with this design, eight were positive and eleven were negative. These patterns of findings led most reviewers to conclude that under specific conditions latent learning does occur (Hilgard and Bower 1956; Thistlethwaite 1951).

Before simulating this controversy, let me acknowledge three reasons why the psychological community may have not adopted the cognitive-learning theory of Tolman and maintained the necessity of reinforcement in learning. One reason is that, at that time, Skinner's operant psychology was amassing a large amount of evidence supporting the effectiveness of schedules of reinforcement in modifying behavior. From the perspective of Laudan's (1977) and Lakatos's (1970) philosophy of science, Skinner's theory was more progressive than the theories of Tolman and Hull. Therefore, it may have been rational for psychologists to pursue the operant research programme. Indeed, Skinner's operant psychology dominated psychology throughout the 1960s and early 1970s. Second, in discussing the place-versus-response controversy that also separated the Hullians and Tolmanians, Amundson (1985) has argued that deep epistemological differences made it difficult to resolve the differences between these theories. A third reason is sociological. Freedman and Smith (1985b) have examined the differences in the training styles of Tolman, Hull, and Spence. Tolman encouraged autonomy and independence in his graduate students while Hull and Spence were strict disciplinarians who encouraged their students to pursue research programs very similar to theirs. Freedman and Smith concluded that Tolman had a substantial, albeit indirect, influence on the development of modern cognitive psychology in that his students (e.g., Garcia) conducted important research in other areas that eventually led to the demise of behaviorism. With these three caveats, let us turn to Thagard's theory of explanatory coherence.

2. Explanatory Coherence by Harmany Optimization

Thagard (1989a) assumes that hypotheses cannot be evaluated in isolation. Instead, a hypothesis is evaluated in terms of the entire set of hypotheses within which it is embedded. Moreover, the evaluation of a particular hypothesis must consider the total sum of evidence that the hypothesis explains. According to Thagard (1989a), theory change in science is assumed to be a function of explanatory coherence. Explanatory coherence is the degree to which a set of propositions or hypotheses cohere. In ECHO, scientific knowledge consists of a set of hypotheses, pieces of evidence, and the relations among them. Explanatory coherence rests on seven principles, several of which are given in List 1. The most important principle is *explanation*, which states: If a set of propositions (e.g, hypothesis P_1 through hypothesis P_m) explains another proposition (e.g., evidence Q), then each of these hypotheses coheres with evidence Q. The second part of the explanation principle establishes connections between hypotheses that explain the same evidence. The complementary principle is the *contradiction* principle, which states: If evidence Q contradicts hypothesis P, then hypothesis P and evidence Q incohere. The fourth principle, *data priority*, assumes that evidence has an acceptability of its own. ECHO is a data-driven system because it gives priority to the evidence when determining the coherence of the system. The explanation and contradiction principles are used in applying the seventh principle, the principle of explanation or system coherence. *System coherence* is a function of the pairwise local coherence of the set of propositions.

List 1
Principles of Explanatory Coherence

Principle 1. Symmetry.
  (a) If P and Q cohere, then Q and P cohere.
  (b) If P and Q incohere, then Q and P incohere.
Principle 2. Explanation.
  If $P_1 \ldots P_m$ explain Q, then:
  (a) For each P_i in $P_1 \ldots P_m$, P_i and Q cohere.
  (b) For each P_i and P_j in $P_1 \ldots P_m$, P_i and P_j cohere.
Principle 4. Data Priority.
  Propositions that describe the results of observation have a degree of acceptability on their own.
Principle 5. Contradiction.
  If P contradicts Q, then P and Q incohere.
Principle 7. System Coherence.
  The global explanatory coherence of a system S of propositions is a function of the pairwise local coherence of those propositions.

Thagard's (1988) ECHO program (Explanatory Coherence by Harmany Optimization) is a Common LISP program that applies connectionist algorithms. In some connectionist models (that is, parallel-distributed representations), the nodes often have no content. ECHO is similar to connectionist models because the explanatory coherence of the system is assumed to be a function of the global pattern of connections among the units. However, it differs from distributed representational models because each unit reflects either a hypothesis or an evidence.

When ECHO is applied to a set of propositions, each proposition will have two values associated with it: an activation and a weight. The activation level can have values ranging from +1 to −1, reflecting complete acceptance to complete rejection of the proposition. It is analogous to the amount of acceptance or rejection of the hypotheses or evidence. The weight is the connection between propositions, and it expresses the explanatory strength of the connections between the propositions. In other words, it provides an indication of how strongly some hypothesis explains or contradicts some data or some other hypothesis. When the programmer specifies that some proposition explains another proposition, an excitatory (i.e., a positively valued) link is set up between the two propositions. Likewise, if some proposition contradicts another proposition, then an inhibitory (i.e., a negatively valued) link is constructed. When ECHO is run, activation spreads from the data to the explanatory hypotheses. The system continues to adjust the weights and activation levels in order to maximize the system coherence of the entire set of propositions. Activation spreads to hypotheses that explain more of the evidence. ECHO also prefers hypotheses that explain more than their competitors. The acceptability or activation of some hypothesis depends on the other hypotheses with which it is associated. If a hypothesis is part of a set of hypotheses that explains the majority of the available evidence, then it will be less likely to be rejected. Simpler explanations are preferred; activation spreads to those hypotheses that can explain evidence with fewer cohypotheses. ECHO continues to cycle until the system reaches a maximum level of coherence, that is, until further adjustments produce no additional increase in explanatory coherence. Basically, ECHO finds the best set of activations and weights that reduces the disharmony in the system. The number of cycles that it takes to asymptote can be viewed roughly as an indication of the amount of effort it takes to maximize the coherence in the system. Again, the coherence of the system is the sum of each of the weights multiplied by the activations of the propositions associated with a weight (see below). As the system coherence increases, positively activated hypotheses can be conceived of as providing a superior explanation of the available evidence. System coherence is computed according to the following formula:

System Coherence $= \sum w_{ij} * a_i * a_j$

As will be discussed below, Thagard (1989a) has specified a number of parameters that can be manipulated. These parameters are analogous to the cognitive constraints that an individual utilizes in the formation of mental representations. Adjusting these parameters results in changes in the mental representation produced by ECHO.

Thagard (forthcoming) has recently updated ECHO (ECHO.2) to capture the competitive way in which rival hypotheses are tested. Thagard (1989a) originally viewed hypotheses as competing only if they contradicted each other. However, in ECHO.2, if two hypotheses explain the same evidence, they are assumed to be in direct competition. ECHO.2 sets up an inhibitory link between hypotheses that explain the same evidence. For example, both Tolman and Hull predict that reinforcing each response will produce a gradual response change. For Tolman, this occurs through cognition formation, while, for Hull, response change results from drive reduction and an increase in habit strength. In ECHO.2, excitation of a particular hypothesis inhibits the competing hypothesis. Because considering the competitive way in which the cognitive and behavioral theories were tested may shed some further light on the way in which cognitive processes affect the resolution of scientific controversies, the latent-learning controversy will also be simulated within ECHO.2.

The hypotheses that Hull and Tolman proposed can be translated into sets of propositions (see List 2). Tolman's views about the nature of conditioning and the role of reinforcement are translated into hypotheses T1 through T9. Central to this debate are Tolman's hypotheses T4 and T5, which concern the necessity of reinforcement and the distinction between learning and performance. Hull's general hypotheses about conditioning are included (e.g., H2: Habit strength reflects the number of reinforcements; and H4: Reinforcement involves drive reduction) as well as specific hypotheses relevant to latent learning (e.g., H5: Learning occurs only when relevant incentives are present). The findings of the experiments are depicted in List 3. The evidence, E1, E2, and E3, reflects the findings of Tolman and Honzig's (1930) experiment. Latent-learning experiments that were designed to examine the relevance of hunger drive during training are depicted by evidence E4 and E5. Notice that these findings are contradictory. Evidence E6 and E7 reflect the findings of experiments that examined the effects of running rats with an irrelevant drive during the initial trials. The next step in using ECHO is to set up explanatory links between the hypotheses and evidence. List 4 illustrates some of the explanatory and contradictory connections among hypotheses and between hypotheses and evidence. A total of ninety-two connections were specified for the simulation of the latent-learning controversy. The complete set of inputs into ECHO is listed in the Appendix.

List 2
Input Hypotheses for Tolman and Hull
for the Latent-learning Studies

Tolman's Hypotheses

T1 Conditioning involves cognition formation.

T2 Learning involves acquiring spatial representation.

T3 Cognition formation requires repetition.

T4 Reinforcement is not necessary.

T5 Learning is different than performance.

T6 Learning occurs with experience.

T7 Reinforcement produces response change.

T8 Learning is unrelated to drive state.

T9 Performance occurs only when appropriate demand is present.

Hull's Hypotheses

H1 Conditioning involves a change in habit strength.

H2 Habit strength reflects the number of reinforcements.

H3 Habit strength produces stimulus-response association.

H4 Reinforcement involves drive reduction.

H5 Learning occurs only when relevant incentives are present.

H6 Drive reduction increases response strength.

H7 Reinforcement is necessary.

3. Results

Figure 2 shows the network produced within ECHO.1 depicting Tolman's hypotheses and Hull's hypotheses and the evidence surrounding the latent-learning studies. Tolman's theory provides many more explanatory connections to the evidence, whereas Hull has a greater number of inhibitory connections. Contrary to the conclusions of most behaviorists, Tolman's hypotheses obtained greater coherency of the evidence. The explanatory coherence of the system is 4.53, which is mentioned only to provide an anchor point. Within ECHO, the coherence of a system is dependent on the number of propositions and the

---

List 3
Input Evidence for the Latent-learning Studies

Latent-learning Evidence

E1 Hungry rats in maze, always reinforced produces gradual response change.

E2 Hungry rats in maze, never reinforced produces no response change.

E3 Hungry rats in maze, reinforced after nonreinforcement produces dramatic response change.

E4 Satiated rats in maze, reinforced after nonreinforcement produces no response change.

E5 Satiated rats in maze, reinforced after nonreinforcement produces dramatic response change.

E6 Hungry rats in maze with irrelevant goal, reinforced after nonreinforcement produces no response change.

E7 Hungry rats in maze with irrelevant goal, reinforced after nonreinforcement produces dramatic response change.

---

number and type of connections among them. It should be noted that a number of the connections among Tolman's hypotheses were made during the running of ECHO. ECHO establishes connections among hypotheses if they tend to co-occur in explaining some evidence. Figure 3 depicts the activation levels for Tolman's hypotheses, Hull's hypotheses, and the evidence. Tolman's hypotheses remain highly activated (i.e., accepted) while Hull's hypotheses are negatively activated (i.e., rejected). In order to maximize the explanatory coherency of the system, the two failures to support the latent-learning phenomenon (i.e., E4 and E6) are rejected. As ECHO cycled, the activation levels of evidence E4 and E6 were initially positively activated, but as activations of Tolman's hypotheses increased the activations of this evidence decreased. Thus, ECHO shows that Tolman's cognitive theory provides a more coherent explanation of the findings of the latent-learning studies than Hullian theory.

Again, the latent-learning controversy was also simulated in ECHO.2. In ECHO.2, fourteen competing hypotheses were identified (see Appendix). ECHO.2 produces a similar pattern of activation levels with

List 4
Examples of Input Explanations and Contradictions for the Latent-learning Studies

Input Explanations and Contradictions

Explanations

T4 Reinforcement is not necessary.
  Explains
E5 Satiated rats in maze, reinforced after nonreinforcement produces dramatic response change.

H2 Habit strength reflects the number of reinforcements.
  Explains
E1 Hungry rats in maze, always reinforced produces gradual response change.

Contradictions

H5 Learning occurs only when relevant incentives are present.
  Contradicts
T8 Learning is unrelated to drive state.

H7 Reinforcement is necessary.
  Contradicts
E7 Hungry rats in maze with irrelevant goal, reinforced after nonreinforcement produces dramatic response change.

Tolman's hypotheses manifesting strong positive activations and Hull's hypotheses manifesting strong negative activations. In addition, a similar pattern of excitatory and inhibitory connections among the hypotheses and evidence is generated. However, the explanatory coherence of the system decreases to 1.4 (from 4.52 in ECHO.1), and it takes sixty-two cycles for the system to settle (versus forty-seven cycles). When ECHO.2 is implemented with the same default settings as with ECHO.1, the coherence is reduced to 0.90. Thus, when Hull's and Tolman's theories are run in a more competitive situation, Tolman's theory still provides a better account of the results of latent-learning studies, but the coherence with which Tolman can account for the findings of the latent-learning studies is reduced. The pattern of connectivity from Evidence 3 (see Figure 4), that is, the weights from that piece of evidence to the hypotheses that explain or contradict it, shows that, in ECHO.2, the explanatory connections between Tolman's hypotheses and this evidence are slightly

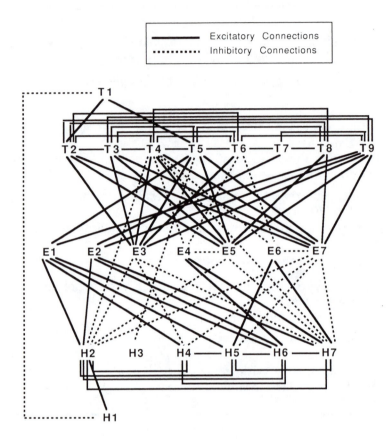

Figure 2. Network representing Tolman's hypotheses (T1–T9), Hull's hypotheses (H1–H7), and the latent-learning evidence (E1–E7).

less than with ECHO.1, and the inhibitory connections between Hull's hypotheses and this evidence are less in ECHO.1 than in ECHO.2. Thus, Tolman's explanation of latent learning is relatively weaker and the contradiction with Hull's hypotheses is less when these theories are tested competitively. Additionally, it can be seen that the positive activations for Tolman's hypotheses and negative activations for Hull's hypotheses are reduced when ECHO.2 is applied to this controversy.

The differences between ECHO.1 and ECHO.2 provide some insights into why the psychological community did not adopt Tolman's theory. Part of the answer lies in the competitive way in which these theories were tested, but this is an incomplete answer because in both contexts

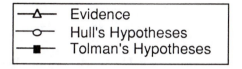

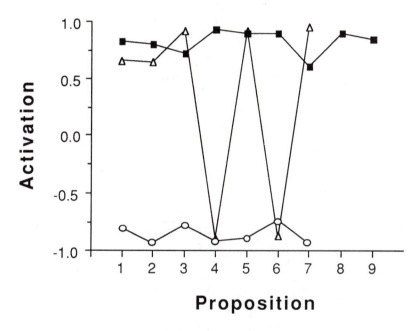

Figure 3. Activation levels for Hull's hypotheses, Tolman's hypotheses, and the latent-learning evidence.

Tolman's hypotheses remain accepted and Hull's hypotheses remain rejected. The central question remains: Why, then, didn't the psychological community adopt Tolman's theory? In order to understand this controversy, several parameters that ECHO uses to determine the coherence of the system were examined.

One way to affect the coherence of the system is to change the initial excitatory and inhibitory weights among the propositions. These parameters determine the a priori strength of the theoretical explanations of the data provided the hypotheses. If the excitatory links are set high

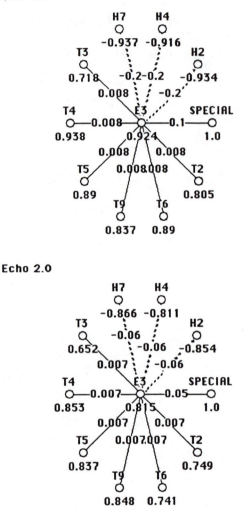

Figure 4. Pattern of connectivity for evidence E3 for ECHO.1 and ECHO.2.

relative to the inhibitory ones, it becomes difficult for contradictory evidence to overcome or refute the hypotheses with strong excitatory links. The absolute ratio of the excitatory to inhibitory settings determines the tolerance of the system. When the tolerance is high, the system will accept competing or incompatible hypotheses. When the tolerance is increased in ECHO.1, the coherence of the system decreases from the baseline of 4.52 to 0.54. It also takes the system longer to settle, requiring 124 cycles rather than forty-seven. Comparing the weights among the hypotheses and evidence under the default settings when tolerance is increased (see Figure 5), the positive activation of Tolman's hypotheses decreases and the negative activation of Hull's hypotheses decreases when the tolerance increases in ECHO.1. In ECHO.2, the coherence of the system decreases from the baseline of 1.4 to 0.95 and it takes 101 cycles for ECHO to settle. Thus, Tolman's hypotheses are less accepted and Hull's hypotheses are less rejected. Additionally, the evidence that fails to support latent learning (i.e., E4 and E6) is no longer rejected. Thus, the strength of Tolman's explanation of the latent learning is reduced when tolerance is high. If the tolerance is increased even more, both Hull's and Tolman's hypotheses remain positively activated (see Figure 6). When the tolerance is very high, the coherence drops in ECHO.2 to 0.654 and it requires 123 cycles for the system to settle. Thus, a tolerance for competing hypotheses results in a situation where both Tolman's hypotheses and Hull's hypotheses are accepted.

The relative importance of a particular piece (or pieces) of evidence can also be manipulated. ECHO allows for the fact that not all data are of equal reliability. It could be the case that Hullians did not treat the evidence supporting latent learning as being as important as the evidence that refutes latent learning. Consequently, the relative importance of the evidence E3, E5, and E7 that supports latent learning was reduced in half (0.5). Reducing the relative importance of these three pieces of evidence had no appreciable reduction in the explanatory coherence or the activation levels in ECHO.1. However, it takes relatively longer for the system to reach its state of maximal coherence (fifty-nine cycles vs. forty-seven cycles). However, unlike in ECHO.1, Hull's hypotheses remain positively activated and Tolman's hypotheses are negatively activated when the weight of evidence E3, E5, and E7 is reduced to 0.5 in ECHO.2. If the importance of the evidence that demonstrates a dramatic response change when reinforcement is introduced is reduced from 1.0 to 0.3 in both ECHO.1 and ECHO.2, a reversal in the activations results with Hull's hypotheses manifesting positive activations and Tolman's hypotheses manifesting negative activations. Figure 7 presents the activation levels when the weight of evidence E3, E5, and E7 is reduced to 0.3 in ECHO.2. Additionally, the evidence supporting latent learning is now rejected. Thus, reducing the relative importance of the evidence supporting the existence of latent learning, in a competitive environment, produces a reversal in the outcome, in that, Hull's hypotheses are accepted and Tolman's hypotheses are rejected.

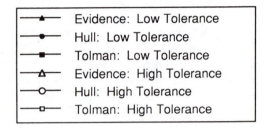

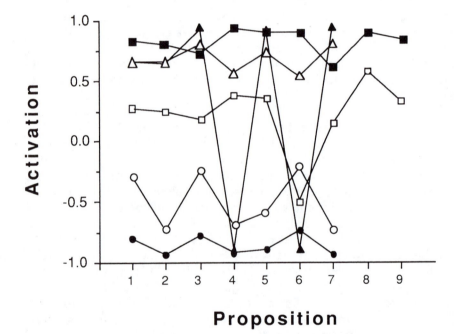

Figure 5. Activation levels for evidence, Hull's hypotheses, and Tolman's hypotheses for low- and high-tolerance conditions.

Yet the problem may not be in the evidence but may reflect the quality of the explanations of the findings. This dimension can be captured by reducing the strength of the explanatory connections between specific hypotheses and evidence. In particular, the strength of the connections between Tolman's hypotheses and the evidence supporting latent learn-

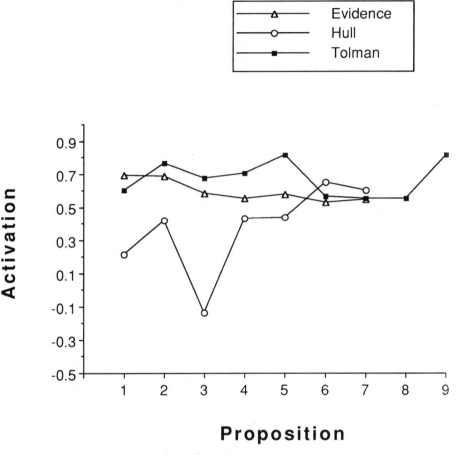

*Figure 6. Activation levels for Hull's hypotheses, Tolman's hypotheses,
and the latent-learning evidence for very high levels of tolerance.*

ing was reduced. In both ECHO.1 and ECHO.2, the modification of the
explanatory strength did not have a significant impact on the coherence
of the system but it did require seventy-two cycles to settle in ECHO.1
and seventy-three cycles to settle in ECHO.2. Thus, it takes longer to
resolve the inconsistencies in the evidence when either the evidence or
the explanations of the evidence are reduced.

The decay function, which is the rate at which activation levels of
propositions are changed, can also be adjusted. According to Thagard
(1989a), this parameter is analogous to skepticism because it reflects

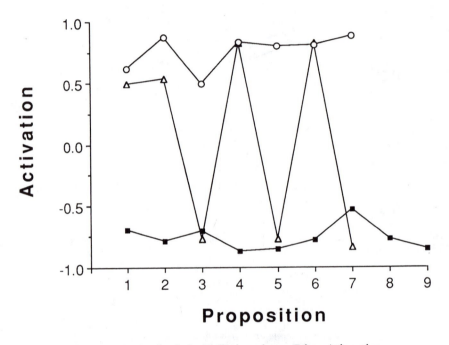

Figure 7. Activation levels for Hull's hypotheses, Tolman's hypotheses, and the latent-learning evidence when simulated in ECHO.2 with evidence #3, 5, and 7 weighted at a 0.3 level.

the amount of data activation that is required to adjust the activation of hypotheses. When the skepticism is increased, additional activation from the data is required to increase the activation of the hypotheses. Whereas tolerance deals with the treatment of contradictory hypotheses, skepticism deals with the treatment of all hypotheses. As the skepticism is increased, the coherence of the system decreases. Figure 8 depicts that increased skepticism also substantially reduces the acceptance of

Tolman's hypotheses and the rejection of Hull's hypotheses as reflected by activations that are closer to zero. Again, when the level of skepticism is high, neither theory is strongly rejected or accepted.

4. Discussion

The present application of ECHO demonstrates that modifying several cognitive constraints has a substantial impact on the activation levels, weights, and explanatory coherence of the latent-learning studies. When it assumed that Hull's and Tolman's hypotheses were tested competitively, ECHO.2 showed the coherence of the system is reduced. Acceptance of Tolman's hypotheses is reduced and the rejection of Hull's hypotheses is reduced when tolerance is increased. Moreover, when tolerance is sufficiently high, it is possible to generate a mental representation of latent-learning studies in which both Hull's and Tolman's hypotheses are accepted. Explanatory coherence decreases as the skepticism increases. Most importantly, if one reduces the importance of the findings that show that introduction of rewards after maze experience produces a dramatic response change, a mental representation of the latent-learning studies is generated in which Hull's hypotheses are accepted and Tolman's hypotheses are rejected. Thus, it is possible to generate distinct mental representations of the same hypotheses and evidence. The changes in the explanatory coherence and activation levels when these cognitive constraints are modified provide a partial explanation of the controversial nature of this historical episode. The Hullians could have weighted the evidence favoring latent learning less than the evidence refuting latent learning before adjusting the activations that are explained or contradicted by this evidence. Another possibility is that as the explanatory coherence decreases, scientists can make radically different conclusions about the findings of the available evidence.

Even though the present essay shows that different mental representations will be formed when scientists' cognitive constraints are modified, cognitive scientists still need to explain why scientists modify these constraints. It may be the case that "hot" cognitive variables (e.g., emotional and motivational variables) cause scientists to modify their cognitive constraints. Thagard (1989b) has proposed that "hot" variables could be integrated into computational approaches. According to Thagard's motivated-coherence principle, if a proposition, P, is consistent with some goal, G, then P and G cohere. Thagard suggests that motivational factors affect explanatory coherence by spreading activation to those hypotheses that are relevant to the scientist's goals. Similarly, Freedman (1990) has argued that the main goal of scientists is to enhance their

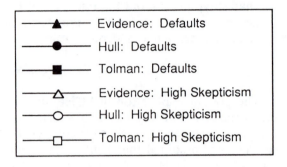

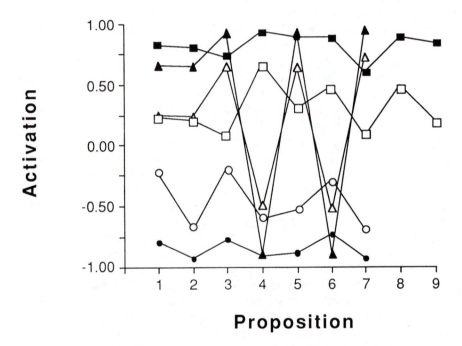

Figure 8. Activation levels for Hull's hypotheses, Tolman's hypotheses, and the latent-learning evidence when skepticism (decay = 0.25) is high and when the default settings in ECHO.1 are set.

mental representations by increasing their explanatory power. Scientists may therefore modify the types of parameters described by Thagard in order to maximize the explanatory coherence of their theories. At present, ECHO provides no mechanism for determining when the default settings should be changed. In future versions of ECHO or similar programs, the settings could be adjusted by examining specified goals. For example, if the goal is to provide an explanation of a scientific debate from one of the theoretical perspectives and disconfirming evidence exists, then ECHO could reduce the weight of this evidence. Alternately, ECHO-like programs could be programmed to search for the settings that maximize either the explanatory coherence or the activations of certain hypotheses.

From a slightly different computational perspective, Pazzani and Flowers (1990) have recently suggested that an understanding of the types of argument structures that develop during scientific debates may be useful in explaining theory formation. Argument heuristics are rules for dealing with contradictory evidence or hypotheses. When contradictions are observed, an argument heuristic could either challenge the validity of a hypothesis or evidence, develop alternative hypotheses, or accept the contradiction. These heuristics are used to resolve the differences among competing theories. Scientific debates may become controversial when the argument heuristics fail to resolve the differences among competing theories. Incorporation of argument heuristics in a computational account of scientific controversies embodies the role of debate discussed by the sociologists.

Again, the present approach is not meant to imply that sociological factors do not influence the development and resolution of scientific controversies. It may be the case that when explanatory coherence is low, the influence of sociological factors on scientific decisions may increase. Sociological factors may serve to reduce the possible dissonance created by the lack of explanatory coherence. Sociological factors may also influence the types of cognitive constraints that scientists apply to the interpretation of evidence. For example, the training styles of Hull and Tolman may have led Hull's students to have been more skeptical while Tolman's students were more tolerant of competing hypotheses. Although Thagard proposed that ECHO was meant to simulate the mental representations of individual scientists, as Wetherick (1989) asserts, Thagard's theory of explanatory coherence may be sociological to the extent that it accounts for the processes (i.e., cognitive processes) by which individual members of a scientific community accept or reject a theory.

The present simulation was also not intended to suggest that the entire psychological community or any particular member of it changed their representations based upon the mechanisms described here. Nor is the present simulation intended to show that Tolman's theory was inevitably correct or that the members of the Hullian camp were ultimately irrational. Instead, it can be argued that the types of cognitive constraints described above determined the conclusions reached by individual members of the psychological community. It is very likely that particular members of the psychological community applied these cognitive constraints differently. Applying ECHO to the latent-learning controversy demonstrates that the same cognitive mechanisms that allow scientists to make progressive theory choices can also be used to account for scientific controversies. Again, the simulation of a scientific controversy does not refute the strong programme within the sociology of knowledge, as Slezak has argued. Instead, the present simulation demonstrates that a cognitive approach can provide an alternative explanation of scientific controversies.

There are, nonetheless, several limitations of this approach. First, the data were treated as categorical; that is, either the studies were assumed to demonstrate a dramatic response change when reinforcement was introduced or they were not. In fact, the results were continuous in that there were varying degrees of response change. However, the categories used to cluster evidence together are the conclusions proposed by several psychologists (Hilgard and Bower 1956; Thistlethwaite 1951). Second, we cannot incorporate the social dimension that was a central feature of this controversy. Clearly, social factors influence scientists' mental representations. For example, Tolman's training style may have encouraged his supporters to be more tolerant of contradictory theories. Future research will need to integrate social processes into computational models. Similar to the implementation of motivational and rhetorical (i.e., argument heuristics) dimensions, it is clearly possible to include social processes in future simulations. A *social principle* that could be implemented in future versions of ECHO is: If some evidence, E, is produced by a rival theoretical camp, then reduce the weight of that evidence. Third, it could be argued that the decision to accept or reject either cognitive or behavioral theory does not depend entirely on the hypotheses and evidence that surrounded the latent-learning experiments but rather depends on the entire theoretical framework and evidence that these learning theorists developed. Both the Tolmanians and Hullians were pursuing other research programmes besides the latent-learning task. As Thagard (1989a) notes, the major limitation of ECHO is that it is dependent on the inputs provided by the programmer. In the future, I plan on including additional hypotheses and non-latent-learning evidence to investigate how this factor influences explanatory coherence. Fourth, all of the evidence from the latent-learning controversy is provided to ECHO simultaneously. However, the latent-learning controversy raged over a twenty-year period. One way to simulate this historical aspect of scientific controversies is to enter certain evidence at specific cycles of the running of the program. Finally, the present essay has not dealt with the influence of the ad hoc hypotheses that were proposed to explain the latent-learning phenomenon. Indeed, a number of ad hoc hypotheses were offered. A Guthrian hypothesis that was proposed to explain the slight although not insignificant reduction in errors on the nonreinforced trials was that removal from the maze served to protect the last response. Therefore, if the last response reduced the number of errors, this response would be more likely on future trials. Hull and Spence developed the concept of incentive motivation, the magnitude of reinforcement of reinforcement, partially in response to the findings of the latent-learning task. However, because ECHO penalizes ad hoc hypotheses, inclusion of these hypotheses would probably not help to understand why psychologists formed mental representations consistent with Hull's theories.

Besides these limitations, several other issues need to be answered before a complete computational model of scientific controversies can be developed. First, additional research is necessary to establish under what conditions these cognitive constraints are modified by scientists. Furthermore, it will need to be determined how modifiable these constraints are. Future research should also focus on the role of discourse in the scientific explanation. Computational approaches (e.g., Pazzani and Flowers 1990), like constructivist approaches (e.g., Mulkay, Potter, and Yearly 1983), presume that scientific discourse plays a central role in controversies. From a computational perspective, scientific discourse is assumed to mediate the construction and modification of scientists' representations (Freedman 1990). As scientists read the research of scientists with opposing theories, this discourse must be integrated into the scientists' mental representations. Thagard (1989a) assumes that discourse is comprehended as a function of its explanatory coherence. A complete computational model of scientific controversies will have to account for how discourse mediates the resolution of controversies. Still, to the extent that cognitive science can provide explanations of the variety of practices in which scientists engage, this research programme will be profitable.

Appendix: Inputs to ECHO.2
for the Simulation of the Latent-learning Controversy

Evidence

E1 stands for: Hungry rats in maze, always reinforced produces gradual response change.

E2 stands for: Hungry rats in maze, never reinforced produces no response change.

E3 stands for: Hungry rats in maze, reinforced after nonreinforcement produces dramatic response change.

E4 stands for: Satiated rats in maze, reinforced after nonreinforcement produces no response change.

E5 stands for: Satiated rats in maze, reinforced after nonreinforcement produces dramatic response change.

E6 stands for: Hungry rats in maze with irrelevant goal, reinforced after nonreinforcement produces no response change.

E7 stands for: Hungry rats in maze with irrelevant goal, reinforced after nonreinforcement produces dramatic response change.

Tolman's Hypotheses

T1 stands for: Conditioning involves cognition formation.

T2 stands for: Learning involves acquiring spatial representation.

T3 stands for: Cognition formation requires repetition.

T4 stands for: Reinforcement is not necessary.

T5 stands for: Learning is different than performance.

T6 stands for: Learning occurs with experience.

T7 stands for: Reinforcement produces response change.

T8 stands for: Learning is unrelated to drive state.

T9 stands for: Performance occurs only when appropriate demand is present.

Hull's Hypotheses

H1 stands for: Conditioning involves a change in habit strength.

H2 stands for: Habit strength reflects the number of reinforcements.

H3 stands for: Habit strength produces stimulus-response association.

H4 stands for: Reinforcement involves drive reduction.

H5 stands for: Learning occurs only when relevant incentives are present.

H6 stands for: Drive reduction increases response strength.

H7 stands for: Reinforcement is necessary.

Contradictions

T1 contradicts H1.

T8 contradicts H4.

T4 contradicts H2.

T4 contradicts H7.

T5 contradicts H3.

T8 contradicts H5.

T6 contradicts E4.

T6 contradicts E6.

T4 contradicts E4.

T4 contradicts E6.

H2 contradicts E3.

H4 contradicts E3.

H7 contradicts E3.

H2 contradicts E5.

H7 contradicts E5.

H2 contradicts E7.

H5 contradicts E7.

H4 contradicts E7.

H7 contradicts E7.

E4 contradicts E5.

E6 contradicts E7.

Explanations

(T1) explains T2.

(T1) explains T5.

(T2) explains T4.

(T5) explains T6.

(T5) explains T9.

(T7) explains T9.

(T5 T9) explain E1.

(T7 T9) explain E2.

(T3 T4 T5 T9 T6 T2) explain E3.

(T3 T4 T5 T8 T9 T2) explain E5.

(T3 T4 T5 T8 T9 T2) explain E7.

(H1) explains H2.

(H4) explains H6.

(H2 H4 H5 H6 H7) explain E1.

(H2 H6 H7) explain E2.

(H5 H7) explain E6.

(H6 H7) explain E4.

Data are: (E7 E6 E5 E4 E3 E2 E1).

Competing hypotheses

H2 competes with T9 because of (E1 E2).

H2 competes with T5 because of (E1).

H2 competes with T7 because of (E2).

H4 competes with T5 because of (E1).

H4 competes with T9 because of (E1).

H5 competes with T9 because of (E1).

H5 competes with T5 because of (E1).

H6 competes with T5 because of (E1).

H6 competes with T9 because of (E1 E2).

H6 competes with T7 because of (E2).

H7 competes with T9 because of (E1 E2).

H7 competes with T5 because of (E1).

H7 competes with T7 because of (E2).

T7 competes with T5 because of (T9).

Symmetric inhibitory links created for competition: 14.

Note

Portions of this essay were presented in a symposium jointly sponsored by the Society for the Social Studies of Science and the Philosophy of Science Association, Minneapolis, October 1990. A special set of thanks to Paul Thagard, who provided me with copies of ECHO, helped with debugging ECHO for use on the Macintosh, and provided insightful comments on earlier versions of this essay. Thanks also to David Gochfield for assisting me with ECHO.2.

References

Amundson, R. 1985. Psychology and epistemology: The place versus response controversy. *Cognition* 20:127–53.

Collins, H. M. 1983. An empirical relativist programme in the sociology of scientific knowledge. In K. D. Knorr-Cetina and M. Mulkay, eds., *Science observed: Perspectives on the social study of science*, pp. 85–114. Beverly Hills, Calif.: Sage.

De Mey, M. 1981. *The cognitive paradigm*. Boston: Reidel.

Freedman, E. G. 1990. Representational enhancement: The role of discourse in scientific explanations. Paper presented at the 98th Annual Meeting of the American Psychological Association, Boston, Mass., August.

Freedman, E. G., and L. D. Smith. 1985a. Implications from cognitive psychology for the philosophy of science. Paper presented at the 93rd Annual Meeting of the American Psychological Association, August.

————. 1985b. Tolman's legacy to modern cognitive psychology: A reassessment. Paper presented at the 93rd Annual Meeting of the American Psychological Association, August.

Gholson, B., E. G. Freedman, and A. C. Houts. 1989. Cognitive psychology of science: An introduction. In B. Gholson et al., eds., *Psychology of science: Contributions to metascience*, pp. 267–74. New York: Cambridge University Press.

Giere, R. N. 1987. The cognitive study of science. In N. J. Nersessian, ed., *The process of science*, pp. 139–59. Dordrecht: Martinus Nijhoff.

Hilgard, E. R., and G. H. Bower. 1956. *Theories of learning.* 2d ed. Englewood Cliffs, N.J.: Prentice-Hall.

Kulkarni, D., and H. A. Simon. 1988. The process of scientific discovery: The strategy of experimentation. *Cognitive Science* 12:139–75.

Lakatos, I. 1970. Falsification and the methodology of scientific research programs. In I. Lakatos and A. Musgrave, eds., *Criticism and the growth of knowledge,* pp. 91–196. Cambridge: Cambridge University Press.

Langley, P., et al. 1987. *Scientific discovery: Computational explorations of the creative process.* Cambridge, Mass.: MIT Press.

Laudan, L. 1977. *Progress and its problems: Towards a theory of scientific growth.* Los Angeles: University of California Press.

Mulkay, M., J. Potter, and S. Yearly. 1983. Why an analysis of scientific discourse is needed. In K. D. Knorr-Cetina and M. Mulkay, eds., *Science observed: Perspectives on the social study of science,* pp. 171–203. Beverly Hills, Calif.: Sage.

Pazzani, M. J., and M. Flowers. 1990. Scientific discovery in the layperson. In J. Shrager and P. Langley, eds., *Computational models of scientific discovery and theory formation,* pp. 403–35. Palo Alto, Calif.: Morgan Kaufmann.

Pinch, T. 1985. Theory testing in science — the case of solar neutrinos: Do crucial experiments test theories or theorists? *Philosophy of the Social Sciences* 15:167–87.

Popper, K. 1972. *Objective knowledge.* New York: Oxford University Press.

Quine, W. V. O. 1961. *From a logical point of view.* 2d ed. New York: Harper.

Slezak, P. 1989. Scientific discovery by computer as empirical refutation of the strong programme. *Social Studies of Science* 19:563–600.

Thagard, P. 1988. *Computational philosophy of science.* Cambridge, Mass.: MIT Press.

———. 1989a. Explanatory coherence. *Behavioral and Brain Sciences* 12:435–502.

———. 1989b. Scientific cognition: Hot or cold? In S. Fuller et al., eds., *The cognitive turn: Sociological and psychological perspectives on science,* pp. 71–82. Boston: Kluwer.

———. Forthcoming. The dinosaur debate: Explanatory coherence and the problem of competing hypotheses. In J. Pollock and R. Cummins, eds., *Philosophy and AI: Essays at the interface.* Cambridge, Mass.: MIT Press.

Thagard, P., and K. Holyoak. 1985. Discovering the wave theory of sound: Induction in the context of problem-solving. *Proceedings of the Ninth International Joint Conference on Artificial Intelligence,* pp. 610–12. Los Altos: Morgan Kaufmann.

Thistlethwaite, D. L. 1951. A critical review of latent learning and related experiments. *Psychological Bulletin* 48:97–129.

Tolman, E. C. 1932. *Purposive behavior in animals and men.* New York: Appleton-Century-Crofts.

Tolman, E. C., and C. H. Honzig. 1930. Introduction and removal of reward, and maze performance in rats. *University of California Publications in Psychology* 4:257–75.

Wetherick, N. E. 1989. Psychology, or sociology of science? *Behavioral and Brain Sciences* 12:489.

PART III

MODELS FROM NEUROSCIENCE

A Deeper Unity:
Some Feyerabendian Themes
in Neurocomputational Form

1. Introduction

By the late 1960s, every good materialist expected that epistemological theory would one day make explanatory contact, perhaps even a reductive contact, with a proper theory of brain function. Not even the most optimistic of us, however, expected this to happen in less than fifty years, and most would have guessed a great deal longer. And yet the time has arrived. Experimental neuroscience has revealed enough of the brain's microphysical organization, and mathematical analysis and computer simulation have revealed enough of its functional significance, that we can now address epistemological issues directly. Indeed, we are in a position to reconstruct, in neurocomputational terms, issues in the philosophy of science specifically. This is my aim in what follows.

A general accounting of the significance of neural-network theory for the philosophy of science has been published elsewhere (Churchland 1989a, 1989b). My aim here is to focus more particularly on five theses central to the philosophy of Paul Feyerabend. Those five theses are as follows.

1. Perceptual knowledge, without exception, is always an expression of some speculative framework, some *theory:* it is never ideologically neutral (Feyerabend 1958, 1962).

2. The common-sense (but still speculative) categorial framework with which we all understand our mental lives may not express the true nature of mind nor capture its causally important aspects. This common-sense framework is in principle *displaceable* by a matured materialist framework, even as the vehicle of one's spontaneous, first-person psychological judgments (Feyerabend 1963b).

3. Competing theories can be, and occasionally are, *incommensurable*, in the double sense that (1) the terms and doctrines of the

one theory find no adequate translation within the conceptual resources of the other theory, and (2) they have no logical connections to a common observational vocabulary whose accepted sentences might be used to make a reasoned empirical choice between them (Feyerabend 1962).

4. Scientific progress is at least occasionally contingent on the *proliferation* and exploration of mutually exclusive, large-scale conceptual alternatives to the dominant theory, and such alternative avenues of exploration are most needed precisely when the dominant theory has shown itself to be "empirically adequate" (Feyerabend 1963a).

5. The long-term best interests of intellectual progress require that we proliferate not only theories, but research *methodologies* as well (Feyerabend 1970).

In my experience, most philosophers still find these claims to be individually repugnant and collectively confusing. This is not particularly surprising. Each claim is in conflict with common sense and with a respectable epistemological tradition as well. Taken in isolation, and against that background, each one is bound to seem implausible, even reckless. But taken together, they form the nucleus of an alternative conception of knowledge, a serious and far-reaching conception with major virtues of its own.

Those virtues have been explored by a number of writers, most originally and most extensively by Feyerabend himself, but it is not my purpose here to review the existing arguments in support of these five themes. My purpose is to outline an entirely new line of argument, one drawn from computational neuroscience and connectionist artificial intelligence (AI). Research in these fields has recently made possible a novel conception of such notions as mental representation, knowledge, learning, conceptual framework, perceptual recognition, and explanatory understanding. Its portrayal of the kinematics and dynamics of cognitive activity differs sharply from the common-sense conception that underlies orthodox approaches to epistemology. The mere existence of such an alternative conception, one grounded in the brain's microanatomy, is sufficient to capture one's general interest. But this novel conception is of special interest in the present context because it strongly supports all five of the Feyerabendian themes listed above. It provides a unitary explanation of why all five of them are jointly correct.

The claim being made here is a fairly strong one. Just as Newtonian mechanics successfully reduced Keplerian astronomy, so does a connectionist account of cognition reduce a Feyerabendian philosophy of science. Not everything in Kepler's account survived its Newtonian reduction, and not everything in Feyerabend's account survives its neurocomputational reduction. But in both examples the parallel of principle is sufficiently striking to make the claim of intertheoretic reduction and explanatory unification appropriate. And as with the case of Kepler and Newton, the cross-theoretic parallels serve to vindicate the principles reduced, at least in their rough outlines. I begin with a summary account of the kinematical and dynamical ideas that support this explanatory reduction.

2. Neural Nets: An Elementary Account

A primary feature of neuronal organization is schematically depicted in the "neural network" of Figure 1A. The circles in the bottom row of the network represent a population of sensory neurons, such as might be found in the retina. Each of these units projects a proprietary axonal fiber toward a second population of neuronlike units, such as might be found in the lateral geniculate nucleus (LGN), a midbrain structure that is the immediate target of the optic nerve. Each axon there divides into a fan of terminal branches, so as to make a synaptic connection with every unit in the second population. Real brains are not quite so exhaustive in their connectivity, but a typical axon can make many thousands or even hundreds of thousands of connections.

This arrangement allows any unit at the input layer to have an impact on the activation levels of all, or a great many, of the units at the second or "hidden" layer. An input stimulus such as light produces some activation level in a given input unit, which then conveys a signal of proportional strength along its axon and out the end branches to the many synaptic connections onto the hidden units. These connections stimulate or inhibit the hidden units, as a function of (1) the strength of the signal, (2) the size or "weight" of each synaptic connection, and (3) its polarity. A given hidden unit simply sums the effects incident from its many input synapses. The global effect is that a pattern of activations across the set of input units produces a distinct pattern of activations across the set of hidden units. Which pattern gets produced, for a given input, is strictly determined by the configuration of synaptic weights meeting the hidden units.

The units in the second layer project in turn to a third population of units, such as might be found in the visual cortex at the back of the brain, there to make another set of synaptic connections. (In real brains this pattern typically branches, and is iterated through many

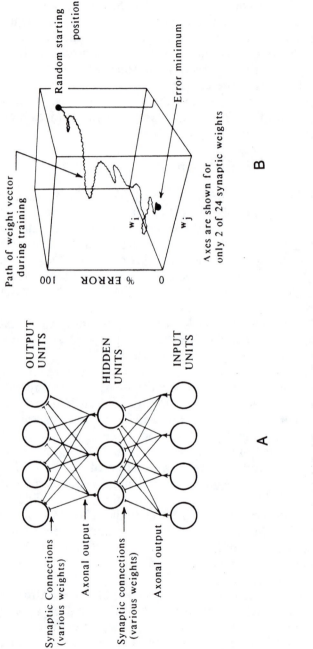

Figure 1. **A** *A simple feed-forward network;* **B** *Weight/error space: the network's journey during learning.*

layers — roughly, $5 < n < 50$ — before the chain concludes in some population of motor or other "output" neurons. Real brains also display recurrent or "feedback" pathways not shown in Figure 1A. But for purposes of illustration here, a nonbranching feed-forward network of just three layers will suffice.) In this upper half of the network also, the global effect is that an activation pattern across the hidden units produces a distinct activation pattern across the output units. As before, exactly what pattern-to-pattern transformation takes place is fixed by the configuration of synaptic weights meeting the output units.

All told, this network is a device for transforming any one of a great many possible input vectors (i.e., activation patterns) into a uniquely corresponding output vector. It is a device for computing a specific function, and exactly which function it computes is fixed by the global configuration of its synaptic weights.

Now for the payoff. There are various procedures for adjusting the weights so as to yield a network that computes almost any function — that is, any general vector-to-vector transformation — that we might desire. In fact, we can even impose on it a function we are unable to specify, so long as we can supply a modestly large set of examples of the desired input-output pairs. This process is called "training up the network."

In artificial networks, training typically proceeds by entering a sample input vector at the lowest layer, letting it propagate upwards through the network, noting the (usually erroneous) vector this produces at the topmost layer, calculating the difference between this actual output and the desired output, and then feeding this error measure into a special rule called the generalized delta rule (Rumelhart, Hinton, and Williams 1986a, 1986b). That rule then dictates a small adjustment in the antecedent configuration of all of the synaptic weights in the network. This particular learning procedure is the popular "back-propagation" algorithm. Repeating this procedure many times, over the many input-output examples in the training set, forces the network to slide down an error gradient in the abstract space that represents its possible synaptic weights (Figure 1B). The adjustments continue until the network has finally assumed a configuration of weights that does yield the appropriate outputs for all of the inputs in the training set.

To illustrate this technique with a real example, suppose we want the network to discriminate sonar echoes of large metallic objects, such as explosive mines, from sonar echoes of large submarine rocks. The discrimination of such echoes poses a serious problem because they are effectively indistinguishable by the human ear, and they vary widely in character even within each class. We begin by recording fifty different mine echoes and fifty different rock echoes, a fair sample of each. We then digitize the power profile of each echo with a frequency analyzer, and feed the resulting vector into the bank of input units (Figure 2A). We want the output units to respond with appropriate activation levels (specifically, {1, 0} for a mine; {0, 1} for a rock) when fed an echo of either kind.

The network's initial verdicts are confused and meaningless, since its synaptic weights were set at random values. But under the pressure of the weight-nudging algorithm, it gradually learns to make the desired distinction among the initial examples. Its output behavior progressively approximates the correct output vectors. Most gratifyingly, after it has mastered the echoes in the training set, it will generalize: it will reliably identify mine and rock echoes from outside its training set, echoes it has never heard before. Mine echoes, it turns out, are indeed united by some subtle weave of features, to which weave the network has become tuned during the training process. The same is true for rock echoes (see Gorman and Sejnowski 1988).

Here we have a binary discrimination between a pair of diffuse and very hard-to-define acoustic properties. Indeed, we never did define them! It is the network that has generated an appropriate internal characterization of each type of sound, fueled only by examples. If we now examine the behavior of the hidden units during discriminatory acts in the trained network, we discover that the training process has partitioned the space of possible activation vectors across the hidden units (Figure 2B). (Note that this space is not the space of Figure 1B. Figure 1B depicts the space of possible synaptic weights. Figure 2B depicts the space of possible activation vectors across the middle layer.) The training process has generated a *similarity gradient* that culminates in two "hot spots" — two rough regions that represent the range of hidden-unit vector codings for a *prototypical* mine and a *prototypical* rock. The job of the top half of the network is then just the relatively simple one of discriminating these two subvolumes of that vector space.

Some salient features of such networks beg emphasis. First, the output verdict for any input is produced very swiftly, for the computation occurs in parallel. The global computation at each layer of units is distributed among many simultaneously active processing elements: the weighted synapses and the summative cell bodies. Hence the expression "parallel distributed processing." Most strikingly, the speed of processing is entirely independent of both the number of units involved and the complexity of the function executed. Each layer could have ten units, or a hundred million; and its configuration of synaptic weights could be computing simple sums, or second-order differential equations. It

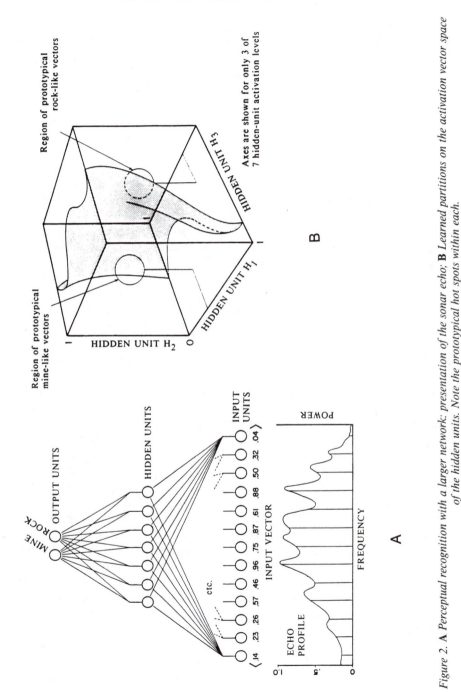

*Figure 2. A Perceptual recognition with a larger network: presentation of the sonar echo; **B** Learned partitions on the activation vector space of the hidden units. Note the prototypical hot spots within each.*

would make no difference. Speed is determined solely by the number of distinct *layers* in the network. This makes for very swift processing indeed. In a living brain, where a typical information-processing pathway has something between five and fifty layers, and each pass through that hierarchy takes something between ten and twenty milliseconds per layer, we are looking at overall processing times, even for complex recognitional problems, of between one-twentieth of a second and one second. As both experiment and common knowledge attest, this is the right range for living creatures.

Second, such networks are functionally persistent. They degrade gracefully under the scattered failure of synapses or even entire units. Since each synapse contributes such a tiny part to any computation, its demise makes an almost undetectable difference. In living creatures, the computational activity at any layer is essentially a case of multiplying an input vector by a very large matrix, where each synaptic weight represents one coefficient of that matrix (Figure 3). Since the matrix is so large — typically in excess of $10^5 \times 10^3$ elements — it might have hundreds of thousands of positive and negative coefficients revert to zero and its transformational character would change only slightly. That loss represents less than one-tenth of 1 percent of its functional coefficients. Additionally, since networks learn, they can compensate for such minor losses by adjusting the weights of the surviving synapses.

Third, the network will regularly render correct verdicts given only a degraded version or a smallish part of a familiar input vector. This is because the degraded or partial vector is relevantly *similar* to a prototypical input, and the internal coding strategy generated in the course of training is exquisitely sensitive to such similarities among possible inputs.

And exactly which similarities are those? They are whichever similarities meet the joint condition that (1) they unite some significant portion of the examples in the training set, and (2) the network managed to become tuned to them in the course of training. The point is that there are often many overlapping dimensions of similarity being individually monitored by the trained network: individually they may be modest in their effects, but if several are detected together their impact can be decisive. Here we may recall Ludwig Wittgenstein's famous description of how humans can learn, by ostension, to detect "family resemblances" that defy easy definition. Artificial neural networks recreate exactly this phenomenon.

Finally, such networks can learn functions far more complex than the one illustrated, and make discriminations far beyond the binary example portrayed. In the course of learning to produce correctly pronounced

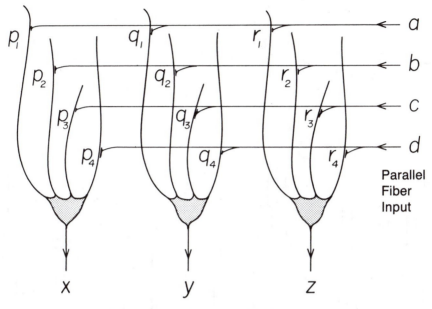

Figure 3. Schematic emphasizing the matrixlike array of synaptic connections between distinct cell populations. The input vector < a, b, c, d > is transformed into the output vector < x, y, z >.

speech (as output) in response to printed English text (as input), Rosenberg and Sejnowski's NETtalk (1987) partitioned its hidden-unit vector space into fully seventy-nine subspaces, one for each of the seventy-nine letter-to-phoneme transformations that characterize the phonetic significance of English spelling. Since there are seventy-nine distinct phonemes in English speech, but only twenty-six letters in the alphabet, each letter clearly admits of several different phonetic interpretations, the correct one being determined by context. Despite this ambiguity, the network learned to detect which of several possible transformations is the appropriate one, and it does so by being sensitive to the contextual matter of which other letters flank the target letter inside the word. All of this is a notoriously irregular matter for English spelling, but a close approximation to the correct function was learned by the network even so.

As in the mine-rock network, an analysis of the behavior of the hidden units during each of the seventy-nine learned transformations reveals an important organization. For each letter-to-phoneme transformation, of

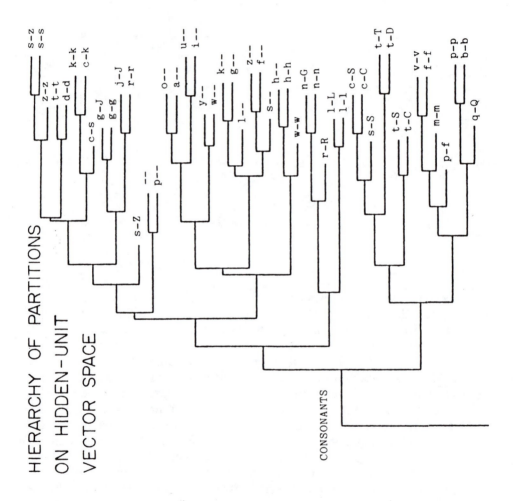

HIERARCHY OF PARTITIONS
ON HIDDEN-UNIT
VECTOR SPACE

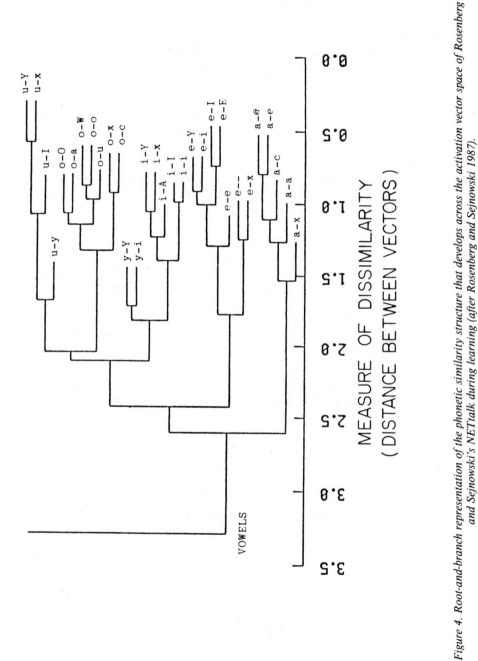

Figure 4. Root-and-branch representation of the phonetic similarity structure that develops across the activation vector space of Rosenberg and Sejnowski's NETtalk during learning (after Rosenberg and Sejnowski 1987).

course, the hidden layer displays a unique activation vector: a total of seventy-nine vectors in all. If one examines the similarity relations between these vectors in the trained network, as judged by their Euclidean proximity in the abstract activation vector space (see again Figure 2B), one discovers that the learning process has produced a treelike hierarchy of types (Figure 4). Similar sounds are grouped together and a global structure has emerged in which the deepest division is that between the consonants and the vowels. The network has spontaneously recovered, from the text on which it was trained, the phonetic structure of English speech!

Such revealing organization across the hidden-unit vector space is typical of trained networks in a great many contexts, and is a provocative feature of these machines. They partition that space into useful and well-organized *categories* relative to the functional task that they are required to perform.

Other networks have learned to identify the three-dimensional configuration and orientation of curved surfaces, given only flat gray-scale pictures of those surfaces as input. That is, they solve a version of the classic shape-from-shading problem in visual psychology (Lehky and Sejnowski 1988; 1990). Still others learn to divine the grammatical elements of sentences fed as input, or to predict the molecular folding of proteins given amino acid sequences as input, or to categorize olfactory stimuli into a hierarchical taxonomy, or to guide a jointed limb to grasp perceived objects, or to predict payment behavior from loan-application profiles. These networks perform their surprising feats of learned categorization and perceptual discrimination with only the smallest of "neuronal" resources — usually much less than 10^3 units. This is less than one-hundred-millionth of the resources available in the human brain. With such powerful cognitive effects being displayed in such modest artificial models, it is plausible that they represent a major insight into the functional significance of our own brain's microstructure.

Let us briefly contrast this approach with the rule-governed symbol-manipulation approach of classical AI. Unlike standard serial-processing, programmable computers, neural nets typically have no representation of any rules, and they do not achieve their function-computing abilities by following any rules. They simply "embody" the desired function, as opposed to calculating it by recursive application of a set of rules listed in an externally imposed program. Moreover, since neural nets perform massively parallel processing, they can be many millions or even billions of times faster than serial machines on a wide range of problems, even though they are constructed of vastly slower physical components.

A further contrast concerns the manner of information storage. In neural networks, acquired knowledge is stored in a distributed fashion: specifically, in the intricate permutational structure of the global configuration of synaptic weights, which number at least 10^{14} in a human brain. The relevant aspects of that vast store are instantly accessed by the input vectors themselves, since the weights have been configured by the learning process precisely so as to produce the appropriate activation patterns in the layer receiving that input vector. This constitutes a form of "content-addressable" memory. Given the very high-dimensional representations employed by neural nets (namely, activation vectors across large cell populations), even smallish nets can be exquisitely sensitive to subtle and hard-to-express similarities among their perceptual inputs, and to the intricate contextual features that they may contain.

This welcome feature allows a network to activate the appropriate prototype vector at the hidden layer even when the input vector is only a partial or degraded version of a typical input. The prototypical "hot spots" in the activation space of the trained hidden layer function as "attractors" into which a wide variety of partial or degraded inputs will "fall." This phenomenon allows a well-trained network to recognize instances of its categorial system even in novel or noisy circumstances, and given only partial information. In the language of philosophical theory, this means that a trained network will regularly make an ampliative "inference" to the best available "explanation" of the input phenomena. And it will do so in milliseconds.

Finally, neural nets can learn a desired function and generate a categorial system adequate to compute it, even where its makers and trainers are ignorant of both. All it needs is sufficient examples of the relevant function. These are some of the more striking rewards we gain from our modest attention to the brain's empirical architecture.

3. Epistemological Issues in Neurocomputational Guise

Let us now turn away from smallish artificial networks and refocus our attention on a large-scale biological network: the human brain. The suggestion to be explored below is that cognitive activity in the brain follows the same pattern displayed in the artificial networks. Knowledge is stored in the global configuration of the brain's synaptic weights. Learning consists in the modification of those synaptic weights according to some adjustment procedure that is somehow sensitive to successful or erroneous performance by the network. Successful configuration of the weights yields a complex and hierarchically organized set of partitions across the various subpopulations of "hidden units" scattered throughout the brain. That is, it yields one of many possible categorial or conceptual frameworks. Conceptual change consists in reconfiguring the synaptic weights so as to produce a new set of partitions across the relevant population(s) of neurons.

In humans, such categorial frameworks can clearly be of remarkable complexity, since the human brain boasts something like 10^3 neural sub-populations ("layers of hidden units") at a minimum, each of which has something like 10^8 distinct neurons. A coding vector with 10^8 elements in it can code the contents of a very large book, so we may expect the prototypes involved to characterize intricate things such as "stellar collapse" and "economic depression," as well as simple things like "raven" and "black."

Perceptual recognition consists in the activation of an appropriate prototype vector across some appropriate population of postsensory neurons. The achievement of explanatory understanding consists in exactly the same thing, although here the occasion that activates the vector need not always be sensory in character. Perceptual recognition is thus just a special case of explanatory understanding.

The preceding begins to evoke the range of epistemological material we can reconstruct in neurocomputational terms. (A more detailed and far-reaching account can be found in Churchland 1989b.) We are now prepared to address the claim that motivated this essay, the claim that five salient themes of Paul Feyerabend's philosophy of science are a natural consequence of the neurocomputational perspective.

3.1. On the Theory Ladenness of All Perception

The argument here is about as brief and as decisive as it could be. Perception is of course more than mere peripheral transduction: it is a cognitive achievement. But on the model of cognition outlined above, no cognitive activity whatever takes place without the relevant input vectors passing through the complex filter of a large set of synaptic weights (see again Figure 1A). Most importantly, any configuration of synaptic weights dictates a specific set of partitions on the activation space of the postsensory neurons to which they connect. And that set of partitions constitutes a specific conceptual framework or theory, one of many millions of possible alternative frameworks.

Any activation pattern produced across the relevant population of hidden units is thus a point in an antecedently existing space, a space with antecedently prepared similarity gradients and antecedently prepared partitions having an antecedently prepared significance for subsequent populations of neurons in the processing hierarchy. That antecedent framework and the configuration of weights that dictates it represent whatever "knowledge" the network has accumulated during past training. That framework may be well trained and finely tuned, or it may be uninstructed and inchoate. But whichever it is, no cognitive activity takes place save as the input vectors pass through that *speculative* configuration of synaptic connections, that *theory*. Theory ladenness thus emerges not as an unwelcome and accidental blight on what would otherwise be a neutral cognitive achievement, but rather as that which makes processing activity genuinely *cognitive* in the first place.

From this perspective it is evident that the process of learning about the world is not just the process of learning which general beliefs to embrace, as guided by our neutral perceptual judgments. It is also a process of learning how most usefully and penetratingly to *perceive* the world, for there is just as much room for conceptual variation and conceptual exploration at the perceptual level as there is at any other level of knowledge.

The basic point to emphasize is that, since there are almost endlessly many different possible observational frameworks (that is, hidden-layer weight configurations), where the choice between *them* is *also* an epistemic decision, there can be no question of grounding all epistemic decisions in some neutral observation framework. There is no such framework, and epistemic decisions are not made by reference to its contents in any case. One can certainly regard the unprocessed activation vectors at the sensory-input layer as theoretically neutral, but those epithelial activation vectors are not themselves propositional attitudes, they are not truth-valuable, and they stand in no logical relations to anything. Their impact on subsequent activity is causal, not logical. Human knowledge thus has *causal* "foundations," but it has no *epistemic* foundations.

3.2. On Displacing Folk Psychology

Given the model of cognition outlined above, any conceptual framework whatever is a speculative attempt to process incoming vectors in a way that is useful to the network, and it is subject to modification or replacement as a function of whatever pressures are exerted by the network's learning algorithm. The system of partitions that constitutes one's "folk" conception of mental reality is no exception. It is a learned framework whose purpose is to render intelligible both the introspectible reality of one's own case and the continuing behavior of people in general. A suitable regime of training should be able to produce any one of a large variety of alternative conceptions (indeed, even the "folk" conception is nonuniform across cultures, and across individuals).

The idea of embracing an alternative to folk psychology was never very compelling so long as we could not even point toward a plausible alternative conception. But now we can. The neurocomputational framework of the preceding pages portrays cognitive representations as high-dimensional activation vectors, rather than as sentential or propositional attitudes. And it portrays cognitive activity as the synapse-driven transformation of vectors into other vectors, rather than as the rule-governed drawing of inferences from one proposition to another. It presents a fundamentally novel kinematics and dynamics of cognitive activity. It is not yet sufficiently developed for a general transfer of allegiance to take place. But it does hold promise of being descriptively and explanatorily superior to current folk psychology, by a wide margin, and it already presents real opportunities for first-person use.

One class of such opportunities concerns the various subjective sensory "qualia" that have so often been held up as paradigm examples of what materialism can never hope to explicate. A specific color quale emerges as a specific activation vector in a three (or four) dimensional space whose axes correspond to the three types of retinal cones (and perhaps also the rods). A taste quale emerges as a four-element activation vector in a four-dimensional space whose axes correspond to the four types of taste sensors in the mouth. Auditory qualia emerge as more variable vectors whose elements correspond to the places on the cochlea whose natural frequency corresponds to one element in the complex incoming sound. The dimensionality of these qualia is relatively low, and thus their internal structure is potentially learnable and reportable in detail, just as the structure of musical chords is learnable and reportable. As in the musical case, there is also an increased insight into the structure of and the relations within the apprehended domain. One has therefore mastered more than just an esoteric set of labels: one has increased one's understanding of the phenomena.

Qualia are peripheral phenomena, to be sure, and complexity goes up as we ascend the processing hierarchy. It remains to be seen how the story will go in the case of cognitive processing at the level of systematic linguistic activity. Perhaps the familiar propositional attitudes will be smoothly reduced by the computational structures we find there, and perhaps they will simply be eliminated from our scientific ontology because nothing of dynamical importance in the brain answers to them. But whichever is their fate — reduction by something superior, or elimination by something superior — the categories of folk psychology remain displaceable in favor of some more penetrating categorial framework. The only real question is how large the doctrinal and ontological gap will turn out to be, between the triumphant new framework and its poorly informed historical predecessor.

3.3. On Incommensurable Alternatives

Consider a typical brain subpopulation of something like 10^8 neurons. Its abstract activation space will have 10^8 dimensions. Clearly a space of such high dimensionality can support an extraordinarily intricate hierarchical system of similarity gradients and partitions across that space. Equally clear is the commensurately great *variety* of such partitional configurations possible with such generous resources. Now the demand that all possible conceptual frameworks must be somehow translatable into our current conceptual framework is just the demand that each and every one of the billions of possible configurations just alluded to must stand in some equivalence relation to our current configuration. But there is not a reason in the world to think that there is any such relation that unites this vast diversity of frameworks: not in their internal structure, or in their relations to the external world, or in the input-output functions they sustain. On the contrary, they are all in competition with one another, in the sense that they are mutually incompatible configurations of the same activation vector space.

The prospect of widespread incommensurability is unsettling to many philosophers because it threatens to make a reasoned empirical choice between competing theoretical frameworks impossible. The real threat, however, is not to the possibility of rational empirical choice, but to a deeply entrenched *theory* of what rational empirical choice consists in. That superannuated theory requires a relevantly neutral observation vocabulary, among whose sentences the competing theories at issue have different logical consequences. So long as one embraces that superannuated theory, one will perceive incommensurability as a threat to reason and objectivity. But once one puts that theory aside, one can get down to the serious business of exploring how empirical data *really* steer our theoretical commitments.

On the model of cognition here being explored, ongoing learning consists in the continual readjustment of the value-configuration of one's myriad synaptic weights. Exactly what factors drive such readjustments in the human brain is currently the focus of much research, but the familiar philosopher's story about sets of sentences being accepted or rejected as a function of their logical relations with other sentences plays no detectable role in that research, and no detectable role in the brain's activity either. Instead, synaptic change appears to be driven by such factors as local increases in presynaptic or postsynaptic activity (post-tetanic potentiation), by temporal correlations or anticorrelations between the activity reaching a given synapse and the activity reaching other synaptic connections onto the same postsynaptic neuron

(Hebbean learning), by the mutual accommodation of synaptic values under specific global constraints (Boltzmann learning), perhaps by the return distribution of conflict messages (back propagation), and by other decidedly preconceptual or subconceptual processes.

None of this precludes the possible relevance of sentential and logical factors for some cases of learning at some high level of processing, but it does undermine the parochial view that all or even most of human learning must be captured in those terms. And it therefore frees us forever from the short-sighted objection that incommensurable alternatives would make objective learning impossible. Learning then proceeds, as it *usually* does, by other than "classical" means. This is good, because incommensurable alternatives are both possible and actual. They are also welcome, since "commensurability" is just a measure of the similarity between alternative frameworks, and sometimes what the epistemic situation requires is a profoundly different perspective on the world. Which brings us to the next theme.

3.4. On Proliferating Theories

Feyerabend's argument (1963a) for the wisdom of proliferating theories is a very striking one. He points out that important empirical facts can often be quite properly dismissed, as unrevealing noise or intractable chaos, when viewed from within one conceptual framework, while those "same" empirical facts appear as tractable, revealing, and as decisively incompatible with the first framework when they are viewed from within a second conceptual framework. His illustrative example is the empirical phenomenon of Brownian motion, which constitutes a perpetual motion machine of the second kind and is thus incompatible with the second law of classical thermodynamics. However, its status as such was not appreciated, and could not be fully appreciated, until the relevant details of Brownian motion were brought into clear focus by the new and very different kinetic theory of heat. What had been an exceedingly minor and opaque curiosity then emerged as a major and unusually revealing experimental phenomenon, one that refuted the classical second law.

Feyerabend has been criticized for overstating the case here. Laymon (1977) insists that Brownian motion could have been, and to some extent actually was, recognized as a problem for classical thermodynamics in advance of its successful analysis by the kinetic theory. Laymon may have a point, but I think his resistance here is a quibble (see also Couvalis 1988). Whatever minor worries might have been brewing in a few isolated breasts, the fact remains that the kinetic theory transformed our conception of Brownian motion, and made salient certain of its experimentally accessible features that otherwise might never have risen to consciousness. It is not true that all empirical facts are equally accessible, nor that their significance is equally evident, independently of the conceptual framework one brings to the experimental situation. This is all one needs to justify the proliferation of theories. And the case of Brownian motion remains a striking example of this important lesson.

From a neurocomputational perspective, this lesson is doubly clear. Anyone who has spent idle time watching a Hinton diagram evolve during the training of a neural network will have noticed that networks often persist in ignoring or in outright misinterpreting salient data until they have escaped the early and relatively benighted conceptual configuration into which the learning algorithm initially pushed them. They persist in such behavior until they have assumed a more penetrating conceptual configuration, one that responds properly to the ambiguous data. (A Hinton diagram is a raster-like display of all of the synaptic-weight values of the network being trained. These displayed values are updated after each presentation of a training example and consequent modification of the weights. Accordingly, one can watch the weights evolve under the steady pressure of the training examples.) If the proper final configuration of weights happens already to be known from prior training runs, one can even watch the "progress" of the weights as they collectively inch toward their optimal configuration.

What is striking is that for some problems (the exclusive OR function, for example), some specific weights regularly start off by evolving in exactly the *wrong* direction — they become more and more strongly negative, for example, when their proper final value should be strongly positive. One may find oneself yelling at the screen "No! This way! Over this way!" as the early network persists in giving erroneous outputs, and the wayward weights persist in evolving in the wrong direction. These frustrations abate only when the other weights in the network have evolved to a configuration that finally allows the network's learning algorithm to appreciate the various examples it has been "mishandling," and to pull the miscreant weights toward more useful values. The proper appreciation of some of the training data, to summarize the point, is sometimes impossible without a move to a different conceptual configuration.

This example illustrates that the moving point in weight-error space (see again Figure 1B) is often obliged to take a highly circuitous path in following the local error gradient downwards in hopes of finding a global error minimum. That path may well go through points that preclude both a decent output from the network and a proper lesson from the learning algorithm for at least some of the student weights. Only when the network reaches a subsequent point can these defects be repaired.

A more dramatic example of this empirical blindness occurs when, as occasionally happens, the evolving weight-space point gets caught in a purely "local minimum," that is, in a cul-de-sac in weight-error space in which the network is still producing somewhat erroneous outputs, but where every relatively small change in the synaptic weights produces an *increase* in the error measured at the network's output layer. For any learning algorithm that moves the weight-space point in small increments only, the network will be permanently stuck at that point. So far as it is concerned, it has achieved the "best possible" theory.

In order to escape such an epistemic predicament (and occasional entrapments are inevitable), we need a learning algorithm that at least occasionally requires the network to make a relatively *large* jump: a jump to a significantly different portion of synaptic weight space, to a significantly different conceptual configuration. From that new weight-space point, the network may then evolve quickly toward new achievements in error reduction.

It is evident that for some global minima and some starting points, *you can't get there from here*, at least not by small increments of instruction. This is a clear argument for the wisdom of a learning strategy that at least occasionally exploits multiple starting points, or discontinuous shifts, in the attempt to find a descending path toward a genuinely global error minimum. It may be difficult to achieve such diversity in a single individual (but it is certainly not impossible; see Churchland 1989b, chap. 11). But it can readily be done with different individuals in the same scientific community. And of course it is done. That is the point of different "schools."

These considerations do not resolve the essentially political conflict between Feyerabend and Thomas Kuhn concerning how *much* of our resources to put into proliferation and how much into pursuing a single but highly progressive "paradigm." But it does mean that a wise research policy must recognize the need for striking, and endlessly restriking, a useful balance between these two opposing tensions. Proliferation is a desideratum that will never go away, because the prospect of a false but compelling local error minimum is a threat that will never go away, and because complacency is endemic to the human soul.

3.5. On Proliferating Methodologies

The Feyerabend I have in mind here is of course the Feyerabend of "Against Method" (1970), in which he recommends an opportunistic anarchism, constrained only by the innate organization of the human nervous system, as a more promising policy in guiding our scientific behavior than is any of the methodological straitjackets so far articulated by scientists and philosophers of science. In a climate of methodological stories benighted by their formulation in logico-linguistic terms, this is certainly good advice. But it need not always be good advice: someday, perhaps, our acquired methodological wisdom may equal or surpass the innate wisdom of a healthy nervous system, because we have figured out how the nervous system works, and can see how to make it work even better.

This is not a vain hope. Guided by a variety of nonclassical learning algorithms, artificial neural networks have recently proved capable of some astonishing feats of knowledge acquisition, feats that represent a quantum leap over any of the classical logico-linguistic achievements. A new door has opened in normative epistemology, and it concerns the comparative virtues and capabilities of alternative learning algorithms, algorithms aimed not at adjusting sets of propositions so as to meet certain criteria of consistency or coherence, but aimed rather at adjusting iterated populations of synaptic weights so as to approximate certain input-output functions or certain dynamical behaviors. What is striking, even at this early stage of exploration, is that the space of possible learning algorithms is enormous. In the newly developed research program called "connectionist AI," almost as much research time is spent on critically exploring the diverse properties of various existing learning algorithms, and on devising and exploring new ones, as is spent on the properties of trained networks themselves (see Hinton 1989).

This is a healthy situation, and such proliferation should be encouraged. There are at least two major reasons for this. The first concerns the relatively limited aim of trying to understand how the human brain conducts its epistemic affairs. We need to explore the space of possible learning algorithms until we discover which specific place in it corresponds to the brain's mode of operation.

The second reason is deeper. Even supposing we succeed in identifying the brain's place in that space, there is no reason to suppose that our biologically innate learning algorithm is the best possible algorithm, or even that there *exists* a uniquely best learning algorithm. Perhaps they just get better and better, ad infinitum, which means that we must explore them indefinitely. Or perhaps they radiate, along diverse dimensions of distinct virtue, to be explored as our changing needs dictate. The proliferation of learning algorithms is a virtuous policy of long-term science for much the same reasons that proliferation is a virtue in the case of theories. The alternatives are certainly there, and we will not appreciate their virtues unless we explore them.

This may place unreasonable demands on the human nervous system, since presumably it is insufficiently plastic to participate directly in this exploration. Its learning algorithms may be hopelessly hard-wired into its structure. Methodological proliferation may therefore show itself only in artificially constructed brains designed specifically to do novel kinds of scientific exploration on our behalf. But this changes the philosophical point not at all.

The preceding defense of the proliferation of methodologies does not justify exactly the position that Feyerabend outlined in "Against Method." He is there reacting to the shortcomings of an old tradition in methodological research, rather than anticipating the possible virtues of a new tradition. But that is all right. The bottom line is that proliferating methodologies is still a very good idea, and for reasons beyond those urged by Feyerabend.

4. Conclusion

Philosophers are not always so fortunate as Feyerabend appears to be, in respect of finding a systematic vindication of their ideas through an intertheoretic reduction by a later and more penetrating theoretical framework. One must be intrigued by the convergence of principle here, and one must be impressed by the insight that motivated Feyerabend's original articulation and defense of the five theses listed. It seems likely that each one of these important theses will live on, and grow, now in a neurocomputational guise.

References

Churchland, P. M. 1989a. "On the Nature of Theories: A Neurocomputational Perspective." In W. Savage, ed., *Scientific Theories*. Minnesota Studies in the Philosophy of Science, vol. 14. Minneapolis: University of Minnesota Press.

———. 1989b. *A Neurocomputational Perspective: The Nature of Mind and the Structure of Science*. Cambridge, Mass.: MIT Press.

Couvalis, S. G. 1988. "Feyerabend and Laymon on Brownian Motion." *Philosophy of Science* 55:415–21.

Feyerabend, P. K. 1958. "An Attempt at a Realistic Interpretation of Experience." In *Proceedings of the Aristotelian Society*, n.s. Reprinted in P. K. Feyerabend, *Realism, Rationalism, and Scientific Method: Philosophical Papers*. Vol. 1. Cambridge: Cambridge University Press, 1981.

———. 1962. "Explanation, Reduction, and Empiricism." In H. Feigl and G. Maxwell, eds., *Scientific Explanation, Space, and Time*. Minnesota Studies in the Philosophy of Science, vol. 3. Minneapolis: University of Minnesota Press.

———. 1963a. "How to Be a Good Empiricist." In B. Baumrin, ed., *Philosophy of Science: The Delaware Seminar*. Vol. 2. New York: Interscience Publications, pp. 3–19. Reprinted in H. Morick, ed., *Challenges to Empiricism*. Belmont, Calif.: Wadsworth, 1972.

———. 1963b. "Materialism and the Mind-body Problem." *Review of Metaphysics* 17:49–66. Reprinted in C. V. Borst, ed., *The Mind-brain Identity Theory*. Toronto: Macmillan,

1970, pp. 142–56. Also reprinted in P. K. Feyerabend, *Realism, Rationalism, and Scientific Method: Philosophical Papers*. Vol. 1. Cambridge: Cambridge University Press, 1981.

———. 1970. "Against Method: Outline of an Anarchistic Theory of Knowledge." In M. Radner and S. Winokur, eds., *Analyses of Theories and Methods of Physics and Psychology*. Minnesota Studies in the Philosophy of Science, vol. 4. Minneapolis: University of Minnesota Press.

Gorman, R. P., and T. J. Sejnowski. 1988. "Learned Classification of Sonar Targets Using a Massively-parallel Network." *IEEE Transactions: Acoustics, Speech, and Signal Processing* 36:1135–40.

Hinton, G. E. 1989. "Connectionist Learning Procedures." *Artificial Intelligence* 40:185–234.

Laymon, R. 1977. "Feyerabend, Brownian Motion, and the Hiddenness of Refuting Facts." *Philosophy of Science* 44:225–47.

Lehky, S., and T. J. Sejnowski. 1988. "Network Model of Shape-from-shading: Neural Function Arises from Both Receptive and Projective Fields." *Nature* 333 (June 2):452–54.

———. 1990. "Neural Network Model of Visual Cortex for Determining Surface Curvature from Images of Shaded Surfaces." *Proceedings of the Royal Society of London B* 240:251–78.

Rosenberg, C. R., and T. J. Sejnowski. 1987. "Parallel Networks That Learn to Pronounce English Text." *Complex Systems* 1:145–68.

Rumelhart, D. E., G. E. Hinton, and R. J. Williams. 1986a. "Learning Internal Representations by Error Propagation." In D. E. Rumelhart and J. L. McClelland, eds., *Parallel Distributed Processing: Explorations in the Microstructure of Cognition*. Vol. 1. Cambridge, Mass.: MIT Press, pp. 318–62.

———. 1986b. "Learning Representations by Back-propagating Errors." *Nature* 323:533–36.

PART IV

BETWEEN LOGIC AND SOCIOLOGY

Answers to Philosophical and Sociological Uses of Psychologism in Science Studies: A Behavioral Psychology of Science

One major task for psychologists interested in developing the psychology of science is to engage in fruitful dialogue and debate with others in the broad field of science studies (Houts 1989). For at least the last two decades, this interdisciplinary field has been dominated by philosophers, historians, and sociologists. With a few recent exceptions (Fuller 1989; Giere 1988), traditional practitioners of these latter disciplines have consistently opposed incorporation of psychological perspectives into the field of science studies. For example, philosophers of science from Carnap (1936, 1937) to Popper (1970, 1974) and from Lakatos (1970, 1973) to Laudan (1977) have sought to keep epistemology separate from psychology. Even more recent calls to naturalize epistemology (Goldman 1986) have contained elements of the traditional gerrymander to keep psychology and epistemology separated (Heyes 1989). Traditional historians of science have treated the history of science as part of the general history of ideas according to which ideas are disembodied abstractions that have some life of their own quite incidental to the individual scientist who might have given them concrete expression (Sarton 1927–48). Sociologists, especially those influenced by the sociology of knowledge thesis of the early Marx, the latest incarnation of which are the various social constructivists (Brannigan 1981; Collins 1983; Collins and Pinch 1982; Gilbert and Mulkay 1984; Knorr-Cetina 1981, 1983; Latour 1987; Latour and Woolgar 1986; Pickering 1984; Woolgar 1981, 1986, 1988), have opposed psychological contributions to science studies on the grounds that appeals to cognitive psychological accounts obscure rather than clarify the material and social conditions they take to be causes of scientific knowledge production.

At the present time within the "official" community of science studies, the psychology of science is faced with the task of steering a course between the Scylla of rationalistic epistemologies and the Charybdis of sociological reductionisms. One possible course through these obstacles is to approach the psychology of science along the lines of radical behaviorism and the experimental analysis of verbal behavior as presented by Skinner (1953, 1957). In order to see how radical behaviorism meets the objections of both philosophers and sociologists, it is first necessary to outline those objections in some detail. Historians' objections to psychology of science are not separately addressed, because these have generally been the same as those offered by various philosophers, and the reader can extend the arguments presented here to those cases.

After setting the problem context by reviewing key philosophical and sociological objections to psychology of science, we show how a behavior-analytic psychology of science can successfully and consistently answer such objections. We present a version of naturalized epistemology based on radical behaviorism and show how traditional epistemological norms of scientific problem solving (e.g., logic and methodology) proposed by philosophers of science can be given a behavioral interpretation as rule-governed behavior. According to this view, the legitimacy and authority of so-called cognitive norms of science are based on naturally occurring contingencies of reinforcement that operate in everyday scientific practice rather than on appeals to special human faculties that intuit the a priori truth of various logical operations. Also, a radical behavioral version of naturalized epistemology handles many of the objections that recent sociologists have made against psychology of science by eliminating appeals to "cognitive" processes as a way to account for knowledge production.

1. The General Problem and the Psychologism Objection

Traditional objections to the psychology of science rest on various formulations and reformulations of an old argument (charge, indictment, accusation) known as psychologism. From the philosophical side, the psychologism objection follows from certain metaphysical assumptions about the privileged status of logic and logical truths. From the sociological side, the psychologism objection rests on ontological assumptions about what sorts of things can enter into an account of how scientific knowledge is produced. Although traditional philosophers of science and recent sociologists of science stand in dialectical opposition with regard to the status of classical epistemology and correspondence theories of truth, both seem to be equally opposed to psychological accounts of knowledge production in science. The charge of psychologism has been used to cover both quite different types of objection to psychology of science.

Before considering the details of these different versions of the psychologism objection, it is instructive to see how the objection functions as a verbal response in the science studies community. This functional analysis of "psychologism" will serve as a brief illustration of how radical behaviorism deals with verbal behavior, a subject that is discussed in more detail later in this essay. According to this functional analysis approach to verbal responses, the meaning of a term (e.g., the verbal response, "psychologism") is given by specifying the conditions under which using the term (making the objection) will be reinforced by listeners. As the sound of the expression suggests, the verbal response "psychologism" is functionally equivalent to verbal responses such as "bad," "stupid," "dumb," "fallacious," "unsophisticated," "anti-intellectual," etc. As a verbal response made by science studies scholars, "psychologism" is likely to occur on occasions when psychologists claim that their discipline has something unique to offer science studies over and above contributions already available from philosophers, historians, and sociologists. (This essay may be one such occasion.) The "psychologism" response is strengthened or made more likely to occur because of various contingencies of reinforcement that operate in the science studies community of speakers and listeners. These contingencies of reinforcement may take the form of positive social support or relief from aversive states of affairs. For example, a group of philosophers of science may all smile and nod in agreement when a defender of the tradition points out that cognitive psychologists who study human reasoning must have made use of elementary logic to design their studies. In this manner, the priority of logic and philosophical analysis can be comfortably reasserted in science studies. On other occasions the "psychologism" response may be reinforced by a self-congratulatory relief that follows upon asserting that one may safely get on with the project of studying science without having to consider psychology, because psychology is nothing more than mentalistic introspection or psychoanalytic obscurantism.

As a verbal response made in the science studies community, "psychologism" and its equivalents have a long history that bears careful examination. It is, after all, consistent with the approach of radical behaviorism to suppose that verbal responses are maintained and either strengthened or weakened as a function of the history of the community of speakers and listeners in which the response gets made. For this reason, it is important to see how philosophers and sociologists have used the psychologism objection.

1.1. Philosophical Usages of Psychologism

A first step in meeting philosophical objections to psychology of science is to examine what philosophers have meant when they have invoked the charge of psychologism as a response to various efforts to naturalize epistemology and the norms of science. Notturno (1985) and Aach (1987) have provided comprehensive and detailed studies of philosophical debates about psychologism, and much of the following synopsis is taken from their work. Notturno (1985) summarized the philosophical position against which psychologism was the charge as follows: "As a first approximation toward such an understanding [of psychologism], we will denote by 'psychologism' a family of views, all tending to deprecate or deny distinctions between epistemology and metaphysics on the one hand and psychology on the other" (p. 19).

Historically speaking, the term "psychologism" first appeared in the context of philosophical debates that took place in German philosophy during the first half of the nineteenth century. Psychologism was introduced into philosophical discourse as a derisive epithet to describe the implicit and often under-articulated claim of those philosophers who held, against the prevailing Hegelian philosophy of their day, that introspection and subjective experience were the foundations of all philosophical and scientific knowledge (Abbagnano 1967). Those opposed to the thesis of psychologism, most notably followers of Kant, recognized that if the thesis were correct, then the foundationalist aim of establishing universal, transhistorical, and objective truth would have to be abandoned. In short, if the claims in favor of psychologism were to be accepted, then there would be no possibility of grounding knowledge on clear and distinct truths of absolute certainty. At least since Descartes, philosophers had set about attaining a goal of justifying knowledge claims as either true or false with reference to some objective standard. In his transcendental deduction Kant had proposed to locate truth and objectivity not in experience, but rather in certain a priori truths that he took to be the conditions for the possibility of having any experience in the first place. By invoking a distinctly nonpsychological form of intuition, Kant proposed to rescue philosophical truth from the vagaries of introspection and to ground human reason on some objective realm apart from experience.

1.1.1. The Objections of Frege and Husserl

At the turn of this century, the chief proponents of this Kantian objection to psychologism were Frege ([1884] 1960) and Husserl ([1911] 1965, [1900–1901] 1970, 1977). What Frege and Husserl objected to was the claim by J. S. Mill and others that the truths of logic and mathematics were derived from experience by means of inductive inference. Against such claims, Frege and Husserl sought to preserve the Kantian notion that the truths of logic and mathematics had a compelling quality of certainty that did not come from experience but were prior to experience. Husserl (1977) stated his position this way:

> "Prior" to the givenness of a fact and a factual world, we could be a priori unconditionally certain of the fact that, what we state as logicians or as arithmeticians must be applicable to everything that may be encountered as the corresponding factual reality. . . .
>
> Irrespective of whether there is a world or not, the truth that $2 + 2 = 4$ subsists in itself as a pure truth. It does not contain in its sense the least information about real facts. The same holds good of the law of non-contradiction and other such laws. Pure ideal truths are "a priori," and are seen to be true in the unconditioned necessity of their generality. (pp. 198–99)

About Husserl's project to secure some objective foundation for logic, Kolakowski (1975) made the following observation: "Husserl was sure that psychologism ended in scepticism and relativism, that it made science impossible, and that it devastated the entire intellectual legacy of mankind" (p. 17). Although it may sound hyperbolic, this is a fair characterization of the threat to philosophical epistemology that Husserl perceived in psychological formulations of logic. What was troubling for Husserl about proposals to ground logic in psychology was that philosophical theories of knowledge would lose their footing and authority and that the normative side of epistemology would disappear only to be replaced by descriptive empirical studies of human reasoning. According to this view, such descriptive accounts of human reasoning might describe how we come to think as we do but could never tell us whether or not our thinking was valid or true.

Whereas Husserl opposed psychologism because of the disastrous consequences he perceived that it entailed for classical epistemology, Frege's opposition to psychologism was more metaphysically focused. What the logician Frege sought to defend was the Platonic idea that there is some objective realm of knowledge that exists quite apart from our experience of it (Dummett 1981). Notturno (1985) summarized Frege's opposition to psychologism as follows:

> If the objectivity of knowledge requires justification by purely logical methods, then the laws of logic cannot themselves be based upon empirical observation (sense perceptions) — for this would render them subject to revision and undercut the objectivity of the very means of justification. (p. 153)

For Frege as for Husserl, the task of epistemology was *not* to describe what causes us to believe as we do. Instead, the task of epistemology was to provide and secure the objective standards and norms against which we might judge the truth or falsehood of our beliefs irrespective of descriptive accounts about how we acquired those beliefs. By opposing psychological accounts of logic, Frege sought to establish that in logic and mathematics we have instances of objective certainty that do not depend on experience. These instances of objective certainty counted for Frege as indications that there is actually some realm beyond experience where objective, certain, and eternal truths reside. On such a view, then, the task of philosophy is to apprehend and point out these objective truths.

In much of the philosophical discourse about logic and philosophy at the turn of this century, psychology was seen as a threat to the philosophical project of securing the foundations of knowledge and science. This separation of psychology from epistemology became part of the canonized wisdom of logical positivist philosophy of science of the 1930s. The Husserlian and Fregean type of division between psychology and epistemology was recapitulated in Reichenbach's insistence on a distinction between the context of discovery and the context of justification. Reichenbach coined the phrase "context of discovery" to distinguish the aims of philosophy of science from what he took to be the aims of psychology of science.

> We might say that [a rational reconstruction] corresponds to the form in which thinking processes are communicated to other persons instead of the form in which they are subjectively performed.... I shall introduce the terms *context of discovery* and *context of justification* to make this distinction. (Reichenbach 1938, pp. 6, 7)

According to this analysis, the thought processes of the individual scientist were unimportant so far as philosophy of science was concerned, because "epistemology is only occupied in constructing the context of justification" (Reichenbach 1938, p. 7). For Reichenbach the philosophical problem of knowledge and hence of science was the problem of justifying knowledge by appeal to some objective standards.

1.1.2. The Objections of Popper

With the publication of Karl Popper's *Logic of Scientific Discovery* ([1934] 1959) and its eventual wide dissemination, the project of justifying knowledge claims in terms of some absolute and certain standard was seriously questioned. Although Popper rejected the idea that knowledge claims could be justified and favored instead a program of criticism and falsification, he nevertheless retained a notion of absolute standards based in logic (Musgrave 1974). He further rejected as psychologism any proposals that attempted to use psychological explanations to account for theory change and development in science. For example, when Kuhn (1970) posed the controversial question "Logic of Discovery or Psychology of Research?" Popper (1970, p. 58) responded by ridiculing Kuhn's appeal to psychology and social science as "a regress to these often spurious sciences," which he described as containing "a lunatic fringe." For Popper (1974), to use psychology to account for scientific theory change was tantamount to claiming that science is irrational.

The basis for this response to Kuhn was clarified in Popper's later work. In his later writings, Popper (1972a, 1972b, 1972c; Popper and Eccles 1977) introduced his three-world typology to further bolster his objections to basing epistemology on psychology, as he understood it. According to this later scheme, World 1 comprises the material, physical world, and World 2 comprises the domain of individual private experiences including thought processes. Echoing Frege's preference for an objective realm of knowledge that is independent of knowing subjects, Popper claimed that scientific knowledge belonged in the domain of World 3, or objective knowledge. Simply put, World 3 is the totality of scientific theories and arguments that may be found in the libraries of the world, and also those not yet written down anywhere. Moreover, Popper claimed that the truth or falsehood of these statements belonging to World 3 is decidable without reference to subjective experiences of scientists. Like Reichenbach, Popper (1972a) drew a sharp distinction between epistemology and psychology by confining epistemology to the study of the products of scientific work as opposed to the study of the production processes that led to those products.

> Contrary to first impressions, we can learn more about production behaviour by studying the products themselves than we can learn about the products by studying the production behavior. This third thesis may be described as an anti-behaviouristic and anti-psychologistic thesis....
>
> In what follows I will call the approach from the side of the products — the theories and the arguments — the "objective" approach or the "third-world" approach. And I will call the behaviorist, the psychological, and the sociological approach to scientific knowledge the "subjective" approach or the "second-world" approach. (p. 114)

In an uncharacteristic departure from traditional canons of logic, Popper (1972a) defended his stance against psychological (and sociological) accounts of knowledge production by invoking an argument of *post hoc ergo propter hoc*. He noted that the appeal of psychological explanations of scientific theories is due to the fact that such naturalistic explanations are causal. In other words, such naturalistic accounts proceed on the assumption that knowledge is an effect produced by human behavior. Popper faulted such accounts because they rested on reasoning from causes to effects rather than from effects back to causes, the latter being what he took to be characteristic of scientific reasoning! Haack (1979) has commented that Popper's assiduous efforts to keep psychology out of any explanatory role in scientific knowledge production are based on his insistence that logic occupies a privileged place in theories of knowledge. That is, unlike empirical science and even mathematics, which Popper took to be byproducts of human language use, logic could not be a mere byproduct of human language use. Thus, in Popper's presentation of the psychologism objection, there is a full circle return to the metaphysics of pure thought (now called World 3) envisioned by Frege.

1.2. Summary and Conclusions about Philosophical Objections

What is perhaps most interesting about these philosophical objections to a role for psychology in science studies is the extent to which the objections have persisted without taking into account differences of approach within the field of psychology. To put it another way, the charge of "psychologism" has become a type of overgeneralized negative response to anything psychological. The psychology to which Frege and Husserl objected was the empirical program of introspection of the late nineteenth century. What was objectionable for them was not introspection per se, but rather the proposal that empirical evidence obtained through psychological methods of introspection as opposed to philosophically intuited (introspected) logical truths might serve as a foundation for knowledge (Aach, in press). The fact that psychologism as an objection could be defended even when introspectionist psychology had been rejected is evident in both Carnap ([1932] 1959) and Reichenbach (1951). Both advocated that psychology take a behaviorist turn away from introspection, yet neither ever developed the implications this might have for their attempts to segregate epistemology from psychology. In order to avoid the problem of infinite regress in justifying knowledge claims, their justificationist agenda for the philosophy of science mandated the need for "non-empirical" foundations (logic) in terms of which knowledge claims could be evaluated. The idea of a nonintrospectionist but nevertheless still naturalized epistemology seems to have been foreign to these leading logical positivists of the 1930s.

With Popper's nonjustificationist program for philosophy of science, the tables were turned in the opposite direction so far as psychology was concerned. Against Kuhn, Popper invoked the psychologism objection in order to avoid the possibility that empirical psychological studies of human reasoning and problem solving might be taken as a new type of foundationalism. As Notturno (1985) has noted, Popper's objection to psychologism is complicated by the fact that at times Popper seems to oppose all appeals to absolute standards (especially naturalistic, inductive ones), yet at other times he seems to appeal to standards of logic as if those were absolute. On the one hand, in some of his writings, the psychology to which Popper (1968) objected was Freudian and Adlerian psychoanalysis, and the force of his argument turned on the claim that such psychologies were not amenable to empirical refutation. Similar allusions to psychoanalysis and limited views about what psychology is have persisted in the writings of Lakatos (1970) and Laudan (1977), both of whom have been vigorous opponents of psychological approaches to understanding science. On the other hand, in different discussions about the role of psychology in explaining science, Popper's ([1934] 1959, 1972a) objections seem to have been directed against psychology conceived as the study of immediate, private experience or sensations. In either case, it is worth noting that psychology conceived as a science of behavior along the lines of radical behaviorism is both empirically testable and deliberately designed to be an alternative to subjectivist, sensationist psychologies.

This brings us, then, to the crux of these and similar philosophical objections to psychology of science. The heart of the matter seems to be that Frege, Husserl, and Popper all presupposed the authority of deductive logic in one way or another. In countenancing a complete separation of logic from psychology, however, they overlooked an important paradox: If the laws of logic do not depend on human activity for their authority, then how is it that the laws of logic have an effect on human activity (Aach, in press)? What is needed, then, to meet these objections and deal with this paradox is an account of logic that is naturalistic and psychologically (behaviorally) plausible. Fortunately, this type of naturalistic approach to logic has been anticipated by Dewey (1938) and Wittgenstein (1975) among others (Quine 1960). For example, Dewey anticipated the problem of naturalizing logic as follows:

> The application of the postulate of continuity to discussion of logical subject-matter means, therefore, negatively, that in order to account for the distinctive, and unique, characters of logical subject-matter we shall not suddenly evoke a new power or faculty like Reason or Pure Intuition. Positively and concretely, it means that a reasonable account shall be given of the ways in which it is possible for the traits that differentiate deliberate inquiry to develop out of biological activities not marked by those traits....

> If one denies the supernatural, then one has the intellectual re-
> sponsibility of indicating how the logical may be connected with the
> biological in a process of continuous development. (Dewey 1938,
> pp. 24–25)

For Dewey (1938) the continuity between logic and biological development was to be found in an interactive, environmental analysis of
human behavior, especially human verbal behavior, as that behavior has
been shaped and changed by listeners and speakers. Similarly, Wittgenstein's (1975) sociological naturalization of logic and mathematics as
forms of language use that are a function of social practices and conventions also provides an approach to the problem of how to naturalize
logic. As we will show, the psychology of radical behaviorism pioneered
by Skinner (1957), especially the Skinnerian analysis of verbal behavior, can provide an important and rather specific analysis of logic using
psychological principles of operant conditioning.

Before turning to objections to psychology of science offered by recent
sociologists, a relevant and rather obvious point regarding introspection
in the arguments of Frege and Husserl needs to be reiterated. As an approach to psychology, radical behaviorism does not rely on empirically
based introspection. For reasons more similar to those of Wittgenstein
(see Bloor 1983) than those of Frege and Husserl, Skinner ([1945]
1988b) arrived at the same conclusion: reports of private sense-data
experience are unreliable and therefore cannot serve as the foundations
for scientific psychology, or anything else including epistemology for that
matter. The experimental analysis of behavior specifically avoids assign-
ing explanatory power to private events and provides a causal account of
verbal behavior in terms of observables. How radical behaviorism handles traditional psychology conceived as private experience or Popper's
World 2 is a subject of some importance that can be best presented once
it has been made clear that the chief psychologism objection offered by
recent sociologists is an objection to "cognitive" psychology.

1.3. Sociological Usages of Psychologism

A recent development in sociology of science has been the rise of a
group of sociologists who travel under the banner of "social constructivism." Although there are differences and debates within the social
constructivist movement, most of these sociologists have accepted the
demise of rationalistic justificationism, a conclusion consistent with a
more general trend observed in postpositivist philosophies of science
(Fuller 1988) and postmodernist philosophy (Lyotard 1984). For them,
scientific theories are not only neither true nor false, but they are also

the products or fabrications of the scientific community. What interests these sociologists are the social interactions and practices that surround and lead to knowledge or text production. Thus, their empirical studies employ the methods of ethnography and rhetorical analysis and consist of intensive local observations of the behavior of scientists in laboratories coupled with some rather ambiguous accounts about how those behaviors result in the production of publications.

Curiously enough, the social constructivists who seek a naturalistic rather than a philosophical or epistemological account of science have also opposed psychology of science. Although they have not often used the term "psychologism," they do seem to equate any talk of the psychology of science, especially cognitive psychology of science, with a regress to traditional epistemology based on representationalism and the correspondence theory of truth. This ironic inversion of traditional philosophical objections to psychology of science has taken the form of denouncing anything cognitive. For example, Latour and Woolgar (1986) have stated:

> Perhaps the best way to express our position is by proposing a ten-year moratorium on cognitive explanations of science. If our French epistemologist colleagues are sufficiently confident in the paramount importance of cognitive phenomena for understanding science, they will accept the challenge. We hereby promise that if anything remains to be explained at the end of this period, we too will turn to the mind! (p. 280)

What the social constructivists object to in proposals for a psychology of science are precisely some of the same objections raised about cognitive science more generally by Skinner (1985). Like Skinner (1957), they want to treat knowledge claims as verbal performances, and they are reluctant to accept as explanations of these events various attributions about what may or may not have been happening in the minds of scientists (Woolgar 1987). In a recent essay, Woolgar (1989) has echoed themes of radical behavioral critiques of cognitive psychology in stating that "explanation in terms of cognition may be *needlessly mystifying*" (p. 217) and that "explanation in terms of cognition may also be *superfluous*" (p. 217).

From the perspective of social constructivists, many of whom view their chief accomplishment to have been to offer an alternative to traditional philosophical epistemology, the major liability of cognitive science of science is the latter's failure to take seriously the problems of epistemologies based on representationalism. Only rarely have social constructivists been explicit about the philosophical traditions that inform their critiques of cognitivism (Bloor 1983; Bowker and Latour 1987), and their cognitive critics have continued to talk about representation, either in mentalistic or computer metaphors, as if there were no problem to be addressed (Slezak 1989). This has led both groups to talk past each other and resulted in debate that has often failed to move the issues forward (see Slezak [1989] and responses in *Social Studies of Science*). In this regard, the social constructivists could be vindicated because they could explain why they lost the debate by referring to the social forces in science and technology studies according to which they were outnumbered by cognitivists. In the interest of avoiding that outcome, it is important to clarify some of the philosophical and implicit psychological differences between social constructivists and advocates for cognitive psychological approaches to science and technology studies.

In keeping with the theme of this essay to present the merits of a behavioral psychology of science, the focus of the following reconstruction of social constructivist objections to cognitivism is on those features of their critiques that lead to a reexamination of cognitive approaches in terms consistent with radical behaviorism. What has been omitted from this synopsis, however, are some of the more extravagant claims of social constructivists to the effect that science and technology studies should specifically avoid pursuing traditional social science aims of causal explanation, prediction, and control (Latour 1988; Woolgar 1987). In this regard and from the standpoint of radical behaviorism, we are sympathetic with those critics of the constructivist program who find it difficult to be enthusiastic about a program of research in science and technology studies that at the outset eschews traditional scientific aims. As behaviorists, we, too, are inclined to ask: Where is the science in science and technology studies? Nevertheless, we also think that social constructivist critiques of the turn to cognitive psychology contain important points about the liabilities of shifting the focus of science and technology studies away from scientists' practices and toward an explanation of their abstract products (e.g., theories and discoveries). It is in this sense that a psychology of science based on radical behaviorism may preserve social constructivist critiques of traditional philosophy of science without sacrificing the aim of making science and technology studies more scientific.

1.3.1. Social Constructivist Objections to Cognitivism

One way or another, all proposals for cognitive science of science rest on an assumption that the mind or brain represents states of affairs in the natural world. From the perspective of the social constructivists, this picture of the individual picturing the world is a retreat to classical epistemology, a theory of knowledge they take to have been discredited through postmodernist critiques of the Cartesian tradition based on interpretations of Wittgenstein, Heidegger, and Foucault (e.g., Rorty 1979). Interestingly, in its naive form this account of the mind or brain as mirror of nature was also rejected by Skinner (1953, 1957, 1985) in formulating his radical behaviorism, a point we will return to below. Rather than exhaustively review these various arguments against representationalist epistemology that undergird the alternative epistemology offered by social constructivists, we will list and very briefly discuss the major ones in order to show how such criticisms are consistent with a behavioral approach to science and technology studies.

1. *The Homunculus Problem.* Simply put, this criticism asserts that if what scientists are doing is picturing the world or re-presenting it, then it is fair to ask just who or what is there to picture the picture. This infinite regress problem of cognitivism is not solved by more recent, sophisticated views of representation such as those based on the analogy between human performances and those of digital computers. Cognitive science models of the brain or nervous system based on the computer analogy simply replace the homunculus with a computer programmer. Current proposals to "discharge" the homunculus by postulating dumber and dumber homunculi in the form of parallel distributed neural networks rest on promises that such schemes will eventually achieve more than pattern recognition. In the meantime, the best available computer simulations of scientific problem solving are those that apply rules to symbolic systems, and these systems are still troubled by the homunculus-as-programmer problem. Who decided what information to store in the computer and what rules to tell the computer to follow, and where did this information and these rules come from in the first place? Social constructivists (Brannigan 1989) and social epistemologists (Fuller 1990) have correctly pointed out that computer simulations of "scientific discovery" import the social conventions of the scientific community through the back door.

2. *Representational Systems Are Necessarily Social.* Representations, however abstract, neural, and outside of our awareness, whether in a machine or putatively in the brain, have the meaning that we language users give them. It is we social users of language who manufacture representations and give them their "meaning" in the social practice of talking. A socially neutral, asocial, or "objective" representational system is a figment of our imagination. Representation systems acquire their meaning through action or use in the social context of a community of language users. For example, drawing on Wittgenstein's later works, Bloor (1983)

has argued that such apparently socially neutral representation systems as mathematics are in practice "social all the way down." To take the example earlier cited from Husserl, it can be shown through the following illustrations that "$2+2=4$" is a matter of social convention.

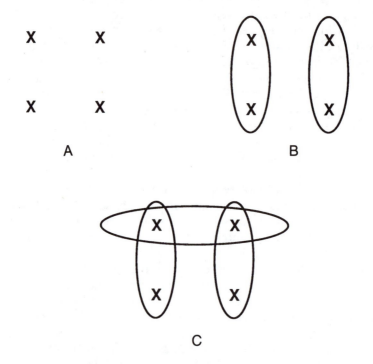

Figure 1. An illustration of how "$2+2=4$" depends on social conventions and rules. Adapted from Wittgenstein as cited in Bloor (1983).

We may show someone Figure 1A and then illustrate "$2+2=4$" is what is meant by Figure 1B. If the individual then responds, "Right, I see how this works," and then draws Figure 1C and says "$2+2+2=4$" we are likely to intervene and explain the rule. Bloor (1983) summarized this point as follows:

> In other words, the compelling force of mathematical procedures does not derive from their being transcendent, but from their being accepted and used by a group of people. The procedures are not accepted because they are correct, or correspond to an ideal; they are deemed correct because they are accepted. (Bloor 1983, p. 92)

The social basis of representation systems, even ones as abstract as mathematics and logic, may be hidden from view, but a naturalistic account of these practices, even a machine implementable simulation of them, requires that they be brought to the forefront (Collins 1987).

3. *Cognitive Science Confuses the Mechanics of Description with Doings of Actors.* Cognitive scientists construct formal descriptions of human problem solving based on observations of human behavior. A mistake occurs when those formal descriptions are then imported from the domain of the observer and placed inside the actor (Heil 1981). For example, cognitive scientists may find it useful to describe the behavior of a pigeon that fails to learn a discrimination task (or of a behaviorist who fails to respond to talk about cognitions) as "an information-processing system with a limited channel capacity," but this description in terms of capacities and capabilities is merely a restatement of their observations and does not explain the causes of the behavior. Similarly, because observers may find it useful to describe scientific problem solving as following rules, it does not follow that a scientist solving a problem is necessarily following a rule. Social constructivists such as Woolgar (1987) and Coulter (1983) have aptly noted that people, including scientists, may do many things such as catching a ball or adjusting a laboratory apparatus without following rules. The fact that current machine simulations of scientific problem solving are rule based does little more than hide from view the social basis of scientific activity. Rule-following computer programs such as BACON are indeed impressive at solving problems that have been well defined by a programmer. What such powerful demonstrations fail to show is how scientists find their problems in the first place. To date, no such computer system has predicted a new scientific discovery.

4. *Cognitive Science Focuses Attention on a Decontextualized Individual Scientist.* By looking to cognitive psychology and cognitive science, scholars in science studies run the risk of excluding from view the social practices of scientists. In this manner the object of study in science studies may become a disembodied and abstracted problem-solving process presumably located in the individual scientist or group mind. This hypostatization of problem-solving processes is clearly evident in computer simulations of scientific discovery, and some proponents of these demonstrations (e.g., Slezak 1989) have argued that their success at "reproducing" scientific discoveries is evidence that social factors do not play a significant role in scientific discovery. Given the vast literature of social psychology, this focus on the decontextualized individual is unfortunate. It is not necessary to focus on the individual scientist as decontextualized information processor in order to apply the methods of

experimental psychology to the aim of understanding scientific activity (Gorman 1989; Gorman, this volume; Fuller, this volume).

5. *Cognitive Psychology of Science Internalizes Knowledge Production.* The chief accomplishment of recent sociology of science has been to focus attention on the production of tangible objects that we call knowledge, namely papers, books, experimental findings, and so on (Latour 1987). Recent sociology of science has called attention to the behaviors that produce these tangible products. In effect, knowledge production has been externalized in the sociology of science. Although some social constructivists have advocated stopping this externalization at the level of merely describing what they record (Woolgar 1987), others have advocated moving further to give a causal account of knowledge production in terms of externalized variables (Bloor 1983). The point of this objection to the cognitive turn is that should the field of science studies adopt the methods and practices of cognitive science, there is a danger that observable events in the knowledge production process will be overlooked because the focus of investigation will be shifted to inferred events going on inside the scientist.

6. *The Vocabulary of Cognitive Science Obscures Social Interactions.* In many respects, the vocabulary of cognitive science is synonymous with the vocabulary of folk psychology. From a sociological perspective, the vocabulary of folk psychology is not well understood, in part because it is so familiar as our everyday talk. By adopting the vocabulary of cognitive science, the field of science and technology studies may inadvertently obscure the social conditions and social interactions that sustain such talk. Uncritical adoption of the vocabulary of beliefs, desires, cognitions, mental models, etc. may lead to overlooking the importance of verbal responses and the extent to which observable conditions determine the initiation and repetition of those responses. In other words, the language of folk psychology could be taken as just more behavior to be explained rather than as an explanation for the activities of scientists. Although advocates of discourse analysis such as Gilbert and Mulkay (1984) have not made this point directly, perhaps because they eschew more traditionally scientific causal accounts of scientific discourse, Giere (1988) and others (Houts 1989) have noted the behavioral implications of discourse analysis.

1.4. Summary and Conclusions about Sociological Objections

The major objections that recent sociologists have raised against psychology of science have targeted various proposals to bring cognitive science and cognitive psychology to bear on the field of science studies. Unlike philosophical usages of "psychologism," which find their rhetorical punch from normative models of rationality, sociological usages of "psychologism" achieve their persuasive edge by insisting that the language of explanation in science studies be confined to observed actions and verbal performances. In various calls for cognitive approaches to science studies, the social constructivists perceive a retreat on two fronts. The first is a retreat to some version of classical epistemology and traditional rationalistic philosophies of science, a move they correctly take to be a direct challenge to their alternative epistemological program of relativism and the social determination of knowledge. From the perspective of social constructivists, the second point of retreat marked by cognitive programs is a retreat to mental events to explain the observed practices of scientists. What is lost in this feature of the cognitive turn are the basic data of the social constructivist program, namely the practical activities of working scientists that have been featured in the empirical work of sociology of science over the last decade.

What is interesting about these objections to the cognitive turn in science studies is just how similar they are to radical behaviorists' objections to the cognitive turn in psychology. Although a few social constructivist sociologists have noted this affinity between their views and those of radical behaviorists (Bloor 1983; Collins 1983), and some of their critics have charged them with endorsing behaviorism as a defunct psychology (Giere 1988; Slezak 1989), only rarely have participants in these disputes suggested that behavioral psychology may be a viable way to advance the sociology of science in the face of the cognitive turn (Fuller 1989, 1990). On the one hand, it is unlikely that some of the more radical social constructivists will welcome a suggestion to turn to radical behaviorism, because radical behaviorism is a version of psychology that is consistent with traditional scientific aims of prediction and control, aims that radical constructivists seek to avoid in favor of description and nothing more. On the other hand, it is also unlikely that the empirical achievements of recent sociology of science will continue to be taken seriously if they cannot be incorporated into the growing effort to make science and technology studies more "scientific."

Social constructivists have noted that there are philosophical and practical liabilities associated with adopting the language of cognitive science for science studies, the chief one being a return to classical epistemology and the Cartesian world picture. The language of cognitive science will send investigators off on a course of construing the activities of scientists as rational picturers of the world whose chief aim is to get the right picture. Cognitive scientists of science will watch scientists at the laboratory bench and infer what is going on in their minds or brains. When they notice that the scientists change some feature of an apparatus, they will infer that the change was made because the scientists' mental models dictated the change. What such a view is likely to miss is the practical activity that is being engaged in and the possibility that this practical activity is controlled by what is happening on the laboratory bench rather than by what is happening in some putative representational system. One way out of this dilemma is to turn to radical behaviorism as an approach to psychology of science.

2. Epistemology from the Standpoint of Radical Behaviorism

Both philosophical and sociological objections to psychology of science in terms of psychologism are based on epistemologies, albeit opposing ones. Based on their rationalistic epistemologies, certain philosophers reject as psychologism any attempts to reduce logic to naturalistic terms, especially the sensationist, subjectivist terms of the old introspective psychology. Based on their social constructivist epistemologies, certain sociologists of science reject as psychologism any attempts to explain knowledge production in terms of cognitive processes said to be occurring in the minds or brains of scientists. Epistemology from the standpoint of radical behaviorism can address both types of objection. In order to see how this may be accomplished, it is necessary to understand how radical behaviorism deals with private experience and rule-governed behavior on the one hand, and with so-called cognitive processes as verbal behavior on the other hand. Before taking up the topics of private experience, rule-governed behavior, and the social basis of verbal behavior, it is first necessary to clarify the basic stance of radical behaviorism as it differs from other types of behaviorism. Confusions between radical behaviorism, logical behaviorism, and methodological behaviorism abound in the literature of philosophical psychology, and these confusions have often led critics to reject behaviorism as if their criticisms found a target in radical behaviorism when, in fact, they have not.

2.1. Radical Behaviorism Is Not Logical Behaviorism or Methodological Behaviorism

Logical behaviorism of the type espoused by Ryle (1949) is not the same as radical behaviorism, even though both types arrive at some of the same conclusions regarding the ontological status of the referents of talk about mental events. The basic thesis of logical behaviorism is that mental predicates may be reduced to behavioral predicates. That is, talk about cognitive processes such as thinking and feeling may be given equivalent meaning in terms of talk about dispositions to behave in certain ways. For example, to say that "John is thinking about buying a new car" is equivalent to making the behavioral observations that John is going around to car dealers and test driving new cars, he is applying for a new car loan at the bank, and he is regularly expressing his dissatisfaction with his old car. Logical behaviorism is a thesis about the logical equivalence of certain forms of expression and a claim that the verification for cognitive forms of expression can be achieved only by referring to overt behaviors. In contrast to logical behaviorism, radical behaviorism is not primarily concerned with the meaning equivalence of cognitive and behavioral forms of expression. For the radical behaviorist, the meaning of "John is thinking about buying a new car" is not given by some equivalent dispositional proposition. Instead, the statement "John is thinking about buying a new car" is treated as a verbal performance of a speaker, and the "meaning" of the expression is given by specifying the conditions that occasion the verbal performance along with the consequences that follow it. In other words, "John is thinking about buying a new car" is a response of a speaker, and the "meaning" of the response is specified by noting the conditions that control its occurrence as a response of the speaker. For example, the speaker of the statement makes the statement on the occasion of having seen John do all of the above acts noted by the logical behaviorist, and making the statement is reinforced by the behavior of listeners, including John, who nod in affirmation. What the logical behaviorist takes to be evidence for a behavioral disposition, the radical behaviorist takes to be discriminative cues that indicate that the verbal performance "John is thinking about buying a new car" will be socially reinforced by members of the verbal community. For the radical behaviorist, logical behaviorists do not go far enough when they substitute talk of cognitive processes with talk of behavioral dispositions. Radical behaviorists go further and analyze the talk of behavioral dispositions as forms of expression that are themselves verbal performances under the control of listeners' responses. The difference between logical behaviorism and radical behaviorism is that the latter treats verbal expressions as behavior to be functionally analyzed in terms of controlling external (social) variables rather than as abstract logical equivalents of "meaning."

Radical behaviorism is also quite different from methodological behaviorism. This distinction is crucial because so many of the canonical criticisms of behaviorism find an easy target in methodological behaviorism, yet totally miss the mark with respect to radical behaviorism. Methodological behaviorism is a psychological thesis based on associationism and stimulus-response (S-R) theories of learning, according to which organisms come to associate certain responses to certain stimuli. In its strictest form, methodological behaviorism asserts that there is no need to introduce unobserved, mediating variables between stimuli and responses provided that one can reliably predict responses from knowledge of the directly observed stimulus conditions. Thus, in this strict form, methodological behaviorism is a methodological prescription to avoid introducing variables that refer to unobserved inner processes of the organism such as cognitions and physiological states, unless it can be shown that by introducing them one gains a degree of predictive accuracy otherwise not attainable. It is ironic that once the problem of predicting behavior was set up in this S-R fashion, the way was made clear for all manner of mediating variables to be introduced on the grounds that an experimenter's knowledge of stimulus conditions alone did a relatively poor job of predicting behavior. This version of S-R associationism led rather easily to the so-called cognitive revolution in psychology. All that was required to overcome methodological behaviorism was to show that by introducing mediating variables that refer to hypothetical states of the organism, one could do a better job of predicting responses to stimuli in the context of laboratory experiments (see interviews with M. Levine and G. Mandler in Baars 1986). Viewed in this way, the cognitive revolution in psychology was probably not a revolution at all but was, instead, a consistent and predictable development from S-R associationist versions of behaviorism (Gholson and Barker 1985).

In contrast to methodological behaviorism, radical behaviorism is *not* an S-R version of psychology; nor is it a form of associationism. S-R psychologies view the control of behavior as belonging on the stimulus side of the so-called S-R bond. In radical behaviorism, behavior is controlled not by the stimuli, but rather by the environmental consequences that behavior produces. This is the basic concept of the operant, that behavior operates to produce changes in the environment. In radical behaviorism, the environment is said to shape the behavior of organisms because there exist regularities in the relationships between an organism's actions and the effects of those actions on the environment. These regularities are called "contingencies of reinforcement." Just as contingencies of survival may select certain genotypes, contingencies of reinforcement may select certain types and forms of behavior. The basic model is one of selection by consequences where the causes of behavior are the consequences that the behavior produces. An experimenter may arrange environments in such a way that a particular stimulus regularly precedes or accompanies certain response-environmental consequence relationships so that responding reliably occurs in the presence of a particular stimulus. In this manner it can be said, though it is perhaps misleading, that the behavior is "under the control" of a discriminative stimulus, but the cause of the behavior is the past consequences that were produced by performing the behavior. In radical behaviorism, organisms do not "associate" stimuli and responses as in methodological behaviorism. The "association" between stimuli and responses occurs in the environment, not in the behaving organism. Organisms do not form associations, except in the derivative sense that an experimenter can arrange environments in such a way as to bring about reliable relationships between stimulus conditions and consequences. Selection by consequences, not association between stimuli and responses, is the causal mechanism in radical behaviorism.

The confusion of S-R associationism with radical behaviorism has led to numerous criticisms of behaviorism that miss their mark, the most famous example being Chomsky's (1959) review of Skinner's (1957) *Verbal Behavior*. Although there have been some critical responses to Chomsky's review that addressed its weaknesses in some detail (Catania 1972; MacCorquodale 1970), the basic fallacy of Chomsky's argument as applied to radical behaviorism is so simple that the fallacy has gone relatively unnoticed. In Baars's (1986) interview with Chomsky, Chomsky clarified his deductive argument against behavioral approaches to verbal behavior as follows:

Major Premise: Finite State Automata Models Cannot in Principle Account for Known Facts of Language Use and Language Comprehension

Minor Premise: All Behavioral Psychologies Are Finite State Automata Models of Human Behavior

Conclusion: Therefore All Behavioral Psychologies Cannot in Principle Account for Known Facts of Language Use and Language Comprehension

Chomsky's main criticism of Skinner's functional analysis of verbal behavior was based on Chomsky's (1957) unargued and still unsubstantiated claim that like S-R associationist theories that rely on serial left-to-right conditioning (chaining) to build up verbal performances, Skinner's approach could be subsumed under Chomsky's minor premise. Yet, this claim as applied to Skinner's approach is *patently false*, because radical behaviorism is not S-R associationism and therefore cannot be subsumed under Chomsky's minor premise. That such molecular chaining of individual responses to other molecular responses was not the case for Skinner's analysis is obvious from the basic units of

Skinner's functional analysis of verbal behavior, illustrations of which Chomsky (1959) himself quoted as containing multiple words and entire sentences. In retrospect, it is astonishing that such misplaced criticism and invalid argumentation have been widely accepted as the basis for canonical dismissals of behaviorism. As MacCorquodale (1970) noted, the functions of verbal behavior that Chomsky assigned to a hidden grammar competence (a homunculus equivalent), Skinner (1957) assigned to the observable actions of speakers and the social consequences delivered through listeners.

With these distinctions between logical behaviorism, methodological behaviorism, and radical behaviorism as caveats for what follows, it can be shown that philosophical objections about subjectivism and the priority of logic can be handled within a naturalistic epistemology based on radical behaviorism. Rather than digress any further to give an introduction to the radical behavioral analysis of verbal behavior, we assume that the reader will be able to follow the analysis as presented in the context of the specific topics that are of interest for this essay. However, there is probably no substitute for actually reading and studying Skinner's (1957) *Verbal Behavior, pace* Chomsky, a book that we consider to be the key to a behavioral approach to issues of psychology of science and science studies more broadly construed.

2.2. An Answer to Subjectivist Psychology: Behavioral Analysis of Private Experience

From the standpoint of radical behaviorism, attempts to anchor knowledge on subjective experience are rejected, not because the sanctity of logic must be preserved, but rather because statements about private experience are not reliable. As Bloor (1983) has noted, Skinner's treatment of the problem of privacy is similar to that of Wittgenstein, only Skinner ([1945] 1988b) gives a more detailed set of reasons as to why appeals to private experience are not trustworthy. The basis of this account of private experience is a functional analysis of ordinary verbal behavior (folk psychology) that employs subjectivist psychology terms.

This naturalistic account of private experience views responses such as "I am in pain" to be lawful verbal responses made in the presence of physical stimuli to which the speaker alone has direct access. However, because private events such as pain are physical phenomenon, there is no reason why an objective psychology cannot consider the social processes through which a language of private experience gets generated. The problem posed by such an analysis is that the scientific community cannot, as in the case of publicly observable stimuli, account for a response occasioned by a private event by pointing to some public event that set the occasion for the response to be reinforced. Moreover, even the antecedents of one's own private cues are often out of the reach of introspection. Thus, the task for a naturalistic account of private experience is to show how verbal responses made in the presence of private experiences come to be controlled by the responses of other people.

Skinner ([1945] 1988b) has suggested that there are at least four ways in which responses occasioned by private experiences come under the control of the speaker's verbal community. First, responses to private events can be shaped by making reinforcement contingent upon public accompaniments of the private stimuli. In the case of pain, the verbal response "I am in pain" may be reinforced in the presence of concomitant tissue damage. The verbal community utilizes an overt stimulus (tissue damage) to shape the individual's verbal response. However, in the course of this shaping by the community of listeners, the verbal response "I am in pain" also may be reinforced in the presence of private bodily cues available only to the speaker. In this way, the private event (a felt bodily state) may set the occasion for the response "I am in pain" to be reinforced by others at later times. Second, reinforcement of a response in the presence of a private experience may be made contingent upon collateral responses to the same experience. In this instance, the verbal community infers the private experience on the basis of collateral nonverbal responses such as facial expression, grasping the bodily area that hurts, or groaning. Third, some responses made in the presence of a private experience may begin as descriptions of public behavior, and then later be performed in the presence of covert accompaniments of that same public behavior. For example, when a child has tissue damage and is crying, an adult may state, "That must hurt" or "Oh, that is painful." At a later time, the verbal response "pain" or "hurt" may be made in the presence of the collateral bodily sensation known only to the speaker. Finally, a verbal response reinforced in the presence of a particular property of a stimulus may naturally generalize to stimulus conditions that are subjectively similar. This metaphorical type of conditioning is called "stimulus induction." The subjective similarity of the private experience of calm and the overt appearance of a calm lake is an example.

Although the above analysis of verbal responses to private events can account for how such responses are possible, it also suggests why such verbal responses are often considered unreliable. Discriminations based on private events are crude because of the difficulties in establishing precise contingencies of reinforcement by the verbal community. That is, the relationship between public and private events may be faulty, the speaker may unknowingly respond to different private sensations with the same verbal response, and the metaphorical link between overt events and private events is necessarily imprecise. In addition, verbal responses about private events may come under the control of deprivations associated with reinforcing consequences rather than actual private experiences, as when one feigns a headache to avoid work. This analysis of the problem of maintaining control over verbal responses made in the presence of private events illustrates further how radical behaviorism is unlike methodological behaviorism. The problem of subjectivism or privacy is not that such talk refers to things that are not directly observed, which is an objection made by methodological behaviorists, and one easily countered by pointing to physicists' behavior of talking about quarks. Instead, the problem of subjectivism is that the verbal community cannot arrange contingencies of reinforcement in such a way as to make talk about private experience useful for scientific purposes of prediction and control. The "success" of the verbal community of high-energy particle physicists is that they have arranged the verbal community in such a way as to bring talk about quarks under the control of contingencies that allow humans to interact with unobserved events in a predictable manner.

It is by way of such a functional analysis of verbal behavior that a behavioral psychology of science arrives at the same conclusion that philosophical opponents of subjectivist versions of psychologism have reached: it is not possible to base knowledge claims on statements that refer to subjective experience. However, the route to this conclusion is quite different. The functional analysis of verbal behavior can also lead to a naturalistic account of logic as a form of rule-governed behavior.

2.3. An Answer to Logical A Priorism: The Analysis of Rule-governed Behavior

In a classic essay on problem solving, Skinner ([1966] 1988a) made a distinction between rule-governed and contingency-shaped behavior. As noted above, contingency-shaped behavior is the basic mechanical notion that behavior is selected and formed by the consequences it produces in the environment. In contrast to contingency-shaped behavior, rule-governed behavior is controlled by verbal descriptions of contingencies, and for that reason rule-governed behavior is something that happens only in organisms that use language. For example, the experienced center fielder acquires the skill of catching long fly balls through a process of successive approximations to catching the ball, and the behavior of catching the ball is shaped through contingencies of reinforcement that govern human behavior with respect to falling objects. The skill of catching a baseball is not generally acquired by first learning the rules of physics that permit one to plot the trajectory of falling objects. In contrast, a ship's captain who is steering the ship to catch a falling satellite must follow rules and plot trajectories in order to arrive at the proper location at the right time. This is due primarily to the fact that, unlike the center fielder, the ship's captain has not had the requisite history for the behavior of catching satellites to be shaped by successive approximations that have been reinforced by natural contingencies.

The practice of formulating verbal rules is what makes it possible for humans to behave effectively with respect to the contingencies of the natural world without having to undergo extensive shaping by natural contingencies of reinforcement. Rules are verbal descriptions of contingencies, and some rules get called "good," "accurate," "valid," etc. when the rule enables the rule follower to respond to the contingencies of the environment that control the rule follower's behavior. From this perspective, the laws of science are rules that enable humans to deal with nature and continue to survive under its contingencies. These rules are especially useful when the actual consequences described in the rule are remote, such as when medical researchers announce that there is a lawful relationship between smoking and the development of lung cancer. The significance of such knowledge is not that it is a rule that nature follows, but that it is a rule that people can follow to bring their behavior under control of delayed consequences. Rules function as discriminative stimuli that people provide for themselves as occasions for performing behaviors that get reinforced by environmental consequences, including the social consequences provided through the verbal community.

According to this approach, the field of logic is a part of the field of verbal behavior that consists of sets of rules that people have devised for manipulating other rules. The origin and continued use of various logical rules can be explained naturalistically as verbal discriminative stimuli that occasion reinforcement by the verbal community of scientists and philosophers and, of course ultimately, by the practical problems that get solved through research practices (Skinner 1957, chap. 18). Logic is a type of second-order rule that has been devised to bring the behavior of formulating rules (e.g., laws of science) under the control of other rules (e.g., laws of logic). Logic is a set of rules adopted by the verbal community to enable speakers to formulate rules according to certain forms and sequences. For example, in such expressions as "It is true that swans are white," the logical expression "It is true" functions as a type of verbal behavior that qualifies the remaining verbal behavior, "swans are white." In a functional analysis, the expression "It is true" produces an effect on the listener. For example, prefacing an assertion ("swans are white") with "It is true" may set the occasion for the listener to pay particular attention or to repeat the assertion "swans are white." The function of the expression can be established only by analyzing actual behavioral episodes of speaking where the expression occurs in the presence of an audience. From the standpoint of radical behaviorism, logical expressions are analyzed in the same manner as other verbal behavior, and their meaning is specified not by their form, but by the conditions in which they occur and the consequences they produce in a community of listeners.

Logical expressions of quantification such as "all" or "none" can be analyzed functionally as verbal behavior about other verbal behavior, and Skinner (1957) introduced the technical term "autoclitic" (from the Greek "leaning on itself") to refer to this general class of functional units of verbal behavior. For example, in the verbal behavior "All swans are white," the quantifying autoclitic "all" performs a function with respect to the listener. "All" performs the same function as "It is always the case that." The expression tells the listener that the response "white" will always be reinforced by the verbal community on occasions when the response "swan" is made and also reinforced in the presence of a particular type of bird. The "truth" of the expression is in the extent to which it describes the actual practices of the verbal community with respect to swans. The quantifier "all" is a rule that specifies the contingencies of reinforcement that the verbal community supplies for verbal behavior. Like the field of grammar, the field of logic is a set of rules that describes the practices of the verbal community with respect to contingencies of reinforcement that apply to the order and form of verbal behavior. Just as people spoke grammatically before the rules of grammar were articulated, people argued logically before the laws of logic were formulated. Once the rules of grammar and the laws of logic were formulated as second-order verbal behavior, they could be followed so that other verbal behavior could be made to conform to such rules.

Such a naturalistic account of logic as verbal rules about verbal behavior vitiates appeals to logical a priorism based on inner faculties of the mind or structures of the brain and shows instead that claims for logical a priorism are actually based on practices and conventions that operate in the verbal community of speakers and listeners. Nevertheless, such an account does not entail that logic is not useful. To the contrary, it suggests that the laws of logic have been formulated precisely because they have enabled and continue to enable people to behave effectively with respect to their own verbal behavior. However, the fact that the laws of logic are not immutable and that they depend finally upon their usefulness has been illustrated with the case of quantum logic where classical laws of distribution are violated in favor of formulating some laws of nature (Putnam 1969). As Skinner ([1945] 1988b, pp. 158–59) summarized his earliest work on verbal behavior: "If it turns out that our final view of verbal behavior invalidates our scientific structure from the point of view of logic and truth value, then so much the worse for logic."

Radical behaviorism as an approach to psychology of science can not only address the chief concerns of philosophical objections of psychologism; the approach also is consistent with sociological objections to the recent cognitive turn in science studies.

2.4. An Answer to "Cognitive Processes": Verbal Behavior Is Social Behavior

From the perspective of radical behaviorism, the attractiveness of recent calls for scholars in science studies to adopt the language of cognitive science (represented by this volume as well as a previous essay by Gholson and Houts 1989) may be understood as a correlate of how behavior is explained in everyday discourse. When asked why we are eating we typically respond, "Because I am hungry," or when someone inquires why we espouse a particular set of beliefs, we may state, "I believe them to be true." This way of talking usually leads us to say that when others eat, they must be hungry, and when others verbalize certain propositions, it is due to their beliefs. At least in folk psychology, these feelings or beliefs are usually thought to reside in a nonphysical realm called the mind. But from the standpoint of radical behaviorism, to stop with explanations of behavior that refer to feelings or states of mind as causes is a problem. To be sure, important parts of a scientist's behavioral repertoire may be covert, such as thinking and problem solving. As noted previously in presenting sociological objections to psychology of science, cognitive psychologies of science attempt to study this aspect of scientific behavior by restating their observations of a scientist's behavior in terms of purpose, intention, or ideas that are moved into the scientist's mind or computerlike brain. In a cognitive account, a scientist's practical activity in the world and the reinforcement history of that activity get merely redescribed in terms of hypothetical cognitive mechanisms. Such cognitive activities are then further abstractly described as undetermined, decontextualized information processing that originates solely within the scientist. Such an account of scientific behavior both fails to explain the behavior of the scientist and effectively halts further inquiry.

In a manner that is consistent with sociological criticisms of appeals to cognitions, radical behaviorism takes the view that language is a system of verbal operant behavior that is a part of a larger set of social behaviors. Verbal behavior has a special character only because it is reinforced by its effect on the verbal community (i.e., people, including the speaker). The practices of the verbal community of which a person is a part determine how a person will speak. Verbal behavior is reinforced as a means of securing primary or nonlinguistic secondary reinforcers from others, as when an author alters a manuscript to get published, to gain tenure, and so forth. However, unlike nonverbal actions, verbal behavior requires no environmental support (other than an occasional audience), because a speaker may reinforce her or his own verbal utterances (Skinner 1957, 1974). Thus, thought, according to this analysis, is verbal behavior within individuals (e.g., having learned how to adjust an apparatus to produce a result, one can repeat instructions and then follow these instructions).

Traditionally, language has been viewed as a thing, something a person acquires and possesses, then uses when necessary. Words and grammar are held to be tools used to express meanings, thoughts, and ideas that are said to be in the mind of the speaker. The radical behaviorist avoids the objectification of words as things or tokens used to transfer ideas to the minds of others. Instead, the "meaning" of a response is found in its antecedent history. Meaning is not determined either by properties of the specific verbal response or the situation, but by the contingencies of reinforcement that determine the topography of the behavior and the controlling stimuli (Skinner 1957, 1974). For example, if an experimenter rings a bell each time before food is presented to an organism, the sound of the bell "means" food to the organism. However, if water is presented contemporaneous with the ringing of the bell, the bell "means" water. The meaning of the bell does not refer to some property of the bell, but to the contingencies of which the bell is a part.

From the standpoint of radical behaviorism and the analysis of talk about private events, events that take place within an organism's skin do not have special properties for that reason. A covert event differs from publicly observable behavior due only to its limited accessibility, not to any special structure or nature (Skinner 1953). Once private events are viewed in this light, they can be linked to their behavioral referents, thus becoming accessible to people who study scientists. This is possible because covert behavior, such as thinking, almost always originates as overt behavior. In the only way that it is available for study, thought is verbal behavior. It is a mistake to assume that this type of verbal behavior is somehow controlled by variables other than the social variables specified in the functional relationships identified by the behavioral analysis of verbal behavior.

According to this analysis, scientific and logical thinking are simply behaviors to be explained in terms of the reinforcement history of the scientist. For example, there is no compelling reason to posit that science involves discovering a priori truths of various logical operations. Rather, subtle and complex reinforcement histories create a special type of stimulus control that enables the scientist to behave effectively with respect to prevailing contingencies of reinforcement found in nature. Thus, the point of science is to analyze the contingencies of reinforcement found in nature and to formulate rules or laws that make it unnecessary to be exposed to the contingencies in order to behave effectively. It is misleading to say that science involves forming ideas or representations of the world. If we become content with talking about scientific concepts, we will never discover what the scientist actually has learned. The referents of scientific concepts or ideas are in the real world, not in the verbal behavior of the scientist (Skinner 1974).

By subsuming verbal behavior under social behavior, radical behaviorism can give an account of the behavior of scientists that does not sacrifice the social determinants of scientific knowledge claims. The approach of radical behavioral psychology of science is, in this respect, consistent with certain sociological analyses. However, this type of behavioral psychology of science is also inconsistent with some of the extreme relativism espoused by radical constructivists. Epistemology from the standpoint of radical behaviorism does reach bedrock in that the constraints on verbal behavior are taken finally to be the same constraints that control the survival of the human species. To find out what those constraints are without having to undergo extinction is the point of science.

3. Summary and Conclusions

In this essay we have presented the rudiments of a radical behavioral analysis of some traditional problems of science studies associated with philosophical and sociological usages of psychologism as objections to psychology of science. First, we argued that some traditional philosophical objections to psychological approaches to science studies either do not apply or are answered by the approach of radical behaviorism. Radical behaviorism deals with the charge of subjectivism through a functional analysis of verbal behavior made in response to private events. Radical behaviorism also offers a functional, behavior-analytic account of logic as a form of rule-governed behavior. Second, we argued that recent developments in sociology of science suggest that behavioral analysis is compatible not only with certain trends toward naturalizing science studies, but also with sociological critiques of cognitive approaches. What cognitivists and traditional philosophers of science have called "cognitive" norms of science may be accounted for as verbal behaviors that function in the verbal community of scientists to control scientific practices. According to a radical behavioral account, acting in accord with scientific methodology may be understood as another instance of rule-governed behavior that is controlled by the verbal community. Third, we argued that traditional "cognitive norms" may be naturalized by treating them as verbal behaviors that may be controlled immediately by the verbal responses (consequences delivered by people) of philosophical and scientific audiences, but also ultimately by the practical outcomes (consequences delivered by nature) of scientific practices.

Although as psychologists we would like to reinforce the verbal behavior of our nonpsychologist colleagues when they introduce psychological talk into the field of science studies, as behaviorists we wish to introduce some cautions regarding what kind of psychological talk is more likely to lead to achieving the aim of making science studies more scientific. To that end, we have suggested that before we all get on board the cognitive train to improve science studies, we should at least take a hard look at some of the costs of playing the language game of cognitive psychology and cognitive science. The perspective of radical behaviorism, especially the functional analysis of verbal behavior, provides one psychological alternative for salvaging the insights of recent sociology of science without at the same time abandoning traditional scientific aims of prediction and control.

Note

C. Keith Haddock's work on this chapter was supported by the Van Vleet Fellowship of Memphis State University. Arthur C. Houts's work on this chapter was partially supported by a grant from the National Institute of Child Health and Human Development.

References

Aach, J. 1987. Behaviorism and normativity: Prospects of a Skinnerian psychologism. Ph.D. diss., Boston University.

———. In press. Psychologism reconsidered: A re-evaluation of the arguments of Frege and Husserl. *Synthese*.

Abbagnano, N. 1967. Psychologism. Trans. N. Langiulli. In P. Edwards, ed., *The encyclopedia of philosophy*. Vol. 6, pp. 520–21. New York: Macmillan.

Baars, B. J. 1986. *The cognitive revolution in psychology*. New York: Guilford.

Bloor, D. 1983. *Wittgenstein: A social theory of knowledge*. New York: Columbia University Press.

Bowker, G., and B. Latour. 1987. A booming discipline short of discipline: (Social) studies of science in France. *Social Studies of Science* 17:715–48.

Brannigan, A. 1981. *The social basis of scientific discoveries.* Cambridge: Cambridge University Press.
——. 1989. Artificial intelligence and the attributional model of scientific discovery. *Social Studies of Science* 19:610–13.
Carnap, R. 1936. Testability and meaning: I. *Philosophy of Science* 3:419–71.
——. 1937. Testability and meaning: II. *Philosophy of Science* 4:2–40.
——. [1932] 1959. Psychology in physical language. In A. J. Ayer, ed., *Logical positivism*, pp. 165–98. New York: Free Press.
Catania, A. C. 1972. Chomsky's formal analysis of natural languages: A behavioral translation. *Behaviorism* 1:1–15.
Chomsky, N. 1957. *Syntactic structures.* The Hague: Mouton.
——. 1959. Review of *Verbal behavior*, by B. F. Skinner. *Language* 35:26–58.
Collins, H. M. 1983. The sociology of scientific knowledge: Studies of contemporary science. *Annual Review of Sociology* 9:265–85.
——. 1987. Expert systems and the science of knowledge. In W. E. Bijker, T. P. Hughes, and T. J. Pinch, eds., *The social construction of technological systems: New directions in the sociology and history of technology*, pp. 329–48. Cambridge, Mass.: MIT Press.
Collins, H. M., and T. J. Pinch. 1982. *Frames of meaning: The social construction of extraordinary science.* London: Routledge and Kegan Paul.
Coulter, J. 1983. *Rethinking cognitive theory.* London: Macmillan.
Dewey, J. 1938. *Logic: The theory of inquiry.* New York: Henry Holt.
Dummett, M. A. E. 1981. *The interpretation of Frege's philosophy.* Cambridge, Mass.: Harvard University Press.
Frege, G. [1884] 1960. *The foundations of arithmetic.* 2d ed. Trans. J. L. Austin. New York: Harper and Row.
Fuller, S. 1988. *Social epistemology.* Bloomington: Indiana University Press.
——. 1989. *Philosophy of science and its discontents.* New York: Westview.
——. 1990. Some hints on how to be cognitively revolting. Unpublished manuscript, Center for the Study of Science in Society, Virginia Polytechnic Institute and State University, Blacksburg, Va.
Gholson, B., and P. Barker. 1985. Kuhn, Lakatos, and Laudan: Applications in the history of physics and psychology. *American Psychologist* 40:755–69.
Gholson, B., and A. C. Houts. 1989. Toward a cognitive psychology of science. *Social Epistemology* 3:107–27.
Giere, R. N. 1988. *Explaining science: A cognitive approach.* Chicago: University of Chicago Press.
Gilbert, G. N., and N. Mulkay. 1984. *Opening Pandora's box: A sociological analysis of scientists' discourse.* Cambridge: Cambridge University Press.
Goldman, A. 1986. *Epistemology and cognition.* Cambridge, Mass.: Harvard University Press.
Gorman, M. E. 1989. Beyond strong programmes: How cognitive approaches can complement SSK. *Social Studies of Science* 19:643–53.
Haack, S. 1979. Epistemology with a knowing subject. *Review of Metaphysics* 33:309–35.
Heil, J. 1981. Does cognitive psychology rest on a mistake? *Mind* 90:321–42.
Heyes, C. M. 1989. Uneasy chapters in the relationship between psychology and epistemology. In B. Gholson et al., eds., *Psychology of science and metascience*, pp. 115–37. Cambridge: Cambridge University Press.
Houts, A. C. 1989. Contributions of the psychology of science to metascience: A call for explorers. In B. Gholson et al., eds., *Psychology of science and metascience*, pp. 47–88. Cambridge: Cambridge University Press.

Husserl, E. [1911] 1965. Philosophy as rigorous science. In Q. Lauer, ed. and trans., *Phenomenology and the crisis of philosophy*, pp. 71–147. New York: Harper and Row.
———. 1970. *Logical investigations*. 2 vols. Trans. J. N. Findlay. New York: Humanities Press.
———. 1977. The task and the significance of the Logical Investigations. In J. N. Mohanty, ed. and trans., *Readings on Edmund Husserl's Logical Investigations*. The Hague: Martinus Nijhoff.
Knorr-Cetina, K. D. 1981. *The manufacture of knowledge: An essay on the constructivist and contextual nature of science*. Oxford: Pergamon.
———. 1983. The ethnographic study of scientific work: Towards a constructivist interpretation of science. In K. D. Knorr-Cetina and M. J. Mulkay, eds., *Science observed: Perspectives on the social study of science*, pp. 115–40. London: Sage.
Kolakowski, L. 1975. *Husserl and the search for certitude*. New Haven: Yale University Press.
Kuhn, T. S. 1970. Logic of discovery or psychology of research? In I. Lakatos and A. Musgrave, eds., *Criticism and the growth of knowledge*, pp. 1–23. Cambridge: Cambridge University Press.
Lakatos, I. 1970. Falsification and the methodology of scientific research programmes. In I. Lakatos and A. Musgrave, eds., *Criticism and the growth of knowledge*, pp. 91–198. Cambridge: Cambridge University Press.
———. 1973. History of science and its rational reconstructions. In R. C. Buck and R. S. Cohen, eds., *Boston studies in the philosophy of science*. Vol. 3, pp. 91–136. Dordrecht: D. Reidel.
Latour, B. 1987. *Science in action*. Cambridge, Mass.: Harvard University Press.
———. 1988. The politics of explanation: An alternative. In S. Woolgar, ed., *Knowledge and reflexivity: New frontiers in the sociology of knowledge*, pp. 155–76. London: Sage.
Latour, B., and S. Woolgar. 1986. *Laboratory life: The construction of scientific facts*. Princeton, N.J.: Princeton University Press.
Laudan, L. 1977. *Progress and its problems: Towards a theory of scientific growth*. Berkeley: University of California Press.
Lyotard, J. F. 1984. *The postmodern condition: A report on knowledge*. Minneapolis: University of Minnesota Press.
MacCorquodale, K. 1970. On Chomsky's review of Skinner's Verbal Behavior. *Journal of the Experimental Analysis of Behavior* 13:83–99.
Musgrave, A. E. 1974. The objectivism of Popper's epistemology. In P. A. Schlipp, ed., *The philosophy of Karl Popper*, pp. 560–96. LaSalle, Ill.: Open Court.
Notturno, M. A. 1985. *Objectivity, rationality and the third realm: Justification and the grounds of psychologism*. Dordrecht: Martinus Nijhoff.
Pickering, A. 1984. *Constructing quarks: A sociological history of particle physics*. Chicago: University of Chicago Press.
Popper, K. R. [1934] 1959. *The logic of scientific discovery*. New York: Harper and Row.
———. 1968. *Conjectures and refutations*. New York: Harper and Row.
———. 1970. Normal science and its dangers. In I. Lakatos and A. Musgrave, eds., *Criticism and the growth of knowledge*, pp. 51–58. Cambridge: Cambridge University Press.
———. 1972a. Epistemology without a knowing subject. In *Objective knowledge: An evolutionary approach*, pp. 106–52. Oxford: Oxford University Press.
———. 1972b. On the theory of the objective mind. In *Objective knowledge: An evolutionary approach*, pp. 153–90. Oxford: Oxford University Press.

———. 1972c. Two faces of common sense. In *Objective knowledge: An evolutionary approach*, pp. 32–105. Oxford: Oxford University Press.

———. 1974. Replies to my critics. In P. A. Schlipp, ed., *The philosophy of Karl Popper*, pp. 1144–53. LaSalle, Ill.: Open Court.

———. 1983. *Realism and the aim of science*. London: Hutchinson.

Popper, K. R., and J. C. Eccles. 1977. *The self and its brain*. London: Springer International.

Putnam, H. 1969. Is logic empirical? In R. S. Cohen and M. Wartofsky, eds., *Boston studies in the philosophy of science*. Vol. 5, pp. 216–41. New York: Humanities Press.

Quine, W. V. O. 1960. *Word and object*. Cambridge, Mass.: MIT Press.

Reichenbach, H. 1938. *Experience and prediction*. Chicago: University of Chicago Press.

———. 1951. *The rise of scientific philosophy*. Berkeley: University of California Press.

Rorty, R. 1979. *Philosophy and the mirror of nature*. Princeton, N.J.: Princeton University Press.

Ryle, G. 1949. *The concept of mind*. London: Hutchinson.

Sarton, G. 1927–48. *Introduction to the history of science*. 3 vols. Baltimore: Williams and Wilkins.

Skinner, B. F. 1953. *Science and human behavior*. New York: Macmillan.

———. 1957. *Verbal behavior*. New York: Appleton-Century-Crofts.

———. 1974. *About behaviorism*. New York: Alfred A. Knopf.

———. 1985. Cognitive science and behaviorism. *British Journal of Psychology* 76:291–301.

———. [1966] 1988a. An operant analysis of problem solving. In A. C. Catania and S. Harnad, eds., *The selection of behavior: The operant behaviorism of B. F. Skinner: Comments and consequences*, pp. 218–77. New York: Cambridge University Press.

———. [1945] 1988b. The operational analysis of psychological terms. In A. C. Catania and S. Harnad, eds., *The selection of behavior: The operant behaviorism of B. F. Skinner: Comments and consequences*, pp. 150–217. New York: Cambridge University Press.

Slezak, P. 1989. Scientific discovery by computer as empirical refutation of the strong programme. *Social Studies of Science* 4:563–600.

Wittgenstein, L. 1975. In C. Diamond, ed., *Wittgenstein's lectures on the foundations of mathematics: Cambridge, 1939*. Chicago: University of Chicago Press.

Woolgar, S. 1981. Interest and explanation in the social study of science. *Social Studies of Science* 2:365–94.

———. 1986. On the alleged distinction between discourse and praxis. *Social Studies of Science* 16:309–17.

———. 1987. Reconstructing man and machine: A note on sociological critiques of cognitivism. In W. E. Bijker, T. P. Hughes, and T. J. Pinch, eds., *The social construction of technological systems: New directions in the sociology and history of technology*, pp. 311–28. Cambridge, Mass.: MIT Press.

———, ed. 1988. *Knowledge and reflexivity: New frontiers in the sociology of knowledge*. London: Sage.

———. 1989. Representation, cognition and self: What hope for an integration of psychology and sociology? In S. Fuller et al., eds., *The cognitive turn: Sociological and psychological perspectives on science*, pp. 201–24. Dordrecht: Kluwer.

Simulating Social Epistemology: Experimental and Computational Approaches

Philosophers of science have made a bewildering variety of recommendations concerning how scientists ought to behave. Popper (1972), for example, tells them to falsify; Kuhn (1970) says that is good advice only during a crisis, but that the rest of the time they should do research within a paradigm or set of exemplars; Feyerabend (1975) tells them to do whatever they want. One could go on. The point is, how does one assess which (if any) of these philosophical prescriptions is right?

One answer would be to simply analyze the way scientists behave, deriving general principles of good conduct. Lakatos's method of "rational reconstruction" serves as an example: reconstruct historical cases of science in the light of what ought to have happened, rationally speaking, allowing the history and principles to inform each other. The problem with this sort of method is that every philosopher can potentially select her favorite cases and read her philosophical prejudices into them.

In this essay, I wish to suggest a complementary approach: simulations of science in which important variables can be controlled and manipulated to assess their effect on reasoning and representations. To illustrate the potential value of such simulations, let us consider a thought experiment. Suppose we could take large groups of scientists in different areas and persuade them to adopt precisely the methods advocated by different philosophers, taking care to insure that the scientists and the problems they focused on were equivalent in all other respects. Then we could compare the results obtained via each philosopher's prescriptions, and determine the best method objectively.

One of the new sociologists of scientific knowledge (e.g., Collins 1985) might object that the criteria for determining success would have to be the results of social negotiations, thereby undermining the objectivity of the whole exercise; after all, at the end, the losers could still redefine what constituted success, and claim that they had won. Fair enough, but one could counter that a giant simulation of the sort just outlined would at least shed important light on philosophical theories, forcing them to be more precise. It would also allow us to look at all kinds of measures besides success, including scientists' representations and rhetoric.

Of course, such a simulation is impossible. But smaller scale simulations are possible, and have in fact been carried on by psychologists for some years now. In this essay, I will review two major types of simulation, experimental and computational, describing research to date and outlining a program of future studies. My colleague Steve Fuller (this volume) will follow with a discussion of how such a program of research can address issues in social epistemology.

1. Two Types of Validity

At the outset, it is important to distinguish between the different types of claims one can make from simulations that fall short of the idealized and unrealistic thought experiment outlined above. Berkowitz and Donnerstein (1982) distinguish between two types of validity, ecological and external. An experiment has ecological validity if both the pool of subjects and the experimental situation are representative of some real-world situation. To have ecological validity, or what Berkowitz and Donnerstein somewhat disparagingly call "mundane realism," experimental studies of scientific reasoning would have to use working scientists, have them work on tasks that bear a close resemblance to actual scientific problems, and set up a working environment that resembles an actual laboratory or laboratories in competition (see Houts and Gholson 1989, for a discussion). This takes us close to the thought experiment outlined above. One could legitimately wonder why it would not be better to take Latour's (1987) advice and simply "follow scientists around."

Of course, one could remind those enamored of thick descriptions that they cannot generalize beyond specific social settings, either. The value of a social constructivist approach cannot be demonstrated by studying a few laboratory situations, because these might be peculiar; it depends either on a potentially infinite range of such studies or on specifying with greater precision what aspects of one thick description can be generalized to another.

The other problem with simply "following scientists around" and generating thick descriptions is there is no way of controlling for the ideology of the investigator. One can construct a thick description according to one's preconceptions: a kind of irrational reconstruction, perhaps, in which the content of the science and the ideas are ignored in favor of social networks.

So, the experimenters and the social constructivists may both be criticized on the grounds of insufficient generalizability. The experimenters have dealt with the problem of generalization via the notion of external validity. An experiment has external validity if it addresses what ought to happen, under ideal conditions. Berkowitz and Donnerstein argue that "artificiality is the strength and not the weakness of experiments" (1982, p. 256). The higher the ecological validity, the more complex the situation, until it becomes impossible to tell what variables are playing a causal role. Similarly, Banaji and Crowder (1989) argue that "ecological validity of the methods as such is unimportant and can even work against generalizability" (p. 1187).

Controlled experiments, therefore, can deliberately isolate variables of theoretical interest and assess their impact under circumstances that eliminate the noise due to other variables. If a theory does not work even under ideal laboratory conditions, it cannot be generalized beyond the laboratory. In mundanely realistic situations, one can always invoke *ceteris paribus* clauses to explain why people did not act in accordance with the theory. These clauses can be invoked in the laboratory as well, but they should be framed in a way that leads to new tests.

Externally valid experiments can also act as hypothesis generators (see McGuire 1989). In other words, they can beg the question of generalizability: people perform actions Y with processes Z under conditions X in the laboratory. Might they not do the same under more mundanely realistic situations? One can use thick descriptions to explore hypotheses generated from experiments, in turn leading to proposals for new experiments and new field studies *ad infinitum*.

2. Two Types of Experimental Research

To clarify this discussion, we need to consider examples of experimental research on scientific reasoning. For the purposes of illustration, I will focus on one of the most "artificial" programs, one involving abstract tasks that purport to simulate scientific reasoning.

2.1. Experiments Using Abstract Tasks

Consider Wason's (1960) 2-4-6 task: college students (and sometimes scientists; see Mahoney 1976) are told that the number triple "2, 4, 6" is an instance of a rule that the experimenter has in mind, and subjects are to guess the rule by proposing additional triples. The experimenter tells them whether each triple is an instance of his rule. Each triple is analogous to an experiment whose results can be viewed as either supporting or contradicting a hypothesis. For example, if one supposes the rule might be "numbers go up by twos," one could propose triples like

8, 10, 12 and 13, 15, 17; if these triples are right, they provide confirmations of one's hypothesis. Alternately, if one of these is wrong, one's hypothesis is disconfirmed.

This is a highly artificial situation, far removed from the daily practices of working scientists. Nonetheless, Wason's task and similar ones were adapted to the study of scientific reasoning. Some of these tasks were far more complex and therefore potentially possessed more ecological validity (see Mynatt, Doherty, and Tweney 1978; Klahr and Dunbar 1988), but let us once again deliberately stick to highly simplified and artificial situations.

Consider two *generalizations* that emerge from experiments on scientific reasoning using the 2-4-6 task:

1. If subjects must find a rule that is more general than initial instances suggest, they can be trained to disconfirm, but will succeed only if their representation of the rule suggests where to look for disconfirmatory information (see Gorman, Stafford, and Gorman 1987; and Klayman and Ha 1987).

2. The possibility of system-failure error significantly disrupts problem solving only when results of the current experiment depend on results of previous ones (Gorman 1986, 1989c).

To understand more clearly how such generalizations emerge, let us discuss part of the chain of experiments that led to (1) above, unpacking one in particular detail. Research by Gorman and Gorman (1984) suggested that giving subjects instructions to disconfirm dramatically improved their performance on Wason's traditional 2-4-6 rule, "numbers must ascend in order of magnitude," thereby supporting the idea that falsification might be a useful heuristic. But Gorman, Stafford, and Gorman (1987) showed that this disconfirmatory strategy did not work for a very general rule: "the three numbers must be different."

To find out why, Gorman, Stafford, and Gorman, following Tweney et al. (1980), reframed Wason's task as a search for two rules. Instead of being classified as correct or incorrect, each triple was classified as DAX or MED. "Three different numbers" was the DAX rule, which meant that the MED rule was "two or more numbers the same." The triple "2, 4, 6" was given as an instance of the DAX rule. A control condition was used as well, in which Ys were used to indicate correct triples and Ns were used to indicate incorrect triples.

Differences between conditions were dramatic. DAX-MED subjects solved the rule significantly more often than control subjects, obtained a significantly higher proportion of MED triples than incorrect triples obtained by control subjects, and proposed significantly more triples (see Table 1). Previous research had established a strong correlation between proportion of incorrect triples and solving a general rule (Gorman and Gorman 1984).

Table 1: A Comparison between DAX-MED and Control Subjects on the "Three Different Numbers" Rule (N = 24 in Each Condition)

	DAX-MED	Control
Number of subjects who solved the rule	21	5
Average proportion of incorrect or MED triples obtained	.29	.09
Average number of triples proposed	18.5	9.5

Tweney et al.'s (1980) effect for DAX-MED instructions was clearly replicated. Tweney et al. offered three explanations for the success of DAX-MED instructions:

1. Telling subjects that some triples are incorrect interferes with their ability to integrate this information. But Gorman et al. (1984) and Gorman and Gorman (1984) have showed that subjects can make effective use of negative information.

2. DAX and MED hypotheses can be processed separately. A good analogy here may be to some of Ryan Tweney's (1985) work on Michael Faraday's discovery of electromagnetic induction. Faraday experimented on two complementary hypotheses separately: one said that electricity could be induced from an electromagnet and the other said that electricity could be induced from an ordinary magnet. Confirmations of the first hypothesis came almost immediately, but the second hypothesis was far harder to confirm. As Tweney noted:

> Faraday seems to have played off an easily verified hypothesis against one that was not so encouraging at the start. So, when things were discouraging on the ordinary magnet side, he switched to electromagnets, to pursue further knowledge about the effect he knew how to produce. As he gained expertise, he was able to refine his attempts with ordinary magnets until he succeeded.... [This] reconstruction is plausible from a cognitive standpoint because similar "dual hypothesis" strategies have been shown to be effective in laboratory studies of problem solving. Working with two closely related hypotheses helps

because disconfirming evidence for one leaves the investigator with the other hypothesis as a focus of effort. (Tweney 1985, p. 204)

The DAX-MED experiment is the "dual hypothesis" research referred to in the above quotation. Tweney sees a similarity between DAX and MED hypotheses and Faraday's "electricity can be induced from electromagnets" and "electricity can be induced from permanent magnets" hypotheses. The DAX and MED rules are mutually exclusive alternatives, whereas Faraday's two hypotheses are complementary. Still, it is possible that both Faraday and the subjects switched between easily confirmed and hard-to-confirm hypotheses to maintain confidence in their research programs.

To see if subjects in the DAX-MED condition were really following a hypothesis-switching strategy, Gorman, Stafford, and Gorman (1987) asked subjects to predict whether each triple they proposed would be a DAX or a MED. A comparison of DAX-MED and control conditions provided some support for Tweney's position: DAX-MED subjects switched from predicting one rule to predicting the other 35 percent of the time, whereas control subjects switched from predicting Y to predicting N 26 percent of the time. This difference is statistically significant, but it is partially explained by the fact that there were significant differences between conditions in terms of proportion of incorrect or MED triples predicted and obtained. In fact, five control subjects proposed only triples they thought would be Y. In other words, control subjects simply proposed fewer triples they thought would be N, and therefore switched from one kind of triple to the other less often than DAX-MED subjects.

3. Tweney et al.'s final explanation is that the DAX-MED manipulation alters the logical structure of the task. Tweney does not spell out the implications of this alteration, but Gorman, Stafford, and Gorman focused on how the DAX-MED design changes the way subjects mentally represent the task.

Wason and Green (1984) argued that performance on the selection task was greatly improved if subjects' mental representations of the task were altered. Similarly, the DAX-MED design appears to highlight or make obvious the role of disconfirmatory information; subjects immediately realize that the key to the problem is discovering the MED rule. As one subject said on a follow-up questionnaire about his strategy, "I tried to change from the first triple in all ways to find out what the MED rule was."

Furthermore, when they discover their first MED instances, they do not stop until they have made some attempt to identify the limits of the MED rule. The subject quoted above found his first MED instance on the tenth triple he proposed, which was 8, 8, 8, and after proposing 1, 1, 1, concluded that the MED rule was "all the elements are the same number." Before he wrote down the DAX rule he paused and tried 1, 1, 2, which he predicted would be DAX but which turned out to be MED. Then he wrote down the correct rules for both DAX and MED.

Thus, Tweney et al.'s third alternative appears to come the closest to being correct: the DAX-MED manipulation changes the way subjects represent the task, from one in which subjects seek to disconfirm a single rule to one in which they try to confirm two mutually exclusive rules. Consider what happened when a DAX-MED subject encountered the triple "0, 0, 0," as two did. Both of these subjects used the fact that this triple was MED as a clue to the identity of the rule, and focused on what happened when three numbers were the same. In contrast, the one control subject who encountered 0, 0, 0 stopped and concluded that it was an exception; his final guess was "three consecutive integers, each multiplied by the same number (but not 0)." He was looking for a single rule that might have exceptions; DAX-MED subjects represented the task as a search for two distinct rules.

Further evidence for this view comes from the fact that no DAX-MED subjects thought the DAX rule could be "any three numbers"; indeed, the one subject in this condition who failed to find a single MED instance realized he had simply failed to find the rule, and was still convinced that there was one. In contrast, six control subjects decided the rule was "any number" because their exhaustive search failed to turn up a single N.

To summarize, DAX-MED subjects represented the task as a search for two exclusive rules, whereas Y-N subjects represented the task as a search for a single rule that might or might not have exceptions. Naturally, there were individual differences, but it is impressive how effective this manipulation was in altering representation, which in turn affected strategy.

Even though the ecological validity of this simulation is open to question, one can derive externally valid generalizations from it, like (1) above: if subjects or scientists must check whether a rule or principle is more general than initial instances suggest, a disconfirmatory strategy is potentially effective, but only if it is coupled with an appropriate task representation.

Tweney (1985), Gorman and Carlson (1990), and Carlson and Gorman (1990) are trying to take generalizations derived in part from the laboratory and use them to shed light on the cognitive processes of scientists and inventors: Tweney focuses on Faraday while Gorman and Carlson focus on three telephone inventors: Thomas Alva Edison, Elisha

Gray, and Alexander Graham Bell. While both Tweney and Gorman use findings from their 2-4-6 experiments, they also borrow generalizations from a wide range of other cognitive studies. This cross-fertilization illustrates the interaction between externally and ecologically valid findings.[1]

2.2. Experiments Using Scientific Problems

Chi (this volume), Larkin (1983), Gentner and Gentner (1983), and others have compared the performance of novices and experts working on problems similar to those that might be included in advanced physics or biology textbooks. For example, Chi, Feltovich, and Glaser (1981) noted that experts use abstract principles of physics like conservation of energy to classify problems, whereas novices focus on surface features of the problems. Larkin (1983) trained subjects to adopt a "physical-representation schema," which corresponds roughly to the expert schemata used by Chi's subjects, and found that such schema significantly improved performance, as compared with a control group.

Similar studies have been done on the role of analogy in solving technoscientific problems. For example, Gentner and Gentner (1983) instructed subjects to use one of two different analogies or mental models on electrical circuit problems. In one, subjects were told to view the passage of electricity in terms of a moving crowd; in the other, they were told to view it in terms of a flowing fluid. The former group performed better on parallel-resistor problems. Clement (1988) asked ten subjects who were either advanced doctoral students or professors in technical areas to solve a problem: What would happen if the diameter of the coils of a spring holding a weight were doubled? He noted that eight of the ten subjects used analogies to solve the problem, and about 60 percent of these analogies led to significant changes in the way they represented the problem.

On the face of it, these experiments and quasi-experiments have more ecological validity than those involving abstract tasks like Wason's, in that most involve actual scientists working on technical problems. However, I would caution that ecological validity depends on how evidence is collected and analyzed. For example, a number of these accounts assume at the outset that the thinking of both subjects and scientists is organized in terms of representations like production rules, which are "if, then" action statements (see Chi et al., in press; Singley and Anderson 1989). One therefore organizes one's data collection and analysis in terms of constructs like production rules. This can lead to a kind of "confirmation bias" in which one sees only what one is looking for. Note that one can still make the externally valid claim that training subjects to use a particular kind of production rule or representation produces superior performance on textbook problems without making the broader ecological claim that experts ordinarily use such rules or representations in their daily scientific activities. Unlike the advocates of thick descriptions, psychologists who infer production rules from protocols can at least check the external validity of their accounts by seeing what happens when students or scientists are trained to adopt their recommendations.

There are still dangers, of course; as McGuire (1989) noted, clever experimenters can usually find a way to confirm their hypotheses. This plays into generalization (1) from the previous section. Just as subjects can carefully select only triples that provide positive evidence for their hypotheses, experimenters can select situations that produce only confirmations. But this is no way to find a rule that is more general than a narrow and restricted version of a hypothesis. To put it another way, production-rule accounts may work well for certain kinds of learning, e.g., mastering LISP syntax, but may not reflect more general learning processes. Externally valid generalizations always come with *ceteris paribus* clauses; these clauses will be very restrictive if the range of experimental situations is narrow.

3. Computational Simulation: An Alternate Approach

Langley et al. (1987) advocate using computers to model scientific discovery. They have written a series of heuristic-based programs like BACON, which "discovers" Kepler's laws by manipulating columns of numbers that happen to correspond to data on the planets (see Gorman [1987] for a more extensive critique). This approach is really no different from putting intelligent mathematics students in a room and giving them a set of heuristics, a calculator, and several columns of numbers. They could discover mathematical relationships in this data, but would have no idea what they meant (see Qin and Simon [1990] for some data on this subject).

As the Kepler example illustrates, these programs have low ecological validity. However, like the experiments cited above, these computational simulations do potentially possess external validity; programs allow one to simulate what might happen under ideal circumstances. BACON's heuristics are sufficient to discover numerical relations in planetary data sets, once what constitutes data has been made absolutely clear and the numbers have been organized. The question of whether scientists actually employ these heuristics needs to be settled by looking at the actual historical cases in detail. Some of the other cases cited by Langley et al. may possess more ecological validity; after all, important discoveries can arise from mathematical manipulations. The problem is that Langley et al.'s use of the term "discover" is ambiguous; it is not clear whether they wish to make an externally or an ecologically valid claim.

Consider a more sophisticated discovery program. Kulkarni and Simon (1988) developed KEKADA to simulate the heuristics used by Krebs in his discovery of the ornithine cycle. KEKADA can propose tests, but results "are supplied interactively by the user" (p. 156). Decisions about what hypothesis to pursue at certain points in the program are also supplied by the user. Kulkarni and Simon claim that KEKADA "constitutes a theory of Krebs' style of experimentation" (p. 171). This is a confusing claim in that it incorporates elements of both external and ecological validity; the former is suggested by their use of the term "theory," the latter by their reference to "Krebs' style."

Kulkarni and Simon highlight Krebs's expertise with tissue slicing as a major feature of his "style," but there is no way in which their program can incorporate this technical skill, which played an important role in Krebs's success. In general, programs cannot duplicate the kind of hands-on, craft knowledge possessed by scientists (see Gooding [1985, 1990] for a discussion of Faraday's procedural skills); this is a limitation on their ecological, but not necessarily external, validity. One externally valid claim might be that scientific heuristics can be organized hierarchically, from very general ones used by a wide range of scientists to domain-specific ones used by chemists to ones used by a single individual, an example of the latter being Krebs's "tissue-slicing" methods (see p. 171). KEKADA provides some evidence that the general/domain-specific distinction may be a useful way of categorizing heuristics, though, as noted earlier, the most important individual heuristics cannot be simulated by the program. Overall, Kulkarni and Simon need to be more precise about what type of validity they are concerned with in KEKADA.

3.1. Artificial Epistemology

Thagard (this volume) wants to extend the range of computational simulations beyond discovery; he uses a program called ECHO to establish that a set of principles he calls "explanatory coherence" is sufficient to establish the superiority of one scientific theory over its rivals. For example, he shows that ECHO can simulate the victory of Lavoisier's oxygen theory over phlogiston theory; unlike KEKADA, it does not try to follow the actual sequence of events by which Lavoisier discovered his theory. Instead, the program is based on a reconstruction of Lavoisier's final arguments.

ECHO uses a connectionist algorithm, which means that instead of representing knowledge as production rules and information, it represents knowledge as a network of connections between hypotheses and data, including excitatory and inhibitory links. ECHO begins with all hypotheses at the same low activation level; as the simulation runs, the oxygen hypotheses acquire higher activation levels, while the phlogiston hypotheses are gradually deactivated. Simon (1989) contrasts ECHO with STAHL, a production-rule program that also simulates the oxygen-phlogiston controversy, and argues that ECHO, rather than being a more sophisticated program, actually requires the programmer to supply a larger number of "givens," including propositions representing the empirical evidence, the hypotheses, and which hypotheses contradict and/or support each other. STAHL, in contrast, is supplied only with data, although — as we saw in our discussion of BACON — these "data" themselves embody a great deal of theory. Simon concludes that, "Both STAHL and ECHO will corroborate the oxygen theory of combustion if given Lavoisier's 'facts' and the phlogiston theory if given STAHL's 'facts'" (p. 487).

Once again, the issue of claims is central. As Giere (1989) has pointed out, Thagard relies on Lavoisier's rhetoric in constructing his simulation, so it is no surprise that the oxygen theory "wins." Therefore, this is not an ecologically valid simulation of the oxygen-phlogiston situation. As Simon has noted, and Thagard himself admits, a different network could be constructed in which the phlogiston theory might emerge as the winner. Therefore, ECHO does not provide externally valid evidence for the superiority of explanatory coherence over alternate accounts of how scientific controversies are resolved. At best, we are left with a weak claim: explanatory coherence is specific enough to be implemented on a connectionist network, and this network does perform according to Thagard's expectations. To put it in Popperian terms, if the network failed to function according to expectations, this would disconfirm Thagard's claims. Or would it, as Kuhn might counter, merely constitute a test of Thagard's programming abilities? The problem is that computational simulations may leave even more room for confirmation bias than experiments.

One partial solution is to add the equivalent of experimental controls to a computational simulation. After all, in an experiment, one at least has to demonstrate that one's treatment has a demonstrable effect in comparison to an equivalent no-treatment condition. ECHO merely shows that it can produce results that mimic the resolution of the oxygen-phlogiston controversy; it does not establish that programs based on other philosophical principles could not do the same. Perhaps a program based on Popperian or Lakatosian principles could perform in a similar manner.

Thagard (1988) does foresee the possibility of using a computer to compare philosophies of science by means of sophisticated parallel-processing systems like the Connection Machine (see Gorman [1989a] for a discussion). In Thagard's analogy, each individual scientist would correspond to one of these processors, and scientific progress would emerge from the interaction between them. There would be a kind of "central executive" that functions like peer review, collecting and communicating information. Unfortunately, this is as far as Thagard goes, and his analogy remains puzzling. In massively parallel systems, each processor is not intelligent; intelligence emerges from the whole. In fact, each processor can be viewed as a kind of "mindless agent"; it is their joint operation that creates intelligence (Minsky 1985). Also, there is no "control executive," just a relatively simple set of algorithms (see Waltz 1988).

Clearly, Thagard's analogy has low ecological validity: to begin to approach this type of validity, each scientist would have to be represented by a complex processing system beyond our current capabilities, not a "mindless agent" whose capacities are less than that of a human neuron. But his analogy might have external validity if he were clearer about what theoretical constructs it was supposed to clarify. Indeed, in his discussion of this system he takes an almost Popperian position: "Perhaps scientific research proceeds best when there is a division of labor between audacious but reckless thinkers, on the one hand, and careful but less original thinkers on the other" (p. 187). This sounds like Popper's conjectures and refutations; perhaps a Connection Machine could be designed to explore the heuristic value of having a division between audacious and critical functions. But at present, Thagard's analogy is too vague for us to be sure which type of validity he wishes to claim.

3.2. Comparing Experimental and Computational Simulations

Both experimental and computational simulations can possess external validity, but while computer simulations merely indicate that a set of rules or heuristics is sufficient to discover relationships in a given knowledge base, experiments indicate that human beings can employ specific heuristics and representations to solve problems. One cannot be certain that results on a computer model actual human thought processes, whereas experiments with humans do model some aspects of human cognition. This means that experimental simulations possess a kind of ecological validity that programs do not possess and potentially have higher external validity if the theoretical constructs being studied make specific claims about what human beings will or will not do. For example, both experimental and computational simulations might be used to make the claim that explanatory coherence or disconfirmation or some other philosophical prescription produced superior performance on a simulation; only experiments could be used to make the further claim that human beings could successfully employ such a strategy.

Note that both kinds of simulations share some of the same weaknesses as well. Consider the classic problem of demand characteristics (Orne 1962), i.e., those characteristics of the experimental situation that seem to mandate a specific response from subjects. As we have seen, this is one of the strengths of the experimental method: it allows us to deliberately set up situations in which certain kinds of representations and behaviors are likely to occur. But from the standpoint of theory testing, this does pose a problem: clever experimenters can often design experiments that will confirm their theories. (Let us remember, however, that a good many of these confirmatory experiments produce results that surprise the researcher and force modification of the theory; see Gorman [forthcoming] for a discussion.) Similarly, computational simulations like Thagard's can be biased in favor of a particular theory; this can be done by clever arrangement of the initial weights or the examples provided to the system, or — in Kulkarni's case — by direct programmer intervention at key phases.

But of course one way to deal with this problem is to design alternate experiments and simulations. For example, Gorman and Gorman (1984) established that disconfirmatory instructions did help performance on a general rule; Gorman, Stafford, and Gorman (1987) established that these same instructions were not helpful on an even more general rule. Similarly, one could search for computational situations on which explanatory coherence fails to predict the "winner" (see the case of the canals on Mars, below, for an example).

3.3. Do Computational Simulations Refute SSK?

One philosopher has recently cited computational simulations as falsifications of the sociology of scientific knowledge (SSK). Peter Slezak (1989) argues that if BACON can discover scientific laws, then one does not need to invoke social negotiations to explain scientific discovery. As we have seen, it is not clear what one means when one says that BACON "discovers." From a sociological standpoint, one could argue that it is given "data" that represent the product of previous negotiations.

Slezak might have used ECHO as further disconfirmatory evidence. After all, if one can account for the result of a scientific controversy based on simple, logical principles like explanatory coherence, then it is not clear why one needs sociology. However, it could be argued that ECHO, like BACON, merely reproduces the results of social negotiations; it is based on Lavoisier's successful rhetoric, which was itself shaped by social factors. In other words, these programs, rather than refuting SSK, may actually illustrate and support some sociological analyses.

But Slezak does raise an important question: Is it possible to refute the new SSK, or is this one of those approaches that explains everything and forbids nothing? First of all, we have to keep in mind that there is no monolithic SSK; instead, there is a wide range of diverse viewpoints that can be loosely grouped under that heading (see Woolgar 1989). But certainly some of the participants make statements that sound unfalsifiable; for example, Woolgar (1989) argues that from an SSK perspective, "'social' no longer refers to extraneous factors which may or may not impinge; instead, it describes the foundational character of all action, thought and behavior" (p. 660). This makes SSK the only perspective from which to view science; it becomes virtually impossible to contradict. It is hard to imagine how experiments and computer simulations could complement SSK and easy to see why people like Slezak feel the former must be placed in opposition to the latter; even worse, the latter might simply ignore the former, and technoscience studies would remain a completely fragmented enterprise.

But, in fact, the situation may be more hopeful than it appears at first glance. What Latour and Woolgar really want to do is eliminate the social/cognitive distinction. Typically, they have made it sound like the social was the only way to view science, but Woolgar (1987) has attacked sociological reductionism as well.

One might turn the tables on apparently cognitive methods like experimental and computational simulations and ask how they could incorporate the concerns of sociologists. One answer is to add "social" variables like group interaction to experimental and computational simulations. Another approach, more relevant to the concerns of SSK, is to use simulations to study the processes of negotiation that lead to socially constructed views of phenomena. Obviously, these two approaches can be linked. The next section will show examples that suggest how this might be done.

4. Experimental Social Epistemology:
Toward a Research Program

In this section, I will suggest how the social/cognitive gap might be bridged by a specific research program. But first, it is necessary to review existing research that points in a promising direction.

4.1. Experimental Simulations of Error

One of the unrealistic features of both the abstract and textbook-problem tasks used in experimental simulations is that they do not incorporate the possibility of error. The computational simulations largely ignore this important aspect of science as well. The standard Popperian view, best expressed by Lakatos (1978), is that errors can be discovered by replication. But sociologists like Collins (1985) have tried to show that what constitutes replication in science is the product of social negotiations.

Gorman (1986, 1989b) created a series of experimental simulations designed to take a step closer to ecological validity by adding the possibility of error to abstract simulations of scientific reasoning. Subjects worked on tasks similar to Wason's original 2-4-6 task, but were told that on anywhere from 0 to 20 percent of the trials, the experimenter's feedback would be erroneous: a triple that was actually correct would be classified as incorrect and vice versa. This meant that subjects could use the possibility of error to immunize hypotheses against disconfirmatory results.

Gorman (1986) added a social aspect to this sort of simulation, running subjects in groups of four. In this study, no actual error was introduced, which meant that subjects had an opportunity to construct error in a manner consistent with their representations of the rule. The inferences subjects could make about what constituted an error were tightly constrained in these experiments; they were told that error would occur on no more than one trial in five, and that when they labeled the result of a trial an error, that meant that an instance that appeared to be consistent with the rule was really inconsistent and vice versa. These restrictions meant that there was little apparent room for social negotiations concerning the meaning of error.

Nonetheless, about 20 percent of these groups actually altered the meaning of error, claiming that one could simply act as though an erroneous triple were a kind of wild card that could be replaced by any other triple. (In fact, these groups worked on a different task with features very similar to those of the 2-4-6 task.) In other words, their construction of error would have permitted these groups to take a sequence of correct triples like 2, 4, 6; 6, 8, 10; 3, 4, 1; 20, 22, 24 and claim the 3, 4, 1 was actually a 10, 12, 14. Other groups assigned errors to potential falsifications, in effect immunizing their hypotheses. Still others spent so much time replicating experiments they thought might be errors that they were unable to solve the task. Overall, far fewer groups discovered the rule when error was possible than when they knew all results were 100 percent reliable, in part because the possibility of error gave groups greater latitude to negotiate what counted as errors and disconfirmations.

Gorman (1989b) tried to determine the exact circumstances under which this possibility of error had its greatest effect, using individual subjects and various modifications of the 2-4-6 task. In a final modification, Gorman used a rule that required a continuous sequence of odd and even integers across the whole set of triples; in other words, if one proposed the triple "4, 9, 6" twice in a row, it would be right the first time and wrong the second because the next correct triple would have to be "odd, even, odd." His conclusion is reflected in generalization (2) (see sec. 2.1., above), which we might restate as follows: Subjects have the maximum opportunity to construct error in a way that immunizes their hypotheses when results of the current experiment or trial depend on the previous one.

Introducing error adds ambiguity to simulations of scientific reasoning, which permits subjects to construct their own sense of the problem space even when what constitutes an error is strictly constrained by the instructions. Obviously, making the source and nature of error even more ambiguous would permit us to explore ways in which subjects can construct different interpretations of error and evidence under a variety of circumstances. Woolgar, in a personal communication, suggested having subjects receive random feedback on the 2-4-6 task. One could do this most effectively if one introduced the possibility of error; subjects would never know which triples were really right and which were wrong. One could then explore how groups constructed what constituted correct and incorrect triples when different variables were added to this simulation.

4.2. Minority Influence

Consider a specific variable that one might manipulate in such a simulation. Moscovici (1974) did a series of experiments in which he established that a minority could affect the perceptual judgments of a majority in an ambiguous situation; he noted that the key to influence was what he called a "consistent behavioral style," in which the influencer argues his or her position adamantly, without making any concessions or modifications. Gorman and Carlson (1989) hypothesized that this sort of consistent rhetorical style might play an important role in advancing scientific hypotheses.[2] One looks, for example, at the success of Watson and Skinner in advancing behaviorism in psychology. Could part of their success have been due to their consistent, highly persuasive behavioral styles? Note that Watson and Skinner were not merely advocating a theory; they were trying to persuade people to construct psychology as being about behavior, not mind.

One could design an experiment to see if a consistent minority could convince a larger majority to adopt its construction of a scientific reasoning task. I would predict that the minority would have little effect unless, following Moscovici, some ambiguity were added to the task. One way to do this would be to add the possibility of error to a moderately complex rule, one involving several potential sources of error. Ideally, minority members would be confederates of the experimenter told to take a hard line or "strong programme" favoring a particular hypothesis and/or interpretation of error. An alternative design would involve running separate majority and minority subgroups (as in Gorman, Lind, and Williams 1977) and having them communicate by means of written or videotaped responses. One could have naive subjects, who had no prior experience with the task, observe the debate. Further tests on these neutral observers would help determine the extent of minority influence.

This experiment would have relatively low ecological validity, but high external validity. The theoretical construct being tested is whether a minority using a consistent behavioral or rhetorical style on a task designed to simulate scientific reasoning can influence a majority and/or a group of neutral observers to adopt its interpretation of the task and rule. If the minority succeeded under certain conditions, e.g, when a particular kind of rhetoric was used and/or error was high, then the ecological validity of these findings could be further explored using actual scientific cases. But one can never evaluate a claim about minority influence from thick descriptions of actual cases alone, because — as in the Skinner and Watson example above — too many other variables interact with the variable of interest.

To take a more social constructivist approach, one could avoid the use of confederates altogether; one could simply divide subjects into majority and minority subgroups and structure their experiences so that they arrived at different views of the task (again, see Gorman, Lind, and Williams [1977] for some preliminary ideas on how this might be done). One could then bring the subjects together in a situation where they could both debate their perspectives and try to persuade some neutral observers. One then might be able to develop some externally valid generalizations about how facts can be constructed under a variety of conditions, e.g., different levels of error and instructions that promoted different task representations.

4.3. Knowledge Transmission

One of the key issues in social epistemology is how these constructions or interpretations are transmitted from one generation in a culture or discourse community to another. To simulate this sort of knowledge transmission, groups could be run in a manner similar to Jacobs and Campbell's (1961) classic experiment on the transmission of social norms in an ambiguous setting. Jacobs and Campbell used a perceptual task called the "autokinetic effect," which involves showing subjects a point of light at the front of a darkened room. Generally, their individual judgments of the amount the light moves vary widely; however, Sherif (1936) found that in a group, individual judgments converged on a common norm. Jacobs and Campbell used confederates to simulate this convergence; as new members were introduced to a group viewing the effect, confederates would agree on a particular distance, and thereby influence others to adopt a similar norm. Gradually, confederates were replaced by new members; once the original group had been entirely replaced, the norm persisted through four or five additional new subjects. This experiment illustrates the way in which the construction of an ambiguous stimulus can persist for at least a few "generations," even in the absence of any social contingencies that reinforce the norm.

Zucker (1977) added an institutional feature to the Jacobs and Campbell design by simulating three conditions: (1) in a personal influence condition, subjects were told that the study involved problem solving in groups; (2) in an organizational condition, subjects were told it involved problem solving in model organizations, which meant that replacements had to perform in the same way as "employees" they replaced; (3) the office condition was identical with the organizational except that the "oldest" member was given control of a switch operating the light. Basically, (3) led to the greatest persistence in norms and (1) to the least. This experiment shows how a simple manipulation of context can radically affect the transmission of socially constructed views of a phenomenon.

Insko et al. (1980) used a similar design to test an aspect of Service's theory of social evolution. Service set up a prototypical situation involving three villages, one of which occupied a central position in both geography and trade. Chiefs in the central village would, according to Service, have the highest ranked descent line. Insko et al. created a laboratory simulation of this prototypical situation by randomly dividing subjects into three groups. Each group manufactured and traded objects, with one group playing the role of the central village, in that negotiations between the other two groups could be carried out only via this central group. Members of all groups were replaced one by one, creating "generations" as in the Jacobs and Campbell design. On a final questionnaire, subjects across generations reported that the central group had adopted a leadership role, even though the other groups sometimes made attempts to circumvent or reduce the importance of the central group. This study provides externally valid evidence for Service's claims: in a laboratory microeconomy, the central village does acquire a leadership role. But it does not, of course, provide ecologically valid evidence; Service's theory may work only in an ideal simulation.

Interestingly, successive generations in the groups evolved a kind of work ethic and adopted a seniority system, in terms of distribution within each group of money gained from trading. These results show the way in which norms can evolve even in artificial laboratory cultures.

None of these studies addresses issues in the philosophy of science, but they suggest a way in which one can follow the advice of Latour, Woolgar, and others and study science as one would study any other form of cultural activity, without simply relying on thick descriptions. One could, for example, add possible or actual error to a more complicated version of the 2-4-6 or a related task, including one of the artificial universes created by Mynatt, Doherty, and Tweney (1978), and use this sort of task to explore how socially negotiated views of a situation were transmitted across generations.

There was no competition between groups in the Jacobs and Campbell and Zucker designs, and only minimal competition in Insko et al.'s simulation. But scientific research "programmes," to use Lakatos's term, are in constant competition with each other. One could add the elements of competition and conversion to these multigenerational designs by allowing replacement subjects to indicate their preference for which of several competing groups they wish to belong to, and assign them accordingly. Therefore, while each group must give up one member, one group might gain two members and another none. Each new member might bring new resources, e.g., the opportunity for additional experiments, increasing the need for converts. Finally, one could allow original group members to defect, if they wanted.

One could combine this sort of design with the minority-influence ideas discussed earlier. Would a consistent behavioral style allow an initial minority to gradually convert new members and transform itself into a majority across generations?

These very general designs indicate the potential for experimental simulations of social epistemology. Again, one cannot claim that these simulations necessarily have ecological validity; instead, they permit us to make externally valid generalizations like: "*Ceteris paribus*, a consistent minority will have the greatest effect on subjects' representations when there is a high possibility of error in results and results of the current experiment are linked to previous ones."

Note that future studies can make this generalization more precise, e.g., by clarifying exactly what sort of linkage between current and previous experiment produces the greatest effect. This additional precision will also help suggest what ecologically valid scientific problems are most likely to produce the same effect. In the next section, I will consider an experimental simulation directed at a specific scientific controversy.

4.4. Simulating a Scientific Controversy

To illustrate the potential relationship between experimental social epistemology and historical studies, let us consider how experimental research could shed light on a specific scientific controversy — deliberately selected to appeal to social constructivists who are suspicious of "winner's accounts" of history (see Collins and Pinch 1982).

4.4.1. Canals on Mars?

Schiaparelli described "channels" (*canali*) on the surface of Mars in 1877. Earlier observers had detected some vague, streaklike features, but Schiaparelli's "were finer, sharper and more systematically arranged on the surface" (Hoyt 1976, p. 6). Percival Lowell, a wealthy American who developed a strong interest in astronomy, interpreted these channels to be canals and used them to make a strong argument for intelligent life on Mars. Other observers disagreed, and a heated debate lasting over a quarter of a century was carried on, both in scientific journals and the popular press. This scientific controversy raises interesting questions about the relationship between theory, perception, and data because some observers saw the canals and others did not. In hindsight, it is apparent that the advocates of canal explanations were victims of an illusion, but this is a classic explanation by "winners" — those who did not see the modern truth were aberrant or deficient scientists, and can be dismissed. In fact, Lowell himself was an important astronomer, and other major astronomers of his day were divided on the issue — he was not merely a fringe quack who was dismissed by the scientific establishment. We need to come up with a better account for the persistence of this controversy.

To begin with, let us analyze the controversy in terms used in the experimental literature discussed earlier in this essay, focusing on falsification and error. Percival Lowell displayed a strong "confirmation bias": "Once his 'general theory' (of the Martian canals) was framed,

neither subsequent observation nor mounting criticism during the ensu-
ing Mars furor caused him to abandon or alter any of its essential points"
(Hoyt 1976, p. 69). This stubbornness ties in well with Moscovici's ideas
about consistency and persuasion. As Mitroff (1981) has shown in his
study of Apollo moon scientists, stubbornness is often the hallmark of
successful scientists:

> The three scientists most often perceived by their peers as most
> committed to their hypotheses...were also judged to be among the
> most outstanding scientists in the program. They were simultane-
> ously judged to be the most creative and the most resistant to change.
> The aggregate judgment was that they were "the most creative" for
> their continual creation of "bold, provocative, stimulating, sugges-
> tive, speculative hypotheses" and "the most resistant to change" for
> "their pronounced ability to hang onto their ideas and defend them
> with all their might to theirs and everyone else's death." (p. 171)

If successful scientists are speculative and stubborn, Lowell certainly
should have been successful. In fact, his views were widely circulated
and dominated popular thinking about Mars for over a decade.

In addition to a wide audience of laypersons, Lowell also persuaded
a number of other astronomers, who independently confirmed the existence of the canals. As more powerful telescopes became available, more
and more astronomers agreed that there were no canals. Still, as late as
1924 at least one astronomer claimed to see canals on Mars (Hoyt 1976,
p. 170). Lowell himself died in 1916, confident that he had confirmed the
existence of canals and of intelligent life on Mars. During his lifetime,
photographic and telescopic evidence was ambiguous, at best.

Personal communication with the members of the Lowell observatory
suggests that Lowell's view of the canals may have resulted, in part, from
his long observation sessions; he would spend up to eight hours at a time
observing Mars, and was therefore indignant when astronomers who devoted less time to the "red planet" criticized his theories. Therefore, his
view of the canals seems to have resulted from a kind of perceptual ambiguity not precisely simulated by existing error experiments. Subjects
working on the 2-4-6 task could clearly see and identify the triple, even
if the feedback was in doubt. A situation closer to Lowell's would be
one in which the object being observed was ambiguous, as in Jacobs
and Campbell (1961).

4.4.2. Proposal for an Experiment

Astronomers in Lowell's time recognized the potential value of experiments in simulating scientific controversies. Douglass, one of Lowell's

team of observers at Flagstaff, Arizona, decided to find out whether Lowell's observations could be caused by some psychological phenomenon:

> I have made some experiments myself bearing on these questions by means of artificial planets which I have placed at a distance of nearly a mile from the telescope and observed as if they were really planets. I found at once that some well known planetary appearances could, in part at least, be regarded as very doubtful. (Hoyt 1976, p. 124)

Douglass's appearance of doubt caused Lowell to fire him, and I have been able to unearth no further details regarding his experiments.

Another astronomer, E. Walter Maunder, asked a group of boys from the Royal Hospital School at Greenwich to copy a picture of Mars; even though there were no lines on the original picture, the boys drew them in. Maunder concluded that the lines drawn by astronomers were no more real than those drawn by the boys: "It seems a thousand pities that all those magnificent theories of human habitation, canal construction, planetary crystallisation and the like are based upon lines which our experiments compel us to declare non-existent" (Hoyt 1976, p. 165). Flammarion ran a similar set of tests on a group of French school boys, none of whom drew imaginary lines, but Maunder's critique caused the British Astronomical Association to reverse its position:

> The members...have themselves seen more than a hundred of the so-called canals during their observations since 1900, yet E. M. Antoniadi, their director and editor, apparently has their concurrence in holding that their eyes were probably deceived and that they really saw something very different from the straight lines they imagined they were looking at. (Hoyt 1976, p. 165)

Details of these early experiments are hard to discover. A modern, follow-up study might be able to address the issue of under what conditions the existence of phenomena like the canals can be constructed out of ambiguous stimuli. One could simply replicate these earlier experiments using planetary disks. But the average person's knowledge of the planets is much greater now, including the fact that there is almost certainly no other life in the solar system, and might influence how people approached any stimulus resembling a planetary disk. But one could construct a more general situation in which people are shown blurred pictures of terrain features and asked to map them. In one condition, these subjects might be given a detailed, persuasive story about the presence of roads or rivers or other terrain features in this area. Another condition would be given no such story, or a different story. If subjects in the first condition tended to draw the roads in, then one could make the externally valid claim that a persuasive story could affect how people map terrain features. Note that one cannot make the ecologically valid claim that Lowell did this, but it certainly raises the possibility.

More specifically, one could investigate the claim about extra observation time. Perhaps those who feel they are beginning to see roads will draw in more and more details, and feel more and more convinced, if they are given more time to study the ambiguous stimulus, moving it on a computer screen and trying to sharpen it. One could then mate this experiment to some of the designs discussed earlier, first seeing if those who observe roads can transmit their expertise to novice subjects and then setting up a confrontation between those who saw roads and those who did not.

Moscovici's research raises another intriguing question: How important was Lowell's consistent, determined behavioral style in persuading others? One could simulate its impact by using a Moscovici design in which a minority argues persuasively for the presence of certain terrain features. Different aspects of the minority's arguments and style could be manipulated. Of particular interest would be whether the minority had its greatest effect on naive outsiders and a lesser effect on other "expert" subjects.

This example indicates how experimental simulations can help settle issues raised by historical cases. Computational simulations might also be employed. For example, one could use ECHO to determine whether Lowell's theory fits the criteria for explanatory coherence. Lowell's canal hypothesis was linked to a network of other hypotheses. He thought that Mars was a dying planet, slowly drying out. The intelligent inhabitants, in order to survive, had built huge networks of canals to irrigate their world, using the water that melted from the polar caps. These canals were surrounded by belts of vegetation, which is why they were visible from the earth. He used this network of hypotheses to successfully predict the appearance of new canals and the presence of water vapor in the Martian atmosphere. He even had hypotheses to account for why some others with more powerful telescopes failed to observe the canals: their observation periods were too short, and they did not have ideal atmospheric conditions. Perhaps Lowell, like Lavoisier and Darwin, used the rhetoric of explanatory coherence to persuade others.

Another kind of computational simulation might involve training a neural net (see Churchland, this volume) to see the roads perceived by one group of subjects, and another net to see the ambiguous stimulus without roads, like control subjects. These nets would give us clues as to how these different ways of looking at the stimulus might be represented in terms of hidden units.

5. Conclusions

Experimental and computational simulations of science have tried to address a number of issues in the philosophy of science, but their relevance depends on the nature of the claims being made. These simulations are powerful tools for making externally valid claims about the value of philosophical prescriptions under ideal, artificial circumstances; before these claims can be generalized to ecologically valid situations, they must be complemented by other studies, which may include further simulations designed to increase ecological validity and the sorts of case studies traditionally favored by philosophers (see Gorman and Carlson [1989] for a discussion). Experiments can also be used to generate ideas for new philosophical prescriptions.

Simulations of science have addressed a wide variety of topics, including how the possibility of error affects falsification (Gorman 1989b), the difference between expert and novice representations of physics problems (Chi, this volume), and whether explanatory coherence can account for the resolution of scientific controversies (Thagard, this volume). These diverse efforts need to be followed by a systematic program of simulation research that addresses concerns in social epistemology (Fuller, this volume). To succeed, such a program will require close collaboration between philosophers, psychologists, sociologists, and historians. This essay has tried to outline how such a program of research might look, setting up a simulation of the "canals on Mars" controversy using a multigenerational laboratory microculture. The point of this example is to inspire others to adopt what Shadish (1989) calls a "critical multiplist" perspective and come up with their own suggestions for collaborative projects.

Notes

1. For reasons of brevity, I have skirted the issue of internal validity in the body of this essay. Basically, this type of validity has to do with whether an experiment or field study is appropriately conducted. For example, one could conduct an experiment that potentially had external validity, but used inappropriate randomization, or controls, or statistics. Similarly, one could conduct a study of inventors that potentially had ecological validity, but relied solely on patent depositions, which are *post hoc* accounts of an invention written to justify an inventor's claim to priority and originality; therefore, any study that relied too heavily on them to reconstruct an inventor's cognitive processes would lack internal validity. W. Bernard Carlson and I are careful to use multiple sources to reconstruct the mental models of early telephone inventors, including notebooks, artifacts, and correspondence as well as patent depositions.

2. Robert Rosenwein does a more thorough job of relating the literature on minority influence to issues in social epistemology; see his "The Role of Dissent in Achieving Scientific Consensus," a paper delivered at the conference on the "Social Psychology of

Science: The Psychological Turn," May 20–21, 1990, Memphis, Tennessee; a version of this paper will appear in Shadish (in press).

References

Banaji, M. R., and R. G. Crowder. 1989. The bankruptcy of everyday memory. *American Psychologist* 44:1185–93.

Berkowitz, L., and E. Donnerstein. 1982. External validity is more than skin deep: Some answers to criticisms of laboratory experiments. *American Psychologist* 37:245–57.

Carlson, W. B., and Michael E. Gorman. 1990. Understanding invention as a cognitive process: The case of Thomas Edison and early motion pictures, 1888–1891. *Social Studies of Science* 2:387–430.

Chi, M. T. H., P. J. Feltovich, and R. Glaser. 1981. Categorization and representation of physics problems by experts and novices. *Cognitive Science* 5:121–52.

Chi, M. T. H., et al. In press. Self-explanations: How students study and use examples in learning to solve problems. *Cognitive Science.*

Clement, J. 1988. Observed methods for generating analogies in scientific problem solving. *Cognitive Science* 12:563–86.

Collins, H. M. 1985. *Changing order: Replication and induction in scientific practice.* London: Sage.

Collins, H. M., and T. J. Pinch. 1982. *Frames of meaning: The social construction of extraordinary science.* London: Routledge and Kegan Paul.

Feyerabend, P. 1970. Consolations for the specialist. In I. Lakatos and A. Musgrave, eds., *Criticism and the growth of knowledge.* London: Cambridge University Press.

———. 1975. *Against method.* Thetford, Norfolk, Eng.: Thetford Press.

Fuller, S. 1988. *Social epistemology.* Bloomington: Indiana University Press.

———. 1989. *Philosophy of science and its discontents.* Boulder, Colo.: Westview Press.

Gentner, D., and G. R. Gentner. 1983. Flowing waters or teeming crowds: Mental models of electricity. In D. Gentner and A. L. Stevens, eds., *Mental models.* Hillsdale, N.J.: Lawrence Erlbaum, pp. 99–129.

Giere, R. N. 1988. *Explaining science: A cognitive approach.* Chicago: University of Chicago Press.

———. 1989. What does explanatory coherence explain? *Behavioral and Brain Sciences* 12:475–76.

Gooding, D. 1985. In Nature's school: Faraday as an experimentalist. In D. Gooding and F. James, eds., *Faraday rediscovered: Essays on the life and work of Michael Faraday, 1791–1867.* New York: Stockton Press.

———. 1990. Mapping experiment as a learning process: How the first electromagnetic motor was invented. *Science, Technology and Human Values* 15:165–201.

Gorman, Michael E. 1986. How the possibility of error affects falsification on a task that models scientific problem-solving. *British Journal of Psychology* 77:85–96.

———. 1987. Will the next Kepler be a computer? *Science and Technology Studies* 5:63–65.

———. 1989a. Artificial epistemology? *Social Studies of Science* 19, no. 2:374–80.

———. 1989b. Beyond strong programmes. *Social Studies of Science* 19, no. 4:643–52.

———. 1989c. Error and scientific reasoning: An experimental inquiry. In S. Fuller et al., eds., *The cognitive turn: Sociological and psychological perspectives on science.* Dordrecht: Kluwer, pp. 41–70.

———. Forthcoming. *Simulating science: Heuristics, mental models, and technoscientific thinking.* Bloomington: Indiana University Press.

Gorman, Michael E., and W. B. Carlson. 1989. Can experiments be used to study science? *Social Epistemology* 3:89–106.

———. 1990. Interpreting invention as a cognitive process: The case of Alexander Graham Bell, Thomas Edison, and the telephone. *Science, Technology and Human Values* 15:131–64.

Gorman, Michael E., and Margaret E. Gorman. 1984. A comparison of disconfirmatory, confirmatory and a control strategy on Wason's 2, 4, 6 task. *Quarterly Journal of Experimental Psychology* 36A:629–48.

Gorman, Michael E., E. A. Lind, and D. C. Williams. 1977. The effects of previous success or failure on a majority-minority confrontation. ERIC Document ED-177-257.

Gorman, Michael E., A. Stafford, and Margaret E. Gorman. 1987. Disconfirmation and dual hypotheses on a more difficult version of Wason's 2-4-6 task. *Quarterly Journal of Experimental Psychology* 39A:1–28.

Gorman, Michael E., et al. 1984. How disconfirmatory, confirmatory and combined strategies affect group problem-solving. *British Journal of Psychology* 75:65–79.

Houts, A., and B. Gholson. 1989. Brownian notions: one historicist philosopher's resistance to psychology of science via three truisms and ecological validity. *Social Epistemology* 3:139–46.

Hoyt, W. G. 1976. *Lowell and Mars*. Tucson: University of Arizona Press.

Insko, C. A., et al. 1980. Social evolution and the emergence of leadership. *Journal of Personality and Social Psychology* 39:431–48.

Jacobs, R. C., and D. T. Campbell. 1961. The perpetuation of an arbitrary tradition through several generations of a laboratory microculture. *Journal of Abnormal and Social Psychology* 83:649–58.

Klahr, D., and K. Dunbar. 1988. Dual space search during scientific reasoning. *Cognitive Science* 12:1–48.

Klayman, J., and Y.-W. Ha. 1987. Confirmation, disconfirmation and information in hypothesis testing. *Psychological Review* 94:211–28.

Kuhn, T. S. 1970. *The structure of scientific revolutions*. Chicago: University of Chicago Press.

Kulkarni, D., and H. A. Simon. 1988. The processes of scientific discovery: The strategies of experimentation. *Cognitive Science* 12:139–75.

Lakatos, I. 1978. *The methodology of scientific research programmes*. Cambridge: Cambridge University Press.

Langley, P., et al. 1987. *Scientific discovery: Computational explorations of the creative processes*. Cambridge, Mass.: MIT Press.

Larkin, J. 1983. The role of problem representation in physics. In D. Gentner and A. L Stevens, eds., *Mental models*. Hillsdale, N.J.: Lawrence Erlbaum, pp. 75–98.

Larkin, J. H., et al. 1980. Expert and novice performance in solving physics problems. *Science* 208:1335–42.

Latour, B. 1987. *Science in action*. Cambridge, Mass.: Harvard University Press.

McGuire, W. J. 1989. A perspectivist approach to the strategic planning of programmatic scientific research. In B. Gholson et al., eds., *Psychology of science: contributions to metascience*. Cambridge: Cambridge University Press, pp. 214–45.

Mahoney, M. J. 1976. *Scientist as subject: The psychological imperative*. Cambridge, Mass.: Ballinger.

Manier, E. 1987. "External factors" and "ideology" in the earliest drafts of Darwin's theory. *Social Studies of Science* 17:581–610.

Minsky, M. 1985. *Society of mind*. Cambridge, Mass.: MIT Press.

Mitroff, I. I. 1974. *The subjective side of science*. Amsterdam: Elsevier.

————. 1981. Scientists and confirmation bias. In R. D. Tweney, M. E. Doherty, and C. R. Mynatt, eds., *On scientific thinking*. New York: Columbia University Press, pp. 170–75.

Moscovici, S. 1974. Minority influence. In C. Nemeth, ed., *Social psychology: Classic and contemporary integrations*. Chicago: Rand McNally, pp. 217–50.

Mynatt, C. R., M. E. Doherty, and R. D. Tweney. 1977. Confirmation bias in a simulated research environment: An experimental study of scientific inference. *Quarterly Journal of Experimental Psychology* 29:85–95.

————. 1978. Consequences of confirmation and disconfirmation in a simulated research environment. *Quarterly Journal of Experimental Psychology* 30:395–406.

Orne, M. T. 1962. On the social psychology of the psychological experiment with particular reference to the demand characteristics and their implications. *American Psychologist* 17:776–83.

Popper, K. R. 1972. *Conjectures and refutations*. London: Routledge and Kegan Paul.

Qin, Y., and H. A. Simon. 1990. Laboratory replication of scientific discovery processes. *Cognitive Science* 14:281–312.

Shadish, W. R. 1989. The perception and evaluation of quality in science. In B. Gholson et al., eds., *Psychology of science: Contributions to metascience*. Cambridge: Cambridge University Press, pp. 383–428.

Shadish, W. R., ed. In press. *Social psychology of science*. New York: Guilford Press.

Sherif, M. 1936. *The psychology of social norms*. New York: Harper.

Simon, H. A. 1989. ECHO and STAHL: On the theory of combustion. *Behavioral and Brain Sciences* 12:487.

Singley, M. R., and J. R. Anderson. 1989. *The transfer of cognitive skill*. Cambridge, Mass.: Harvard University Press.

Slezak, P. 1989. Scientific discovery by computer as empirical refutation of the strong programme. *Social Studies of Science* 19, no. 4:563–600.

Thagard, P. 1988. *Computational philosophy of science*. Cambridge, Mass.: MIT Press.

Tweney, R. D. 1985. Faraday's discovery of induction: A cognitive approach. In D. Gooding and F. James, eds., *Faraday rediscovered: Essays on the life and work of Michael Faraday, 1791–1867*. New York: Stockton Press.

Tweney, R. D., et al. 1980. Strategies of rule discovery on an inference task. *Quarterly Journal of Experimental Psychology* 32:109–23.

Waltz, D. L. 1988. Artificial intelligence. *Daedalus* 117:191–212.

Wason, P. C. 1960. On the failure to eliminate hypotheses in a conceptual task. *Quarterly Journal of Experimental Psychology* 12:129–40.

Wason, P. C., and D. W. Green. 1984. Reasoning and mental representation. *Quarterly Journal of Experimental Psychology* 36A:597–610.

Woolgar, S. 1987. Reconstructing man and machine: A note on sociological critiques of cognitivism. In W. E. Bjiker, T. Hughes, and T. Pinch, eds., *The social construction of technological systems*. Cambridge, Mass.: MIT Press.

————. 1989. A coffeehouse conversation on the possibility of mechanising discovery and its sociological analysis, with some thoughts on "decisive refutations," "adequate rebuttals" and the prospects for transcending the kinds of debate about the sociology of science of which this is an example. *Social Studies of Science* 19, no. 4:658–68.

Zucker, L. G. 1977. The role of institutionalization in cultural persistence. *American Sociological Review* 42:726–43.

Epistemology Radically Naturalized: Recovering the Normative, the Experimental, and the Social

I argue that a radically naturalized study of knowledge would apply the methods and findings of psychology and the social sciences to itself. The results would make epistemology more robustly normative, experimental, and social than it currently is, indeed, more in the spirit of the original American naturalizers of epistemology: Peirce, Dewey, and Mead. I begin by briefly sketching how my own philosophical program, social epistemology, captures this spirit. I then proceed to show the limits to the radicalism of various naturalizers: Churchland's eliminative materialism, mainstream experimental psychology, analytic epistemology, and even the sociological school of ethnomethodology. In the last section, I tackle head on the most hostile opposition to experimentation from within the sociological camp, which leads me to propose an experimental design that even Bruno Latour could love.

1. The Scope of Social Epistemology

Social epistemology (Fuller 1988, 1989) is an interdisciplinary project that mobilizes the empirical resources of the "sociology of knowledge" (understood very broadly to range from cognitive social psychology, through the sociology of science, to the social history of ideas) for the purposes of informing a normative philosophy of science. The project moves on three fronts: (1) as a metatheory, (2) as an empirical research program, and (3) as knowledge policy.

So far, social epistemology has functioned primarily as a metatheory designed to reinterpret what philosophers and social scientists are doing when they study knowledge. All but the last part of this paper continues that job. The social epistemologist's main metatheoretic move is to "naturalize" knowledge, and thereby take seriously that knowledge is *in* the world of which it is about. After the promise of naturalism at the start of this century, normative philosophical accounts of knowledge have grown steadily apart from, if not antagonistic to, empirical social scientific accounts. The fact that recent "postmodernist" philosophers (e.g., Rorty 1979) have felt the need to abandon the normative project in order to pursue the empirical (often specifically historical) project is a testimony to this rift, one which the social epistemologist takes to be the product of a mistaken understanding of the role of philosophy among the sciences. For, properly understood, the naturalization of knowledge should involve not the withering away of epistemology but the dissolution of the boundaries separating epistemology from the social scientific study of knowledge.

But metatheory aside, social epistemology is also an empirical research program and a vehicle of knowledge policy. On the empirical side, the general strategy is twofold. First, to analyze normative philosophical claims about the growth of knowledge (i.e., theories of scientific rationality) in terms of the historiography of science that they presuppose (e.g., from whose standpoint is the history told? See Fuller 1991). Second, to decompose epistemically salient historical episodes into their "working parts" by simulating them in the laboratory, so as to determine the difference that the presence or absence of an epistemic norm made to the outcome of the episode. Section 2.6 of Gorman's contribution to this volume carries out that task, as well as the last section of my own. On the policy side, social epistemology aims to develop a rhetoric for converting its metatheory and empirical research into a means of increasing the public accountability of science (see Fuller 1992).

2. Naturalizing Knowledge and Cognition: Momentum Lost and Regained

A common thread that runs through the naturalistic movement in philosophy — linking thinkers as otherwise diverse as Bacon, Comte, and Dewey — is the belief that epistemology is always playing catch-up with the disciplines at the cutting edge of scientific inquiry. If we take this view seriously, then twentieth-century philosophy, at least in the English-speaking world, has moved two steps forward and one step back. For what had, at the start of the century, distinguished the naturalist most sharply from the classical epistemologist (whose exemplar was usually Descartes or Plato) is nowadays perhaps naturalism's least touted feature. It is the thesis of *metaphysical monism*, the epistemological corollary of which is that knowledge is an integral part of the world of which it is about.

According to Charles Sanders Peirce, John Dewey, George Herbert Mead, and other early twentieth-century American naturalists, the cardinal sin of classical epistemology was to assume that the knower somehow stood metaphysically apart from the knowable world; hence, the problem of the external world. Moreover, this externality was necessary for the possibility of having knowledge. Knowers could represent the world order because they did not need to intervene significantly in its causal processes. It is clear enough, at a common-sense level, what the classical epistemologist had in mind here. If we put aside large laboratories, knowledge appears to arise from an individual's initial sensory contact with the environment, which is then given a socially acceptable verbal formulation. At that level of description, the causal processes involved seem quite unexceptional and self-contained: most of the epistemically relevant changes occur inside the individual and in the puffs of air it takes to exchange words with other individuals. It would seem, then, that the world order is not particularly disturbed by the production of knowledge. This picture of the naturalist's opponent has been variously called "transcendental idealism," the "spectator theory of knowledge," or simply "intellectualism." Although no philosopher ever baldly assents to intellectualism, it may nevertheless be the best explanation for many of the methodological intuitions advanced by philosophers of science today. These intuitions pertain to the nature of *knowledge* and the nature of the *knower*.

In the case of knowledge, when psychology and the social sciences are portrayed as offering a form of knowledge inferior to that of the natural sciences, the difference is often attributed to the degree of intervention that is necessary for social scientists to constitute their objects of inquiry. For example, although even simple experiments in physics are arguably more artificial than ones in psychology, physicists are usually taken to be better than psychologists at presenting their laboratory artifact as the reflection of some natural fact that existed prior to the intervening experiment. It is the mark of intellectualism to value an experiment because it adds to our knowledge without adding to the world. In the case of the knower, intellectualism appears in the parsimony-based appeal to *methodological solipsism* in cognitive science (Fodor 1981, chap. 9): If knowledge is possible because the knower is autonomous from the world, then why not simply study the knower without the world? Unfortunately, given methodological solipsism, what enables knowledge to be had at all also explains why knowledge is not always to be had: If knowers leave no significant trace in the world as they try to represent it, then how does it happen that the world manages to leave a significant trace on their representations? This worry accounts for the re-emergence of skepticism as the signature problem of analytic epistemology in the last thirty years (see Pollock 1986).

Both philosophers and sociologists of science have recently rediscovered the importance of intervention to knowledge production, but the sociologists have been more thoroughgoing in returning intervention to the status it had in the original agenda of naturalism. Philosophers have stressed the role of intervention in standardizing laboratory phenomena upon which more phenomena can then be experimentally grounded (Hacking 1983, pt. 2). However, they have shown remarkably little interest in how scientists use their laboratory findings as a basis for intervening in nonlaboratory settings (an exception is Rouse 1987, chaps. 4, 7; see Fuller 1991, 1992). In other words, philosophers have confined themselves to what psychologists call the "internal," as opposed to the "external," validity of experiments (Houts and Gholson 1989; Fuller 1989, pp. 131–35). One consequence of this self-imposed limitation is that today's epistemology of experiment is practiced in complete isolation from, say, technological ethics, a field that has typically been concerned with interventions of more general social import. Another implication is that the normative dimension of epistemology and the philosophy of science remains limited to the approximation of internally generated standards. Whereas philosophers of science always used to worry about whether theories were chosen according to sound methodological criteria, they are nowadays increasingly concerned with whether phenomena have yielded suitably robust effects across a wide range of laboratory contexts. Thus, the success or failure of an experimental intervention is judged exclusively in terms of the expectations and intentions of the scientists working in the laboratory, not in terms of the extramural consequences to which the experiment gives rise.

By contrast, sociologists of science — often in spite of themselves — are much more in the spirit of the normative mission of naturalism. Consider the extended setting in which some of them examine experimental intervention. If, as Dewey (e.g., 1960) maintained, an action must be judged by the concrete exigencies that arise from it, then the normative naturalist (*pace* Laudan 1987) would want to follow scientists outside the laboratory to the networks in society at large that must be forged before the experimental result constitutes a stable piece of knowledge (see Latour 1987). As Bruno Latour and other actor-network theorists among the sociologists have argued, this process of "manufacturing knowledge" (Knorr-Cetina 1980) involves the interrelation of many otherwise quite unrelated people and things (i.e., "actor-networks"), most of which are displaced from their normal activities, sometimes irreversibly so (Callon, Law, and Rip 1986). The people include not only competing theorists, but also technicians, editors, politicians, financiers, all of whom enable the theoretical competition to take place by their being made to have a stake in its outcome. The experimental result that establishes a victor can then be evaluated according to the types of changes that these "enrolled actors" undergo: Who (or what) is empowered by the result, and who (or what) is not (see Fuller 1989, pp. 42–52)? Why have philosophers been slow to embrace this extended normative sensibility? I now venture a diagnosis that will serve as a springboard for the rest of this essay.

One way to understand the naturalist slogan "knowledge is both in and about the world" is as a call to self-reflection, or reflexivity. Traditionally, the reflexive turn has been portrayed as inhibiting what the inquirer can do. This Russellian view of reflexivity is illustrated in the strategies used to solve the paradoxes of logic and set theory. The self-contradictoriness of a paradox forces the logician to separate and restrict the domains to which the different uses of some problematic term apply. Without this additional structure, the paradoxical claim expresses no determinate proposition and is, therefore, without meaning. However, it is not clear that the neatly divided, well-bounded world of the Russellian is consistent with the naturalist's monism. Suppose we say, instead, that the paradox is indeed meaningful, but that its meaning is concealed in its outmoded expression, such that the paradoxical term needs to be replaced, not simply refined. Let us call this contrasting view of reflexivity Hegelian, as it captures Hegel's *aufgehoben*, an expansion of inquiry (indeed, the model of Dewey's concept of intelligence) that occurs by eliminating and transcending a conceptual impasse. To see the contrast at work, consider the case of someone — say, Paul Churchland (1979, 1989) — who believes that people do not have beliefs. The Russellian would parse this paradoxical claim by immunizing Churchland's own second-order "belief" from the general denial of beliefs implied in his first-order use of the term. The Hegelian would, in contrast, argue that a world in which Churchland's claim makes sense is one in which the ontology that supports beliefs — either his or anyone else's — needs to be replaced. This is, in fact, what I will now argue, in the course of which I reveal the partial character of Churchland's allegedly radical naturalization of cognitive science.

3. Churchland and the Limits of Radical Naturalism

The core tenet of Churchland's position, eliminative materialism, is that the belief-desire psychology used to explain action in both commonsense and most cognitive science settings is probably false, or at least misleading enough to warrant its abandonment by cognitive scientists. But is not Churchland's espousal of eliminative materialism itself a belief? Churchland (1989, pp. 21–22) addresses this reflexive charge by arguing that a variety of semantic theories are available to determine the truth value of a sentence that do not make reference to the status of the sentence as a belief. But the reflexive charge runs deeper. We commonsensically hold Churchland to be the relevant cause of the sentences that come out of his mouth — however their truth values are determined — because we take him to be a system whose transactions with the environment serve to maintain his own well-being and goals in the face of whatever resistance the environment might provide. Indeed, this is just to say that we treat Churchland as an autonomous organism (see Bernard 1951). He is not merely the passive site in which significant causal processes are played out, but rather an active player, whose intervention decisively affects how those processes are manifested. Not surprisingly, then, while Churchland denies the relevance of beliefs and desires to an explanation of his behavior, he does not go so far as to deny the image of the autonomous organism that is commonsensically defined by belief-desire psychology. Whereas in theory Churchland (1989, pp. 20–21) is willing to countenance the possibility of a world brain of which each individual is just a part, in practice he treats each brain-bearing beast as its own self-contained cognitive unit whose principal source of knowledge is its own immediate transactions with the environment. In fact, this beast, whose adherence to belief-desire psychology is held to be radically false, is nevertheless also said to possess scientific "theories" of the sort that have traditionally interested philosophers. How can Churchland's neurocomputational human be so transparent to the world yet so opaque to herself or himself?

This question suggests that Churchland has failed to naturalize consistently. But why the inconsistency? It is clear from the table of contents of A Neurocomputational Perspective (1989) that Churchland is interested in neuroscience, less to transcend the distinction between philosophy and psychology than to provide alternative answers to standing questions in the philosophy of mind and the philosophy of science. However, these questions make sense only when posed against the backdrop of the (human) organism as a well-bounded cognitive unit. By contrast, the neuroscience research on which Churchland (esp. chap. 5) relies assumes a quite plastic brain that is subtly susceptible to changes in the environment, indeed, to the point of the brain-world interface being continually redefined. Thus, Churchland's overriding commitment to the well-bounded organism goes against the ontological assumptions of the neuroscience — and even some of the experimental psychology — that he uses to deny the existence of the beliefs and desires that define such an organism. But why retain the shell of an ontology whose contents have been eliminated? After all, if memory limitations and perceptual biases are said to prevent us normally from understanding the causes of our behavior, as Churchland readily admits, then why shouldn't he also admit research of a similar nature (see Faust 1985; Cherniak 1986) that shows our inability to store, let alone reason with, something having the computational dimensions of a scientific theory? Admittedly, to allow this additional psychological research would be to undermine the epistemic autonomy of the organism. And here Churchland's fixation on the organism is so strong that he retreats from naturalism to the sort of common-sense understandings of "explanation" (as translation to the familiar [pp. 198–99]) and "theory" (as any sort of conceptual framework [pp. 154–59]) that in any other case he would loathe. To be sure, explanations and theories in their folk senses can be neatly parceled out in individual brains with limited capacities, but they are at best based on puns on the words that philosophers of science have tried to explicate since the heyday of logical positivism.

Churchland fails to face the possibility that, left to speak for themselves and not dragooned into pre-existent philosophical debates, the cognitive sciences may invalidate not just this or that theory of the mind but the metatheory that says that the individual mind is the relevant unit for analyzing the most interesting features of knowledge. Churchland should have suspected something from the fact that analytic epistemology and philosophy of science have pursued separate research trajectories since the days of logical positivism. The problem of justifying a scientific theory is not a version of the problem of justifying an individual's belief. In fact, philosophers of science have tended to make a point of resisting an individualist interpretation of their project, for example, by denying any interest in scientific discovery and by making the objects of epistemic appraisal such crypto-social entities as "paradigms" (Kuhn), "research programmes" (Lakatos), and "research traditions" (Laudan). Moreover, this tendency is not limited to the historicists. Back in the 1930s, Reichenbach (1938, pp. 3–16) came close to identifying scientific epistemology with a normative sociology of knowledge, while Popper (1968), as a reaction to the logical positivist tendency to ground knowledge in individual experience, emphasized the conventional, presumptive, and hence publicly accountable character of so-called observation sentences. And even perception has not been immune to the latent sociologism of the philosophers. Polanyi, Hanson, and Kuhn stressed the role of perception in science only to show that each scientist's perception is laden with a *Weltanschauung* that marks him or her as a member of a particular scientific community. Although this *Weltanschauung* imparts a look to the world that it would otherwise not have, it does not follow that the scientist has much voluntary control over the look that is imparted (Suppe 1977, pp. 125–221). Very unlike Churchland's metatheory of the autonomous organism, on this view at least the normal scientist is portrayed as instantiating the collective consciousness of her or his community.

At this point, Churchland might counter that his residual commitment to the autonomous cognizer is benignly consistent with current psychological and neuroscientific research, which, for the most part, still studies individual organisms in laboratory isolation. In addition, he could argue that the evidence against the ultimacy of individuals is by no means as conclusive as the evidence against the existence of beliefs and desires. Yet, all that these responses could show is that the cognitive sciences are as yet just as unreflexive in their naturalism as philosophy. The history of science has repeatedly overturned the idea that the ultimate units of analysis for a given domain are individuals that appear discrete to the naked eye. For, just as midsized objects traveling at midspeed turned out to be inappropriate for conceptualizing mass in physics, and similar-looking organisms turned out to be unsuitable for individuating species in biology, so too (I argue) the solitary cognizer is unlikely to prove adequate for understanding the nature of knowledge (see Hull 1974, p. 48; Fuller 1989, pp. 58–62). From the fact that an individual registers the presence of knowledge in his or her utterances and other behaviors, it does not follow that she or he is the relevant causal source of that knowledge. After all, a meter registers the presence of electric current, but from that we would not infer that the meter is the source of the current. Rather, the meter enables the current to be manifested to some appropriately trained observer. The medieval Scholastics would have said that the "form of the current" was "communicated" from the meter to the observer. In more modern guise, the social epistemologist says that the cognizer is a better or worse *vehicle* (see Campbell 1988, pt. 6) for transmitting knowledge, depending on how the reception of its transmissions by other vehicles stands up against relevant standards of transmission. This is not the place to flesh out this alternative picture (see Fuller 1989, pp. 82–88, 120–24). I just want to drive home the negative point that the reflexive naturalization of the mind would cast doubts on the appropriateness of the individual as the unit of analysis for the cognitive sciences.

It is a truism among historians of philosophy (see Rorty 1979, chap. 3) that when rationalists like Descartes deemed the elements of thought "intuitively clear," they were treating ideas as if they had the clarity of individual objects moving in the foreground of one's visual field. For example, the comprehensibility of an idea corresponded to the viewer's ability to encompass an entire object in his or her line of vision without having to move; the combinatorial properties of the idea corresponded to the viewed object's ability to bring about new situations by interacting with other objects; and so forth. In short, rationalists have exploited the "positive analogy" between thinking and seeing, and philosophers, psychologists, and other social scientists continue to reap the benefits of that exploitation, even after the rationalist project has been officially abandoned. Cognitive psychologists would find this point unremarkable, as it reflects the mind's tendency to capitalize on heuristics, i.e., biases in how we think that play off of our limitations (Wimsatt 1984, 1986). And so, although Descartes and his successors have been always quick to admit that immediate experience is an unreliable guide to ultimate reality, nevertheless they have had no choice but to use experience as a corrigible model for what reality might be. The question for the cognitive psychologist, then, is when does the rationalist's heuristic become more a hindrance than a help to further inquiry.

The heuristic in this case is the tendency to attribute causation to objects that move freely against a background in one's visual field. The heuristic works in everyday life because there we need a concept of causation to coordinate our actions in relation to other things in the immediate environment. The things deemed "causal" are the ones whose movements are likely to make some difference to what we decide to do (see Kahneman 1973). In such settings, we generally do not need to speculate about whether the object's motion is synchronized with the motions of other visually occluded or distant objects. Yet, a speculation of this sort is appropriate once we start wondering whether what we see is all that there is, that is, whether visually discriminable objects are the right units for thinking systematically about reality. Although our intuitive notions about causation are ill-suited for satisfying a metaphysical impulse of this sort, without another intuitively acceptable theory to act as corrective, these notions function as a kind of default theory.

Contrary to what many philosophers seem to think, lack of intuitive acceptability — and not lack of developed alternatives — has been the big cognitive obstacle facing the reinterpretation of the mobile individual as either an instantiation or a part of some greater social process. For example, however much we may intellectually assent to the explanatory potency of, say, socio-economic relations in human affairs, we still do not intuitively see the need to postulate class differences in order to explain a single transaction that might be observed between two scientists in the laboratory. Why not simply invoke the intentions of the specific parties involved, and avoid reference altogether to an occult entity like class? The plausibility of class as an explanatory principle grows with an awareness that many such transactions occur in many places and times that are systematically interconnected by counterfactually realizable situations (e.g., if one party does not conform, then the other party can impose certain sanctions), the entirety of which transcends the awareness of any of the individuals participating in or witnessing a particular transaction. But such awareness has emerged slowly in our understanding of science. If a discovery is made, a scientist is still credited; if fraud is committed, a scientist is still blamed. We may nod sagely that these events are "strictly speaking" the systemic effects of, say, class struggle acting "through" the scientists, but this learned opinion does not undo our bias toward attributing causation to whatever moves something else. And, as far as the eye can see, people move apparatus, but class struggle does not move much of anything in the lab.

4. The Limited Naturalism of Experimental Psychology

I have been arguing that the tendency to treat the individual organism as the generative source of the behavior it manifests is a deeply intuitive one. Historically, this tendency was reinforced by the institution of experiments in psychology, which exaggerated the foregrounding effects of ordinary vision by physically isolating the organism in the artificially sparse setting of the laboratory. It was no accident that virtually all of the original experimental psychology laboratories were founded by professional philosophers and housed in philosophy departments for the first fifty years of psychology's existence (roughly 1870–1920; see Ash and Woodward 1987). Once techniques were introduced in the mid-nineteenth century to determine the increment of physical stimulus necessary for altering a subject's sensory response, philosophers started to believe that they were on the verge of quantifying the mind's grasp of the external world; hence, the name "psychophysics" was given to these techniques that soon formed the core methods of experimental psychology. But almost immediately, every philosopher with an interest in psychology realized that psychophysics alone would not solve the problems of epistemology. However, the philosophers differed over why psychophysics was insufficient.

The two most powerful figures in late nineteenth-century German psychology, Wilhelm Dilthey and Wilhelm Wundt, are nowadays portrayed as exemplars of, respectively, *geisteswissenschaftlich* and *naturwissenschaftlich* approaches to the study of human beings (Fuller 1989, pp. 56–58). Nevertheless, they both agreed that psychophysical methods could not fathom higher-order cognitive processes because these processes, as products of culture, are supraindividual facts that can be accessed by the hermeneutical understanding (*Verstehen*) of texts but not by the controlled introspection of sensations in the psychology lab (Tweney 1989). Notice that, by this time, the experimental method had been identified with the study of isolated individuals, and hence restricted to only those (sensory) features of human experience that are properly regarded as resulting from an individual's direct encounter with the world. This rare moment of agreement between Dilthey and Wundt served to create a stereotype about the limited appropriateness of human experimentation that has now lasted over a century. However, from our standpoint, the more important arguments against the sufficiency of psychophysics to solve the problems of epistemology were made by the now largely suppressed opposition to Wundt and Dilthey, a variegated group of "proto-cognitivists" that included Franz Brentano and Alexius Meinong, who founded the first two psychology laboratories in Austria, and, most influentially at the time, Wundt's research assistant, Oswald Kuelpe (see Lindenfeld 1980). All of these thinkers urged an expanded role for experimentation in the study of human beings, in terms of both objects and methods.

The proto-cognitivists argued that the alleged limitations of the experimental method were an artifact of the epistemology that implicitly informed psychophysics, namely, the radical empiricism of the British associationist tradition. In the typical psychophysics experiment, a specific sensory modality is stimulated, eliciting a response from the subject that is reported as an image consisting of primitive sensory qualities that are directly present to consciousness. Keep in mind that "subject" and "experimenter" usually exchanged roles and were sometimes one and the same person (Danziger 1985, pp. 134–35). Since subjects had to learn the right ways of reporting their experience before they could participate in experiments, they essentially internalized the experimenters' perspectives as their own. Indeed, when subject and experimenter were officially different people, the latter did little more than manipulate instruments. One drawback of this social arrangement, which became evident as psychology moved away from its empiricist origins, was that it did not allow for the distinct third-person perspective that is nowadays associated with the experimenter's standpoint. Such a standpoint enables the systematic study of features of the subject's thought processes that would normally elude her or his own conscious awareness. The thrust of this point may be seen in an example with which both defenders and opponents of an epistemically relevant experimental psychology grappled.

According to a standard reading of Hume, we infer causal relations from nothing more substantial than the regular association of spatio-temporally proximate events. The proto-cognitivists observed that Hume seemed to presume that all that is given in experience are the objects (or "contents") of experience, and not the very act of experiencing. Admittedly, this corresponds to the standpoint of the subjects of psychophysics, who have internalized a precise language for reporting their experience in the "natural attitude," a language that preserves the fact that we normally have only peripheral awareness of the act of experiencing, that is to say, the "intentional character" of the experience. However, the proto-cognitivists argued, what is only indirectly known to the subjects may be more directly known to observing third parties who are able to ask the sorts of questions that will get the subjects to attend more closely to these elusive, yet probably more fundamental, features of their experience (see Humphrey 1951). The proto-cognitivists realized that the sensory and mental images that constitute most of our waking lives are not, *pace* Hume and Wundt, ultimate units of psychic reality, but simply the products of habitualized constraints on how the world is made to appear to us. Change the constraints, and subjects may be able to reveal the conditions that give their everyday images focus and meaning. Here the proto-cognitivists were unwittingly re-enacting Descartes's attempt to methodically distance himself from his ordinary habits of mind. However, by the late nineteenth century, the Cartesian project had been assimilated to an internally prompted form of introspection, instead of a self-administered set of protocols ("rules for the direction of the mind," as Descartes said of his own book) for penetrating the deliveries of introspection (Scharff 1991).

All of this implied a more interventionist role for the experimenter, one requiring the design of innovative tasks and protocols, little of which would be directly tied to apparatus. The use of hypnosis (i.e., "the method of suggestion") during psychiatric treatment had already set a precedent for this practice in the medical community (Danziger 1985, pp. 135–36). Both the psychiatrists and proto-cognitivists agreed that the key determinants of thought normally subsisted at a level removed from the subject's conscious experience. Interestingly, the psychiatrists described this level as *beneath* consciousness, whereas the proto-cognitivists saw it more as a framing device *behind* consciousness. Both, however, understood the unconscious aspect of thought as giving direction to the subject's experience, often to the point of rigidly biasing it. When the bias facilitated problem solving, it functioned as a *mental set*; when the bias impeded this process, it functioned as a *neurosis* (see Humphrey 1951). Moreover, this natural affinity between the proto-cognitivists and the psychiatrists was pursued from both directions. (For a recent revival of this affinity, see "computational psychiatry" in Colby and Stoller 1988.) Not only did Freud study with Brentano, but

Kuelpe campaigned to transfer psychology's academic assignment from philosophy to medicine (Ash 1980).

Although the proto-cognitivists failed to convince most of their contemporaries of the value of experimentally studying the higher thought processes, their impact has been diffusely felt in both philosophy and psychology. For example, Karl Popper, who took his Ph.D. in educational psychology under Kuelpe's student, Karl Buehler, conceived of falsification as a compensatory strategy for mental sets in scientific reasoning (Berkson and Wettersten 1984). And Herbert Simon has gone so far as to draw a direct line of descent from the problem-solving protocols developed in Kuelpe's Wuerzburg school to the ones commonly used in today's cognitive psychology experiments (Baars 1986, pp. 365–66; see Ericsson and Simon 1984). Why, then, did the proto-cognitivists fail? My diagnosis is that they were un-self-conscious of the sociological character of their enterprise.

Consider the following objection to the proto-cognitivists that was also made of psychiatrists: If the experimenter can observe deep features of the subject's psychology that normally elude the subject's consciousness, could not a third party say the same about the experimenter, in which case why trust the experimenter's judgment any more than the subject's? Aside from appealing to the experimenter's special psychological training, the proto-cognitivists had little to say by way of response, which suggests that they felt obliged to answer in terms of the cognitive powers of individuals. After all, they could have said that the objection only shows that there must be communication between at least two people (e.g., experimenter and subject) before attributions of intentionality and the deeper features of cognitive life can be made. In other words, in going beyond Wundt to standardize the two-person experiment, Kuelpe's Wuerzburg school can be seen as having discovered the socially minimal enabling conditions for the emergence of cognition.

Moreover, as the structure of the objection itself implies, the addition of each new potential communicant into the experimental situation enriches the contexts in which cognitive properties can be attributed, until, at the limit, the laboratory group is no longer a simulation, but an actual instance of the sort of "natural" cognition that ultimately interests the psychologist (see Doise 1986). Conversely, when the communicants are removed, the individual internally represents (or, better, re-enacts) the missing parties. This point can be most readily seen in cases of group deliberation, where the sorts of comments that group members make to each other are rehearsed as metacommentary that an individual makes of her or his own thought. Although it has been traditionally held that group interaction minimizes the level of individual reflectiveness, and hence provides a degraded context for the study of cognition, recent experimental work suggests that, rather, the "higher thought processes" associated with reflection are a stylized arrangement of multiple voices abstracted from the group (Clark and Stephenson 1989; see Bazerman 1988, chap. 11).

The failure of proto-cognitivism was also bound up with the introspective method's general loss of credibility within experimental psychology. Instrumental here was J. B. Watson's (1913) *reductio* that if the subjects in each German university could introspect in the professor's preferred categories, then that only demonstrated the ease with which subjects can be conditioned to respond in a desirable manner — an implicit call to behaviorism, if there ever was one! But, clearly, what doomed introspection was not, as Watson had thought, the method's grounding in occult mental entities, but rather the failure of the early psychology labs to agree on the techniques by which data were generated and the language in which they were reported. In short, the introspectionists lacked "data domains" that could be reliably produced regardless of the psychological theory used to account for them (see Ackermann 1985, chap. 4). By contrast, in the seventeenth century, the founding secretaries of the French and British scientific societies, Marin Mersenne and Henry Oldenburg, made a concerted effort against the natural incommensurability of "philosophical systems" by standardizing the means by which scientists presented their observations to each other, namely, by the institution of the scientific journal (Dear 1985, 1988; Bazerman 1988, chaps. 3–5). The goal was to have the community of scientists in, respectively, France and Britain, constitute a literal body of knowledge that transcended their differences of opinion on metaphysics and other contestable issues. Mersenne and Oldenburg succeeded by enabling each scientist to use the other as a potential source of information for his or her own research. In modern psychological terms, the writings of one scientist could function as an "external memory store" for another, who would then be reminded of some other observation or principle that would forge the link between her or his research and that of the first scientist. Together, then, the scientific community embodied one "transactive memory" of the sort commonly seen today in research teams that successfully act as a unit (Wegner 1986). Inveterate philosophers to the end, the introspectionists never had a Mersenne or an Oldenburg, and consequently the students' lab work seemed to do little more than to offer suggestive support for larger speculative points in their major professor's system.

5. The Limits of Naturalism in Analytic Epistemology

In the 1950s, proto-cognitivism yielded two robust disciplinary off-spring. The more obvious one is experimental cognitive psychology. The less obvious one is analytic epistemology, which learned well the lesson of data domains without acknowledging the lesson's sociological character and while forgetting the function of experiments.

Analytic epistemology first laid claim to the legacy of proto-cognitivism when Roderick Chisholm recovered Brentano's and Meinong's work for the English-speaking world just after World War II (see the footnotes in Chisholm 1966, a standard epistemology textbook). The calling card of analytic epistemologists is their supposition that the problem of knowledge is to be solved by eliciting (one's own) intuitions about whether people would have knowledge under a variety of conditions that are designed to test a proposed definition of knowledge, usually a sophisticated version of "justified true belief." Now, as it turns out, the sorts of conditions that end up surviving philosophical intuitions are highly artificial ones, requiring that knowers be in the right state of mind, with just the right background beliefs, observing the objects under just the right ambient conditions, and so forth. While these intuitions are notoriously difficult to elicit from one's undergraduate students, they flow easily and rigorously from the trained epistemologist. The original introspectionists would be very impressed by the degree to which this inquiry has been standardized over the last thirty years, such that even the likes of Chisholm, Keith Lehrer, and Alvin Goldman, philosophers who differ substantially on their proposed definitions of knowledge (roughly, self-evidence vs. coherence vs. reliability as the justificatory ground), nevertheless share a body of intuitions about the paradigmatic states of knowledge for which any adequate definition must account. In this context, the idea of individuals as mobile mnemonics that inform a common memory is especially apt, as analytic epistemologists form a close-knit subculture who, through frequent written and oral rehearsals of the test cases for knowledge, prime each other's intuitions into mutual conformity. As a psychologist of collective memory might expect, the only other times in this century when intuitions (or basic observations) played such an important role as philosophical data were during the heydays of moral intuitionism at Oxbridge and logical positivism at Vienna, each of which involved a well-defined, highly interactive group whose members inferred truths of universal scope from the character of their interaction (see Cohen 1986, esp. pt. 2). The significance that the community of analytic epistemologists have attached to cultivating intuitions explains a still more puzzling feature of their enterprise.

What analytic epistemology gained in standardization over proto-cognitivism, it lost by ignoring any criteria of experimental validity aside from the intersubjective agreement of analytic epistemologists. Happening upon an epistemology conference, an experimental psychologist would most naturally interpret what the philosophers are doing when they are trading intuitions as comparing recipes for producing knowledge in an imaginary laboratory: e.g., if you put someone in this sort of situation, then the belief he or she is most likely to have about this object will constitute knowledge. But having made this interpretation, the psychologist will then be struck by the extent to which the epistemologists neglect the two crucial validity questions, *especially* given their avowed normative aspirations (see Berkowitz and Donnerstein 1982):

1. *Ecological Validity:* What is made of the fact that these contrived situations are highly unrepresentative of the conditions under which people try to make sense of the world? Could it mean that we have a lot less knowledge than we thought, or that we need to have knowledge a lot less than we thought, or maybe that what the epistemological recipe gives us is not knowledge after all?

2. *External Validity:* What steps could be taken to make our ordinary cognitive circumstances more like the artificial one in which knowledge is shown to be produced? Would the results be worth the effort? In short, can a real-world knowledge enterprise — a science — be improved by teaching its practitioners a particular theory of knowledge? If so, how? If not, why not?

In section 1 of his contribution, Gorman has already weighed the relative significance of these two validity claims for an experimental social epistemology. The point that needs to be stressed here is that *both* types of validity are routinely ignored by analytic epistemologists. In effect, the epistemic predicament of the analytic epistemologist is that of someone who grades classroom performance without ever having had any teaching experience (see Fuller 1989, pp. 141–45).

But is the situation any better for the avowed naturalists among the analytic epistemologists? Consider the "ought implies can" principle, the calling card of the analytic naturalist (e.g., Goldman 1986). According to this principle, the requirement to do something implies the ability to do it, or conversely, the inability to do something implies that one is not required to do it. Naturalists invoke this principle when arguing that there are empirical constraints on reasonable norms for human conduct;

specifically, a theory of scientific rationality fails to be compelling unless human beings are cognitively capable of using it reliably. Interpreted in this way, "ought implies can" puts the experimental psychologist in a position to empirically undermine philosophical theories of rationality. Indeed, some claim to have undermined all the standard normative models of inductive reasoning (see Kahneman, Slovic, and Tversky 1982), while an even longer tradition exists that casts aspersions on our abilities as deductive reasoners (see Tweney, Doherty, and Mynatt 1981, pts. 3–4). The most common epistemological response to this line of research is to deny the validity of the experiments in one way or another, but in any case to preserve the presumption that a theory of scientific rationality must apply to an *individual* human being. Philosophers hedge against potential refutations by using the competence-performance distinction to argue that these experiments (or experiments in general) do not tap deeply enough into the competences that tacitly inform an individual's everyday life (see Cohen 1981). In other words, the picture of the autonomous organism exerts a strong enough grip on the thinking of contemporary analytic naturalists that they would rather renege on much of their naturalism — especially the susceptibility of their theories to empirical test — than relinquish their commitment to the individual as the seat of cognitive power. (We shall see, in the next section, that this criticism applies no less to sociologists.)

The social epistemologist prefers another interpretation of "ought implies can," one that preserves the integrity of the experimental method but dislodges it from the grip of the autonomous organism. It starts by taking the "ought" side of the principle somewhat more in Kant's original spirit than analytic naturalists have been inclined to do. Kant (e.g., 1938, p. 290) first proposed "ought implies can" as part of an argument for the existence of a moral faculty that enables us to do the sorts of things required by the categorical imperative. Translated into naturalistic terms, this means that a norm governs a realm of entities, the identity of which is a matter for empirical inquiry to decide (see Fuller 1989, pp. 85–102). Now pair this point with the tendency, especially among philosophers of science, to express epistemic norms as counterfactuals. That is, normative accounts of scientific rationality are typically defended on the grounds that had they been followed, certain epistemically desirable outcomes would have resulted. This explains the great popularity among philosophers of rationally reconstructed histories, which typically define the relevant outcomes in terms of theory choices that would have consistently moved the research trajectory of the past closer to the science of today (see Lakatos 1979). In cases where such claims are plausible, it would be foolish to dismiss a theory of rationality simply because certain historical figures did not in fact use the theory or even because human beings are not cognitively well-disposed to use it (see Fuller 1991). But neither do we need to deny these facts in order to shelter the theory from revision. Rather, the facts should lead us to conclude that perhaps the isolated individual is not the unit of analysis to which the theory applies.

A good case in point, one in which Gorman and his associates have been heavily involved, pertains to Popper's falsification principle (see secs. 2.1 and 2.5 of Gorman's contribution to this volume). Popper held that if scientists had always chosen the theory that had overcome the most attempts at falsification, science would have progressed at a faster rate than it in fact has. Now, someone interested in increasing the rationality of science, or the quality of its knowledge products, needs to ask not whether Popper's assertion is entirely true, but rather whether it is true enough so as to be in the interest of science to manifest more falsificationist tendencies. Notice that the word "science" is operating here as a place holder for whatever unit turns out to be the appropriate one for falsificationism to work. There is a clear role for experimentation in determining the size of this unit. For example, the psychological evidence suggests that individuals are both cognitively and motivationally ill-disposed toward falsifying hypotheses that they themselves generate. However, subjects can learn to become better falsifiers, and thereby improve their performance on rule-discovery tasks that model some of the basic features of scientific reasoning. That already introduces a level of normative intervention that transcends the aspirations of the typical epistemologist.

Of course, it remains an empirically open question whether the learning techniques used in, say, Gorman's lab can be transferred to the workplaces of real scientists with comparable improvements. Yet, at their most prescriptive, philosophers tend to suppose that individual scientists ought to be taught falsificationism (or whatever happens to be the chosen theory of rationality) until it becomes second nature. But if individuals are as resistant to rationality as the experimental evidence suggests, then it is probably more efficient to aim to *compensate* for cognitive deficiencies than to *eliminate* them outright. In other words, where the reasoner cannot be changed, change the reasoner's environment (see Fuller 1989, pp. 141–51). After all, this is how experimental design is supposed to overcome the cognitive deficiencies of the experimenter! In the case of scientists as subjects, relevant compensatory mechanisms may include altering the incentive structure of science so that scientists come to think that trying to falsify their own hypotheses is in their own best personal interest (and not merely in the interest of science as a collective enterprise) or conceiving of falsification as a social strategy whereby each individual is a bold conjecturer but her or his neighbor's rigorous refuter (see Faust 1985). An advantage of the latter idea is that it makes a virtue out of another cognitive adversity that social psychologists have found, namely, the human tendency to underestimate one's own level of error, while overestimating the errors of others (Jones et al. 1987).

6. The Limited Naturalism of Ethnomethodology

As Gorman has already observed in this volume, although artificiality is often counted as a mark against the experimental method, defenders of experiment have traditionally counted artificiality in its favor, since they have realized that the point of experiment is to compensate for the cognitive biases built into normal observation that would normally prevent us from penetrating the phenomena. For example, if I notice that the vast majority of scientists swear by the hypothetico-deductive method, if I am also of an experimental turn of mind, I will then wonder whether that degree of convergence might be an artifact of the conditions that prompted them to issue their opinions. After all, (philosophy) textbook views of scientific method are easy to hold if one tends to talk about method only in contexts far removed from actual research. In that case, a change of conditions might change the tenor of the scientists' responses. The risk that is run by *not* experimenting, especially on human beings, should be clear: namely, a failure to realize the extent to which the routine settings of everyday life constrain the opportunities for people to express all that they might, both individually and collectively.

The strategic advantage afforded by experimental intervention may be seen in a large-scale social phenomenon like science itself, a proper understanding of which requires studying the interaction of "macro" and "micro" perspectives on human behavior (see Knorr-Cetina and Cicourel 1981). The macro level is the systems theorist's bird's-eye view of the regularly governed society, the tables of statistically normal behaviors that testify to the capacity of norms to stereotype the actions of individuals into functionally interdependent role expectations. But if norms are the means by which individuals are mobilized for the benefit of society, they are equally the means by which individuals mobilize societal resources for their own benefit, in the course of which the norms may themselves (unwittingly) change. At this point, we have reached the micro level of the ethnomethodologist, who strategically intervenes to reveal the array of motivations and competences hidden behind normal behavior (Turner 1975). Ethnomethodologists are notorious for maintaining that even the most universally shared forms of behavior do not run very deep in human beings, as they are more the product of the routinized settings that call for such behavior than of some incorrigible feature of the human psyche. And while ethnomethodologists are on record as being among the staunchest critics of experimentation, their opinion on *this* issue is very much one with the experimentalist.

Consider the role of "being a scientist," which is manifested across disciplines both in the rhetorical uniformity of scientific writing and in convergent expectations about the scientist's personal behavior. Can this uniformity and convergence be observed in the same people outside of the settings in which they are typically held to be accountable as scientists? This is a natural question for both the experimentalist and the ethnomethodologist to ask. The difference between them comes in the ethnomethodologist's expectation that the setting would not have to be altered very much for a significantly different behavioral response to be elicited from the scientists under study. This point is usually cast in terms of the radically conventional character of social behavior, such that people behave like scientists *only* in those situations in which such behavior is expected of them. In the privacy of their own labs, say, the scientists become totally different people. For their part, the experimentalists need not presume the truth of such a radical conventionalism, but may choose to treat it as a hypothesis whose level of empirical support will probably vary, depending on the kind of behavior under study.

Despite their obvious affinities with the experimental turn of mind, ethnomethodologists are generally suspicious of experiments. A brief look at the tenor of their objections reveals both blindness and insight into how experiments work. Common to these objections is the observation that experiments are themselves social situations associated with a host of conventions and role expectations: e.g., the subject must respond within the constraints of the experimental protocols, the experimenter is the final authority on the experiment's interpretation, etc. I have already alluded to some of the interesting historical work that has been done on the early battles fought to institutionalize the two-person experiment as a valid means of obtaining knowledge about humans. The implication often drawn from these histories is that experimental situations are typically contrived so as to cause subjects to behave in ways that will enable the experimenter to assert that people generally have certain incorrigible tendencies, even though these tendencies would never have emerged in more natural settings. In one sense, this should count against the ecological, or even the external, validity of the experimental result. Yet, experimentalists have historically been able to turn the tables and use the very elusiveness of these tendencies in everyday life as grounds for increasing the power of psychologists to study and monitor people even more closely, both in and out of the lab (see Morawski 1988). In effect, a methodological adversity has been converted into a political virtue. The experimenter unreflectively (or is it deliberately?) attributes the effect produced in the lab to a tendency in the *subject* rather than to a tendency in the *situation* defined by the interaction of subject and experimenter. By erasing their own presence from the authoritative interpretation they give to this situation, the experimenters implicitly absolve themselves from any causal or moral responsibility for what the subjects have done.

But does this "ethnohistorical" critique argue against the epistemic legitimacy of experiment per se, or, rather, against one very entrenched way of interpreting the outcome of an experiment, namely, as revealing the hidden character of the individual subject? Clearly, I would claim the latter, and, moreover, would trace the persistence of this interpretation to the failure of experimenters to apply their own work to themselves. After all, they were the ones who originally identified the biases in causal attribution in social contexts (e.g., Nisbett and Ross 1980), an instance of which must be the experimenter's own analysis of a subject's behavior.

It is doubly ironic, then, that most ethnomethodologically inspired critics of experiments (e.g., Cicourel 1964; Harre and Secord 1979) turn to so-called natural settings as appropriate for highlighting the capacities of the autonomous individual. In the first place, artifice is hardly eliminated from the natural setting, as may be quickly seen by focusing on the interpretive artifice required to give someone "the benefit of the doubt." Perhaps the most common of all ordinary interpretive principles, it involves imagining that had a natural situation been slightly different, the person's behavior would have more closely approximated some normative standard. Indeed, this ongoing, largely unconscious process of rational reconstruction may explain why our memories tend to be stereotyped (or idealized, as the case may be; see Kruglanski 1989, chap. 3).

Second, and more importantly, the ethnomethodologists are no different from Churchland in failing to challenge the visually biased ontology — the foregrounded mobile organism is still the principal cause of its own behavior — but the background has now changed to something more visually familiar. Notice that this response marks a retreat from the ethnohistorical critique of experiment, since it would seem now that when people are in their native habitats, society merely enables the display of their powers, but does not actually constitute them. The latter, which entails a more thoroughgoing sociologism, would require showing that claims about the existence of innate or fixed human capacities have depended crucially on agreement over paradigm cases of competent performance and standards for measuring the distance of particular performances from those cases, all of which, in turn, have been made possible by the existence of institutions designed to discipline those capacities. It is no accident that the oldest faculties of the liberal arts — logic, grammar, mathematics — should be the ones most often proposed as tapping into such capacities (see Gellatly 1989). Could it be that historically humanity became first acquainted with its most innate and fixed cognitive capacities and then only later with the capacities that are subject to change and convention because they require "specialized training"? Or, is it rather more likely that the disciplines with the longest histories of imprinting and eliciting their forms of knowledge are the ones that turn out to define canonical performances of cognitive competence more generally (see Hoskin and Macve 1986)? State the choice that way, and clearly the latter is the more plausible scenario.

7. Towards an Experimental Constructivist Sociology of Science

It is now time to be more positive in our efforts to define an experimental social epistemology. In sections 2.5 and 2.6 of his contribution to this volume, Gorman has outlined the rudiments of an experimental research program for investigating the issues that most concern social epistemology, namely, the transmission and reception of knowledge (see Fuller 1988, esp. pts. 2–3). His discussion focuses on experiments that were designed to track the reproduction of certain normative standards, an especially pertinent issue given the considerations raised at the end of the last section. Another complement of experiments, performed mostly by European social psychologists, tracks the conditions in which people are regarded as originators, transformers, or simply users of knowledge available to a group (see Wicklund 1989). However, none of these studies truly addresses the concerns of the most radical sociologists of science, the "constructivists," whose rhetoric has been most hostile to any rapprochement with philosophy and psychology in forging a "cognitive science of science" (e.g., Woolgar 1989). In the remainder of this essay, I will meet this hostility with an extended proposal for an experimental framework that incorporates constructivist concerns.

The most common experimental setting for studying science is to have subjects solve problems that share some of the features that are thought to be salient in the problems that scientists solve. Not only does this identification of scientific reasoning with problem solving have a distinguished pedigree in behaviorist, Gestalt, and contemporary cognitive psychology studies of thinking, but it also has the endorsement of philosophers of science as otherwise diverse as Carnap, Popper, Lakatos, Kuhn, and Laudan — all of whom assume that scientific inquiry is best seen as a sophisticated exercise in problem solving. However, constructivists in the sociology of science (e.g., Woolgar 1988) have critiqued this entire approach as begging the question in favor of an "autonomous" or "internalist" conception of science, one in which problems are unambiguously given to scientists as worthy of solution because of the clear place that the problems occupy in the larger edifice of scientific knowledge. Psychologists reproduce this bias when, in designing their experiments, they simplify the real working conditions of scientists, especially when they try (and presume to have succeeded in) presenting subjects with a clear-cut, nonnegotiable task along with "background information," rudimentary cognitive tools such as blackboards and calculators, but no access to other people with potentially useful skills (see Lesgold 1988).

When confronted with this criticism, psychologists may be inclined to respond in a way that serves to reinforce the internal-external (or cognitive-social, as it is nowadays often put) distinction that the constructivists are trying to subvert. The psychologist may take the criticism as simply a call to supplement the laboratory situation with additional factors that bring the subjects' problem-solving task closer to the sort of task that normally faces scientists. This call for greater ecological validity may include motivating the task by giving the subjects a vested interest in their solving the problem within a certain amount of time or before some competitor does. And while the addition of these factors undoubtedly improves the ecological validity of the psychologist's experiment, it does not get to the heart of the constructivist's concerns. For all that the psychologist has admitted at this point is that there are material constraints within which problems must be normally solved. Call these constraints "motivational," "social," or "political," they are nevertheless conceptualized as things external, and hence superadded, to the core cognitive processes proper to problem solving (see Thagard 1989). But in denying the internal-external distinction, the constructivist means to be making the stronger claim that these constraints are actually *constitutive* of the cognitive processes themselves. How can this point be made in an experimental setting?

As might be expected, when the constructivist talks about "problems," it is usually about their *construction* rather than their *solution*. This already sheds some light on how the experiment ought to be designed: namely, subjects should be allowed to construct a problem out of a set of material (and human) resources, rather than be given a problem prepackaged by the experimenter. In other words, it would be better to treat subjects more like Koehler treated his apes than Thorndike treated his cats (Fuller 1988, p. 199). But what motivates this shift in emphasis? For the constructivist, the most interesting feature about "problem solving" in science is not the discovery of a solution but the discovery of a problem. We are led to believe, especially by Callon (1980), that most of the real cognitive work occurs at the planning stages of scientific research, during which the team director determines the best way to mobilize all of her or his resources to make the biggest impact on the scientific community. The most obvious point at which such thinking takes place is in the formulation of research grant proposals. But in the actual course of research, the problem is also subtly reconstructed to fit changes in resources, personnel, the state of competition, and, of course, new data and analyses — though these factors are not as neatly parceled as one might think. For example, data may acquire new significance once a team member has been lured to the competition, which then forces the team to reconstruct the point of their project so as to maximize the remaining resources. Ultimately, a grant renewal deadline turns out to determine the exact nature of the team's accomplishment, since at that point the research director is forced to construct a problem-solving narrative that makes it seem as though the team had taken the most efficient course of action toward some predetermined goal — a goal that the team might not have acknowledged as paramount or even important a few weeks earlier. Once problem solving is seen in this light, it should be clear why constructivists find the usual focus on "solving" misguided, since the problems that scientists explicitly set out to solve are really themselves temporary and corrigible solutions to the ongoing metaproblem of managing laboratory life and the impressions it makes to the outside world (see Lindblom and Cohen 1979).

Staunch internalists might protest that all we have here is an admittedly sophisticated account of external constraints on problem solving that do not touch the cognitive processes that have traditionally interested philosophers and psychologists. To support their suspicions, the internalists could point to the difference in the way they and their constructivist foes describe the laboratory situation. Where the internalist sees theories being tested, the constructivist sees resources being mobilized. But at the meta-level, where the internalist sees two different activities, one involving theories and the other resources, the constructivist sees the same activity under two different descriptions (Fuller 1989, pp. 25–29). If, as psychologists, we want to take seriously the constructivist's claim that the internal-external distinction makes no real difference (see Latour and Woolgar 1986, postscript), then we must design experiments where talk of theories and talk of resources turn out to be alternative ways of characterizing laboratory life. Specifically, subjects should work on something that enables them to alternate between the two types of talk as *they* see fit. This does not mean abandoning experimental protocols, but rather developing protocols that do not bias the subject's verbal reports in one of these two ways. We can then compare the circumstances under which subjects naturally use theory-talk and resource-talk.

Taking a cue from Callon (1980) and his collaborators in actor-network theory, the best circumstances for comparing the two types of talk are when explicit translations need to be effected. For example, Callon points out that one of the first things that research directors must do is to construct the "socio-logic" of the problem that their team is about to tackle. That is, they need to interrelate people and tasks. Now, there are two ways in which research directors might think about this process:

1. They might assign tasks to people (or "map" tasks "onto" people), which involves thinking of the team as a fixed unit divided into parts (people), each of whom may perform one or several of the tasks required for solving the problem.

2. They might assign people to tasks (or "map" people "onto" tasks), which involves thinking of the problem as a fixed unit divided into parts (tasks), each of which can be executed by one or several people in the team.

The subtle difference between (1) and (2) is, in fact, quite profound, as it marks the moment at which theory-talk and resource-talk become complementary ways of viewing the same phenomenon. In (1), the research team itself is seen as setting the ultimate constraints on problem solving. In effect, the scarcity of talent and inclination determines which problems get solved. Clearly, this is to think about things from an "externalist" standpoint, one in which the business of problem solving itself turns out to be a highly negotiable enterprise. By contrast, in (2), the nature of the problem sets the ultimate constraints on how the team's energies are deployed. The relevant "scarcity" here pertains to the limitations imposed on the ways in which the problem may be legitimately solved: e.g., it must be derivable from certain standing theories and compatible with other theories and findings (see Nickles 1980). Clearly, we have here an "internalist" construal of what is at stake, one in which the energies of the team are deployed in whatever ways it takes to get the job done. A vivid way of casting the difference between (1) and (2) is to imagine a research director (1) using a table with the academic track records of the members of his or her team as the basis for deciding which problem to tackle, and (2) using a flow chart of a hypothesized procedure for tackling an unsolved problem as the basis for deciding who should do what.

The difference between (1) and (2) suggests a reason why internalists tend to denigrate sociological considerations, while externalists return the compliment by denigrating philosophical ones. Research directors who model their situation as in (1) presume that people can solve any number of problems, depending on their ability and inclination to work with each other. Consequently, the sorts of problem-solving flow charts endemic to cognitive science research are unlikely to impress these directors unless they can envisage who would be doing the job prescribed by each of the little boxes on the chart. However, research directors in (2) think of matters much differently. They are unlikely to worry as much about the division and coordination of labor because they reckon that if the problem depicted on the flow chart is both soluble and worth solving, then there are a variety of ways in which people can be arranged to work on it (see Prentice-Dunn and Rogers [1989] on the phenomenon of "deindividuated" labor). This difference is replayed, in exaggerated form, when we move from the research directors' offices to the departments of philosophy and sociology. When philosophers of science talk about a theory-saving phenomena or solving problems, they are typically not concerned with the people and circumstances involved in the theory's use, since such information would seem to be incidental to whether the theory "really" saves or solves. And, sociologists rarely dwell on these philosophical matters because they see theories as inherently meaningless tokens that are exchanged in a game of power. Put somewhat more positively, they are prone to think that virtually *any* theory can be shown to work if the right sort of effort is mobilized on its behalf.

Notice that although my arguments have been based on an example drawn from the initial stages of research, it would be misleading to conclude that the research team is permanently locked into the mindset represented by either (1) or (2) — contrary to what defenders of internalist and externalist approaches to science might think. The psychologist should only expect that subjects will alternate between models as the situation demands. Of particular interest will be how the subjects reconstruct the entire problem-solving session after the experiment is over, for it is then that the most translation between models is likely to take place. As they recount what happened during the experiment, do the subjects make it seem as though the cognitive is made to bend to the social, or vice versa? Assuming that the experimenter has also been observing the problem-solving session, the debriefing session may offer considerable insight into the aspects of laboratory life that are more and less susceptible to rational reconstruction, thereby enabling historians to make more informed inferences from a scientist's notebooks and articles to the events putatively described.

Now let us suppose that the psychologist defines the above experimental situation more strictly by dividing up the subjects into two groups, one explicitly instructed to act as the research director does in (1) and the other to act as she or he does in (2). Although each group would still be working on a challenging problem, an epistemologically interesting aspect of the challenge would drop out, namely, the aspect that comes from subjects having to alternate between orthogonal ways of organizing research. When the experiment was more open-ended, subjects were confronted with the mutual interference of cognitive and social factors: e.g., team members who refused to work together, even though each had skills that were necessary for assembling a solution to a problem. However, once one set of factors is portrayed as indefinitely plastic to the fixed needs of the other set, it becomes easy to see science as either socially or cognitively determined, depending on whether subjects are instructed in (1) or (2). Thus, by restricting the subject's considerations to either social or cognitive factors, the experimenter in fact relieves the subject's real-world epistemic burden. In military circles, this lightening of the load would be readily recognized as indicative of the movement, in Clausewitz's (1968) terms, from the standpoint of *Kritik* to that of *Theorie*, a sure-fire way of jeopardizing one's chances of winning a war (Wills 1983). In conclusion, I propose to take seriously Latour's Clausewitzian maxim that "science is politics by other means," as well as the military metaphors that pepper his writings and those of his sociological colleagues (e.g., Callon and Latour 1981), for they may be read as implicit warnings to the experimentalist not to put his or her subjects in a position of merely theorizing about, rather than genuinely organizing, scientific research.

Clausewitz argued that strategists in the past have given tactically poor advice, and even lost entire wars, because they imagined that a rational approach to war required a "military science" whose theoretical perspective would be comparable to that of the other sciences. Clausewitz's diagnosis here can be read as an attempt to locate the limitations inherent in the rationalist heuristic of modeling cognition on vision. The strategist's error consists in taking the metaphor of theories as "mirrors" or "points of view" too literally. As a result, two quite different principles of interrelating objects are conflated, which after Piaget (1971) may be termed *spatio-temporal* and *logico-mathematical*. The former refers to interrelations based on the objects' physical distance from one another, whereas the latter refers to interrelations based on properties that the objects share with other objects not necessarily in the immediate vicinity. While cartographers must explicitly negotiate between these two principles whenever they design maps (see Robinson and Petchenik 1976), most philosophers and sociologists unwittingly conflate the principles in their theorizing.

Consider the difference between *bottom-up* and *top-down* theoretical orientations. The metaphor suggests, respectively, a widening and a narrowing of the theorist's angle of vision (see De Mey 1982, chap. 10). It is as if higher-order concepts, laws, and structures were things that could be actually seen (like an aerial view of a marching band that forms its school's name), if one stood far enough away from an entire domain of phenomena. This is the systems theorist's bird's-eye view, against which the experimentalist and ethnomethodologist were earlier portrayed as joined in common cause. Conversely, lower-order concepts, atomic sensations, and microinteractions are supposed to appear as theorists intensify their gaze on a specific part of the domain (see Capek 1961). Ubiquitous talk of structures as "emergent" and data as "foundational" just reinforces this conflated image. The image is conflated, of course, because the farther back one stands from a field of micro-objects, the less distinct those objects become, but not necessarily because they are yielding to some stable, overarching pattern of macro-objects. If anything, enlarging one's perspective on the field should reveal long-term uncertainties and opportunities that were obscured by the immediacy of day to day combat. This is the method of *Kritik*. In essence, then, Clausewitz's military strategist errs because he or she jumps too quickly from a spatio-temporal to a logico-mathematical orientation, and as a result, fails to realize that a war is won not by seeing one's own plans as the emergent structure of the battlefield, but rather by remaining open to new courses of action that are suggested by seeing the field from different locations.

As a reflexively Clausewitzian thinker, Latour (esp. 1987) both portrays successful scientists as masters of *Kritik*, not *Theorie*, and casts in this light his own success at capturing "scientists in action." In particular, Latour avoids any ontological commitments to macro-objects, at least macro-objects that exist at a level of reality "above" micro-objects. However, this does not mean that he is an ordinary empiricist or positivist. *Pace* Quine, while Latour's world is flat, it is nevertheless densely populated with entities. For example, a scientific law is not something that theoretically hovers in a world above the data, but rather it exists side by side as part of a long and resilient network, the actors in which (producers of data) have their interests served by allowing the law to speak on their behalf — that is, to represent them. (The price one pays, in the Latourian picture, for abandoning the cognition-as-vision metaphor is to assume that the mind and the politician "represent" in the same sense. See Pitkin 1972; Fuller 1988, pp. 36–45.) How are actors enrolled in such a network? Although Latour characteristically eschews anything that would smack of a *theory* of enrollment, some insight may be gleaned from Piaget's account of how the child's cognitive orientation evolves from spatio-temporal principles to logico-mathematical ones.

According to Piaget, children move from a spatio-temporal to a logico-mathematical orientation when they become able to rearrange objects to their liking, which typically involves grouping certain objects together and separating them from others. In this way, primitive set-theoretic relations of inclusion and exclusion emerge. Because the material character of the objects offers little resistance to their manipulative efforts, children can play with the objects without concern for the consequences of their interaction with them. Successful theorists are likewise ones who propose, say, a lawlike generalization that manages to replace the interests of a vast number of actors (i.e., producers of the data subsumed under the law) with their own, as a result of those actors coming to accept (or at least, to not object to) the redescription they receive under the law. In short, the actors offer no resistance, and the disparateness of their original interests can be contained in a Latourian "black box" (see Latour 1986). However, most real-world theorists do not succeed in this task because actors provide interminable resistance, which reminds the theorists that objects exist not only in the logico-mathematical world of theoretical manipulation but in the spatio-temporal world of alternative interests and agendas. The prominence given to this fact in the first formulation of the problem-construction experiment affords it a greater advantage in studying science as a process than the methodologically tighter second version of the experiment, which only succeeds at approximating the conditions needed for theorizing science as determined by either social or cognitive factors exclusively.

Note

The oral version of this essay was subject to a spirited discussion at the Minnesota Studies conference. I would especially like to thank Ron Giere and Dick Grandy for their comments then. Also thanks to Mike Gorman and Art Houts for their long-standing contributions to the social epistemology project. But most of all, I would not have devoted so much energy to purging "cognitive individualism" were it not for the Ph.D. thesis that my research assistant, Stephen Downes (1990), has recently completed on the topic.

References

Ackermann, R. 1985. *Data, Instruments, and Theory*. Princeton, N.J.: Princeton University Press.

Ash, M. 1980. "Wilhelm Wundt and Oswald Kuelpe on the Institutional Status of Psychology." In W. Bringmann and R. Tweney, eds., *Wundt Studies*. Toronto: Hogrefe, pp. 396–421.

Ash, M., and W. Woodward, eds. 1987. *Psychology in Twentieth Century Thought and Society*. Cambridge: Cambridge University Press.

Baars, B. J. 1986. *The Cognitive Revolution in Psychology*. New York: Guilford Press.

Bazerman, C. 1988. *Shaping Written Knowledge*. Madison: University of Wisconsin Press.

Berkowitz, L., and E. Donnerstein. 1982. "Why External Validity Is More Than Skin Deep." *American Psychologist* 37:245–57.

Berkson, W., and J. Wettersten. 1984. *Learning from Error*. La Salle, Ill.: Open Court.

Bernard, C. 1951. *An Introduction to the Study of Experimental Medicine*. New York: Dover.

Callon, M. 1980. "Struggles and Negotiations to Decide What Is Problematic and What Is Not." In K. Knorr-Cetina et al., eds., *The Social Process of Scientific Investigation*. Dordrecht: Kluwer, pp. 197–220.

Callon, M., and B. Latour. 1981. "Unscrewing the Big Leviathan." In K. Knorr-Cetina and A. Cicourel, eds., *Advances in Social Theory and Methodology*. London: Routledge and Kegan Paul, chap. 11.

Callon, M., J. Law, and A. Rip. 1986. *Mapping the Dynamics of Science and Technology*. London: Macmillan.

Campbell, D. 1988. *Methodology and Epistemology for Social Science*. Chicago: University of Chicago Press.

Capek, M. 1961. *The Philosophical Implications of Contemporary Physics*. New York: Van Nostrand.

Cherniak, C. 1986. *Minimal Rationality*. Cambridge, Mass.: MIT Press.

Chisholm, R. 1966. *A Theory of Knowledge*. Englewood Cliffs, N.J.: Prentice Hall.

Churchland, P. 1979. *Scientific Realism and the Plasticity of Mind*. Cambridge: Cambridge University Press.

———. 1989. *A Neurocomputational Perspective*. Cambridge, Mass.: MIT Press.

Cicourel, A. 1964. *Method and Measurement in Sociology*. New York: Free Press.

Clark, N., and G. Stephenson. 1989. "Group Remembering." In Paulus (1989), chap. 11.

Clausewitz, K. v. 1968. *On War*. Harmondsworth, Eng.: Penguin.

Cohen, L. J. 1981. "Can Human Irrationality Be Experimentally Demonstrated?" *Behavioral and Brain Sciences* 4:317–70.

———. 1986. *The Dialogue of Reason*. Oxford: Clarendon Press.

Colby, K. M., and R. J. Stoller. 1988. *Cognitive Science and Psychoanalysis*. Hillsdale, N.J.: Lawrence Erlbaum.

Collins, R. 1989. "Toward a Neo-Meadian Sociology of the Mind." *Symbolic Interaction* 12:1–32.

Danziger, K. 1985. "The Origins of the Psychological Experiment as a Social Institution." *American Psychologist* 40:133–40.

Dear, P. 1985. "Totius in Verba." *Isis* 76:145–61.

———. 1988. *Mersenne and the Learning of the Schools*. Ithaca, N.Y.: Cornell University Press.

De Mey, M. 1982. *The Cognitive Paradigm*. Dordrecht: Kluwer.

Dewey, J. 1960. *The Quest for Certainty*. New York: G. P. Putnam and Sons.

Doise, W. 1986. *Levels of Explanation in Social Psychology.* Cambridge: Cambridge University Press.
Downes, S. 1990. "Cognitive Individualism and the Possibility of a Cognitive Science of Science." Ph.D. diss., Center for the Study of Science in Society, Virginia Polytechnic Institute.
Ericsson, K. A., and H. Simon. 1984. *Protocol Analysis: Verbal Reports as Data.* Cambridge, Mass.: MIT Press.
Faust, D. 1985. *The Limits of Scientific Reasoning.* Minneapolis: University of Minnesota Press.
Fodor, J. 1981. *Representations.* Cambridge, Mass.: MIT Press.
Fuller, S. 1988. *Social Epistemology.* Bloomington: Indiana University Press.
———. 1989. *Philosophy of Science and Its Discontents.* Boulder, Colo.: Westview Press.
———. 1991. "Is History & Philosophy of Science Withering on the Vine?" *Philosophy of the Social Sciences* 21:149–74.
———. 1992. "Social Epistemology and the Research Agenda of Science Studies." In A. Pickering, ed., *Science as Practice and Culture.* Chicago: University of Chicago Press.
Fuller, S., et al., eds. 1989. *The Cognitive Turn: Sociological and Psychological Perspectives on Science.* Dordrecht: Kluwer.
Gellatly, A. 1989. "The Myth of Cognitive Diagnostics." In A. Gellatly et al., eds., *Cognition and Social Worlds.* Oxford: Oxford University Press, chap. 8.
Gholson, B., et al., eds. 1989. *Psychology of Science.* Cambridge: Cambridge University Press.
Giere, R. 1988. *Explaining Science.* Chicago: University of Chicago Press.
Goldman, A. I. 1986. *Epistemology and Cognition.* Cambridge, Mass.: Harvard University Press.
Gorman, M. E. 1990. *Simulating Science: Error, Heuristics, and Mental Models.* Bloomington: Indiana University Press.
Gorman, M. E., and W. B. Carlson. 1989. "Can Experiments Be Used to Study Science?" *Social Epistemology* 3:89–106.
Hacking, I. 1983. *Representing and Intervening.* Cambridge: Cambridge University Press.
Harre, R., and P. F. Secord. 1979. *The Explanation of Social Behavior.* 2d ed. Totowa, N.J.: Littlefield and Adams.
Heyes, C. M. 1989. "Uneasy Chapters in the Relationship between Psychology and Epistemology." In Gholson et al. (1989), chap. 5.
Hoskin, K., and R. Macve. 1986. "Accounting and the Examination: A Genealogy of Disciplinary Power." *Accounting, Organizations & Society* 11:105–36.
Houts, A., and B. Gholson. 1989. "Brownian Notions." *Social Epistemology* 3:139–46.
Hull, D. 1974. *The Philosophy of the Biological Sciences.* Englewood Cliffs, N.J.: Prentice Hall.
Humphrey, G. 1951. *Thinking.* London: Methuen.
Jones, E., et al. 1987. *Attribution.* Hillsdale, N.J.: Lawrence Erlbaum.
Kahneman, D. 1973. *Attention and Effort.* Englewood Cliffs, N.J.: Prentice Hall.
Kahneman, D., P. Slovic, and A. Tversky, eds. 1982. *Judgment under Uncertainty.* Cambridge: Cambridge University Press.
Kant, I. 1938. *The Fundamental Principles of the Metaphysic of Ethics.* Trans. O. Manthey-Zorn. New York: Appleton-Century Crofts.
Knorr-Cetina, K. 1980. *The Manufacture of Knowledge.* Oxford: Pergamon Press.
Knorr-Cetina, K., and A. Cicourel, eds. 1981. *Advances in Social Theory and Methodology.* London: Routledge and Kegan Paul.

Kruglanski, A. 1989. *Lay Epistemics and Human Knowledge.* New York: Plenum Press.
Lakatos, I. 1979. *Methodology of Scientific Research Programmes.* Cambridge: Cambridge University Press.
Latour, B. 1986. "Visualization and Cognition." In A. Pickering, ed., *Knowledge and Society.* Vol. 6. Greenwich, Conn.: JAI Press, pp. 1-40.
————. 1987. *Science in Action.* Milton Keynes: Open University Press.
Latour, B., and S. Woolgar. 1986. *Laboratory Life.* 2d ed. Princeton, N.J.: Princeton University Press.
Laudan, L. 1987. "Progress or Rationality? The Prospects for a Normative Naturalism." *American Philosophical Quarterly* 24:19–31.
Lesgold, A. 1988. "Problem Solving." In Sternberg and Smith (1988), chap. 7.
Lindblom, C., and D. Cohen. 1979. *Usable Knowledge: Social Science and Social Problem Solving.* New Haven: Yale University Press.
Lindenfeld, D. 1980. *Towards a Transformation of Positivism: 1880–1920.* Berkeley: University of California Press.
Morawski, J., ed. 1988. *The Rise of Experimentation in American Psychology.* New Haven: Yale University Press.
Nickles, T. 1980. "Can Scientific Constraints Be Rationally Violated?" In T. Nickles, ed., *Scientific Discovery, Logic, and Rationality.* Dordrecht: Kluwer, pp. 285–315.
Nisbett, R., and L. Ross. 1980. *Human Inference: Strategies and Shortcomings of Social Judgment.* Englewood Cliffs, N.J.: Prentice Hall.
Paulus, P., ed. 1989. *Psychology of Group Influence.* 2d ed. Hillsdale, N.J.: Lawrence Erlbaum.
Piaget, J. 1971. *Psychology and Epistemology.* Harmondsworth, Eng.: Penguin.
Pitkin, H. 1972. *The Concept of Representation.* Berkeley: University of California Press.
Pollock, J. 1986. *Contemporary Theories of Knowledge.* London: Hutchinson and Company.
Popper, K. 1968. *The Logic of Scientific Discovery.* 3d ed. London: Hutchinson and Company.
Prentice-Dunn, S., and R. Rogers. 1989. "Deindividuation and the Self-regulation of Behavior." In Paulus (1989), chap. 4.
Reichenbach, H. 1938. *Experience and Prediction.* Chicago: University of Chicago Press.
Robinson, A., and B. Petchenik. 1976. *The Nature of Maps.* Chicago: University of Chicago Press.
Rorty, R. 1979. *Philosophy and the Mirror of Nature.* Princeton, N.J.: Princeton University Press.
Rouse, J. 1987. *Knowledge and Power.* Ithaca, N.Y.: Cornell University Press.
Scharff, R. 1991. "Monitoring Self-activity: The Status of Reflection before and after Comte." *Metaphilosophy.*
Sternberg, R. J., and E. E. Smith, eds. 1988. *The Psychology of Human Thought.* Cambridge: Cambridge University Press.
Suppe, F., ed. 1977. *The Structure of Scientific Theories.* 2d ed. Urbana: University of Illinois Press.
Thagard, P. 1989. "Scientific Cognition: Hot or Cold?" In Fuller et al. (1989), pp. 71–82.
Turner, R., ed. 1975. *Ethnomethodology.* Harmondsworth, Eng.: Penguin.
Tweney, R. 1989. "A Framework for Cognitive Psychology of Science." In Gholson et al. (1989), chap. 13.
Tweney, R., M. Doherty, and C. Mynatt, eds. 1981. *On Scientific Thinking.* New York: Columbia University Press.

Watson, J. B. 1913. "Psychology as the Behaviorist Views It." *Psychological Review* 20:158–77.

Wegner, D. 1986. "Transactive Memory." In B. Mullen and G. Goethals, eds., *Theories of Group Behavior.* New York: Springer-Verlag.

Wicklund, R. 1989. "The Appropriation of Ideas." In Paulus (1989), chap. 12.

Wills, G. 1983. "Critical Inquiry (Kritik) in Clausewitz." In W. Mitchell, ed., *The Politics of Interpretation.* Chicago: University of Chicago Press, pp. 159–80.

Wimsatt, W. 1984. "Reductionistic Research Strategies and Their Biases in the Units of Selection Controversy." In R. Brandon and R. Burian, eds., *Genes, Organisms, and Populations.* Cambridge, Mass.: MIT Press, chap. 7.

———. 1986. "Heuristics and the Study of Human Behavior." In D. Fiske and R. Shweder, eds., *Metatheory in Social Science.* Chicago: University of Chicago Press, chap. 13.

Woolgar, S. 1988. *Science: The Very Idea.* London: Tavistock.

———. 1989. "Representation, Cognition, and Self." In Fuller et al. (1989), pp. 201–24.

CRITIQUE

Invasion of the Mind Snatchers

Ten years ago I hoped, even expected, that the computational revolution would revive a dying enterprise, philosophy of science, and make it more intelligent and rigorous and insightful and interesting. This book and the recent work that seems to have provoked it have convinced me that my expectations were completely wrong. To judge by this sample, wherever the discipline of philosophy of science has been touched by cognitive science the result has been a zombie — philosophy of science killed dead and brought back to ghoulish, mindless, pseudolife.

"That grumpy guy, at it again," you say. Well, pilgrim, look at what we have from intelligent people who have in the past done good philosophical work:

Paul Churchland likes connectionist computational models of how the brain computes. Three essays in Churchland's recent book, *A Neurocomputational Perspective*[1] ("On the Nature of Theories: A Neurocomputational Perspective," "On the Nature of Explanation: A PDP Approach," and "Learning and Conceptual Change"), seem typical of his recent connectionist crush. They contain accounts of some well-known nonstochastic connectionist systems and accounts of some weight adjustment algorithms. In the first of these three essays Churchland describes some of the reasons why these connectionist models cannot be correct descriptions of how the brain functions if network nodes are identified with cell bodies and network edges with synaptic connections. He even calls for a study of more biologically realistic networks and learning procedures. Unfortunately, he does not propose any, investigate any, or prove anything about any. If he did, I would have no objection to the enterprise. The essays contain no new formal or computational results of any kind. Instead they solve all of the recently popular problems of philosophy of science. What solutions!

What is a theory? Says Churchland: Confused, old-fashioned people think a theory is a body of claims that can be expressed in a language, claims that (normally) are either true or false, and they even foolishly think the business of science is to find the claims that are true, general, and interesting. Equally benighted people think that humans have attitudes toward propositions and even represent propositions locally in the brain. But connectionism reveals that a theory is no such thing: it is an assignment of weights within a neural network (p. 177). In particular, someone's total theory is an assignment of weights within the neural network of that person's brain. *What is a concept?* The same thing (p. 178). Churchland later changes the story and suggests that a concept is a *partition* of weight space — that is, a region of possible values of weights that for all and only input vectors in a specified class will produce a specific output vector — or perhaps an input/output function (p. 234).

One awaits the time when Churchland's opinion will triumph and theorists will routinely present their heads for examination. Here is a view fit for Lavoisier and Ichabod Crane. A few obvious questions spring to mind: If someone's entire theory is just the "weights" of all of his or her synaptic connections, and if no propositions are represented locally, what is a *part* of an entire theory — electrodynamics, say, or thermodynamics, or the theory of evolution? If theories are as claimed, what is a theory that someone considers, reasons with hypothetically, but does not believe? If that is what a theory is, how do people share theories? If a theory is a pattern of weights, then what is *testing* a theory? If that is what a theory is, then what are people doing when they claim to be arguing about theories — are they arguing about the weights in someone's head? And so on. An obvious assessment of Churchland's *nouveau* philosophy of theories suggests itself: the whole business is an elementary confusion between a theory, or a system of claims, and the physical/computational states someone is in when she or he has a system of beliefs. Astonishingly, you will not find a single one of these questions addressed in Churchland's three essays. Not one. Honest.

Some philosophers hold that evidence and observation are *theory-laden*, whatever that means. I thought it meant that there is no level of description of observations such that one and the same objective circumstance (however individuated, whether by the psychologist, the philosopher, or God) will be reported, at that level of description, in the same way by an observer regardless of his or her prior beliefs. *Is observation theory-laden?* Says Churchland: the connectionist picture shows it is indeed, because there are an astronomical number of different input/output functions — hence "concepts" — that a large network can have (p. 190).

You might wonder, as I do, what the connection is between the issue of theory ladenness and Churchland's remark about the multiplicity of concepts that a network can represent. The two seem so, well, *disconnected.*

Churchland's idea seems to be that an *observation* is a particular pattern of values for a specific set of *output* nodes. If someone has sufficiently different weights, she or he will make different observations from the same input pattern. But then it would seem, contrary to Churchland's story, that the content of an observation is indeed locally represented by the values of a specific set of nodes. And, that aside, in would seem that in connectionist systems there is a level of description of observation that is not theory-laden, namely the values of the "input" nodes. But of course Churchland cannot really talk intelligibly about theory ladenness at all, because that issue is about *descriptions*, and in his view (not mine) networks do not describe anything because they do not say or claim anything.

Churchland's three essays contain more like this: for example, a parallel account of "inference to the best explanation." If, however, you ask where methodology fits in — e.g., why and in what sense inference to the best explanation can form part of a reliable method — you will find no answer. You will not even find the question recognized. If you look for a contribution to methodology you will find none; and if you look for any new results about the power and limits of the computational picture for which Churchland is an enthusiast, there too you will find nothing new. This in an area rocking and seething with new ideas and real problems.

In the last twenty years there has been considerable study in psychology of patterns of human irrationality. The studies include limitations on the reliability of human judgments of logical relations, judgments of probability, judgments of causality, and limitations in the reliability of experts of various kinds. The real source of this work lies in the pioneering studies done by Paul Meehl — Ronald Giere's colleague — in the 1950s on the reliability of clinicians. Herbert Simon's analysis of corporate behavior is in the same spirit. According to Simon, corporations do not act like ideally rational agents; they do not maximize profits or expected profits. Instead they try to make enough profit to keep up with the competition, please the directors, obtain bonuses for the management, etc. In Simon's phrase, they *satisfice*.

Giere's *Explaining Science*[2] tries to apply the idea of satisficing to scientific practice. Individual scientists, he claims, do not maximize their expected utility; they satisfice. The trouble with Giere's starting point is that while there is formal structure implicit in the idea of maximizing expected utility, there is scarcely any in the idea of satisficing; it functions as a phrase partly marking the idea that an agent may be unable to do what is ideally best (when cognitive limitations are ignored) and partly marking the idea that an agent may be unwilling to give over the cognitive resources required to determine what is normatively best. The interesting questions are the general forms these irrationalities take, and what can be done to overcome them. The first question is explored by those who follow Meehl through the work of Kahneman, Tversky, Dawes, and others, the second by a wide range of contemporary work in artificial intelligence. So there is plenty of serious work to do along the general line Giere takes. The trouble is he does not do any of it.

The oddity of Giere's book is not in the content but in the speech act. Indiana University, where he worked for many years, has a cyclotron facility, and Giere took advantage of its location to gather evidence of the obvious. One chapter of the book argues that in designing experiments with the cyclotron, physicists act and reason as if particle theory were true. Another long chapter examines the fortunes of the phenomenological Dirac equation in accounts of scattering and, in particular, the work of a particular physicist, Bunny Clark. The philosophical claims are, first, that scientists prefer being right to being wrong, and second, that the choices they make about what to work on are influenced by their "cognitive resources," e.g., by the theories they have experience with, the mathematical methods they know how to use, their ability to write computer programs. No kidding.

This is not quite the same as setting out to gather evidence that $2 + 2 = 4$, but it is close. Shorn of the philosophical trappings Giere's book is sustained by human interest, just as, say, *The Soul of a New Machine* is sustained by human interest. The interweaving of science and human interest is very well done, and so far as I can tell Giere gets the relevant science right. I have nothing against intellectual journalism. I find it more than curious that it should pass as philosophical research.

Paul Thagard's ECHO program is of a genre. I think the genre is dog and pony show. Thagard's computer program, ECHO, is supposed to measure "explanatory coherence." The idea is that propositions cohere with each other and with the data because of their explanatory and analogical relations. Scientists, it is claimed, prefer (ought to prefer? — too subtle a distinction, that) coherent systems of propositions. The principles of coherence Thagard gives are these:

1. Coherence is symmetric and so is incoherence.

2. If P1...Pm explain Q, then Pi and Q cohere and Pi, Pj cohere for i, j from 1 to m, and the degree of coherence is inversely proportional to m.

3. If P1 explains Q1 and P2 explains Q2 and P1 is analogous to P2 and Q1 is analogous to Q2, then P1 and P2 cohere and Q1 and

Q2 cohere; if the circumstances are otherwise the same but P1 is disanalogous to P2, then P1 and P2 incohere.

4. Observational propositions have a degree of acceptability of their own.

5. If P contradicts Q, then P and Q incohere.

6. The acceptability of a proposition P in a system S depends on its coherence with the propositions in S, and if many results of relevant experimental observations are unexplained, then the acceptability of a proposition P that explains only a few of them is reduced.

7. The coherence of a system of propositions is a function of the pairwise coherence of its members.

This is more substance than in the first two examples, but as a piece of philosophical analysis one would object: (1) all the hard questions — what is explanation? what is a better or a worse explanation? what is analogy? how is consistency to be maintained by a computationally bounded system? — have been begged; (2) the account of the dependence of acceptability on coherence is vague; (3) there is no argument for the account as any kind of norm; (4) there is no empirical evidence that the account is correct as a description of anything. The program is supposed to address some of these questions.

Thagard's program takes as *input* a list of "facts" of the following sort: "Q1 is observed to be the case"; "P1 ... Pn explain Q"; "P1 ... Pm explain Pn"; "Pj and Pk are inconsistent"; "Pk and Pm are analogous." The operator, Thagard or his assistant, specifies all of these facts; the system does nothing to determine what explains what, or what is analogous to what, or what is inconsistent with what. Thagard gives a number of examples of historical cases in which he gives his program "facts" about explanation and inconsistency and alternative hypotheses and the program finds the theory that was historically preferred. Copernicus wins out over Ptolemy, Darwin over creationism, etc. These examples are supposed to be the empirical evidence for the theory incorporated in the program.

What is wrong with this work? Here is a start:

If the aim is to describe how humans produce judgments about the best explanation given facts about what explains what, there are no arguments for the particular program. As a description, the ECHO picture assumes that theory comparison has two distinct modules, one for judging what explains what and another for judging "coherence" of separate propositions. There is neither empirical nor historical evidence for such an assumption. The picture assumes that in judging what to believe, all explanations of observations are equally important; all that matters is their numbers. There is no evidence for that claim, either. There is no psychological case at all.

In their paper for this collection, Nowak and Thagard give one argument that is so bad it must be disingenuous. They consider an incredibly simple counting procedure proposed by Jerry Hobbs as an alternative to ECHO, and they claim that Hobbs's count does not give the intuitive answer in some cases. Therefore, Nowak and Thagard conclude, ECHO is "necessary." Grant the premises, and consider the form of the argument, which is worthy of "Star Trek":

If ECHO then P
If Hobbs procedure then ~P
Therefore, ECHO is necessary for P.

(Which reminds me of a story once told me by Hartry Field about an unnamed distinguished philosopher, said to be from Pittsburgh, who was informed in private by a junior colleague that one of the distinguished professor's forthcoming papers turned on a modal fallacy. "Tell me," the distinguished professor is said to have asked his young colleague, "is it a *well-known* modal fallacy?")

Most of what the ECHO program does is simple counting, and one could think of lots of other easy ways to count besides Hobbs's. Any number of alternatives suggest themselves that are much simpler than ECHO and could be implemented on a pocket calculator.[3]

The historical simulations are bogus experiments. What do they test? Thagard and his assistants get to choose the "facts" and count them, and they choose and count them so that ECHO gets the right answer. I assure you that if ECHO were given as a separate fact to be explained that each star does not show parallax, Ptolemaic theory would fare much better; anti-Darwinians could (and did) cite a myriad of facts that Darwin's theory could not explain, but their objections do not show up in Thagard's input. The experiments show nothing about any capacity of the ECHO program to get to the truth reliably, and they show nothing about the unique capacity of the program (as against some simple function whose implementation would scarcely cost military research or the McDonnell Foundation a dime) to reproduce conventional judgments in the examples Thagard uses.

If the aim of this book is to show how philosophy of science can be integrated with cognitive psychology and artificial intelligence, then I think its point is made. There is plenty of work in cognitive science that is more enthusiasm than substance; there is plenty of work in cognitive psychology that tells you no more than your grandmother could, and probably less; there is plenty of work in artificial intelligence that is dog and pony show. Giere, Churchland, and Thagard each present examples of work that could mingle indistinguishably with the worst of cognitive psychology and artificial intelligence. One hopes their examples will not become the standard for the union of cognitive science and philosophy of science. There are philosophers whose work could mingle with the best of cognitive science and computer science. Maybe we need a grant to discover what makes some philosophers immune to the invasion of the mind snatchers.

Notes

A fellowship from the John Simon Guggenheim Memorial Foundation afforded time to write this essay. I am grateful to Ronald Giere for his courtesy and consideration in publishing it.

1. Cambridge, Mass.: MIT Press, 1989.
2. Chicago: University of Chicago Press, 1989.
3. For example, keeping close to the spirit of Thagard's program, one could define:

$E(P1P2)$ = the set of all tuples P1, P2 ... Pm, Q such that P1, P2 ... Pm explain observation Q;

$N(P1P2)$ = the set of all tuples P1, P2 ... Pm, Q such that P1, P2 ... Pm explain not-Q, and Q is observed;

For $s = <$ P1, P2 ... $<$ Pm, Q $>$ in $E(P1P2)$ or $N(P1P2)$, $L(s) = m$

$$\sum_{s \in E(P1P2)} \frac{1}{L(s)} - \sum_{s \in N(P1P2)} \frac{1}{L(s)}$$

And take the "activation" or "acceptance" of a proposition to be the sum of its coherences with other propositions in the system. Clearly a lot of other simple functions are possible; for example, with simple recursion or iteration one could (as Thagard wants) give P1 and P2 some degree of coherence if they explain a hypothesis that explains a hypothesis ... that explains an observation; the degree of coherence contributed by such a relation could be weighted by the number of explanatory steps between P1, P2, and the observation. You could write the particular function illustrated above in a few minutes on a programmable pocket calculator.

REPLIES TO GLYMOUR

Reconceiving Cognition

I think Glymour is right to be upset. The old epistemological ways are dying — cause enough for distress. Worse, their prospective replacements are still rag-tag, unpolished, and strange. Moreover, there is an unsettling uncertainty or shift of focus as to which are the truly important problems to be addressed. But even so, the objective theoretical situation holds more promise than at any time since the 1940s, and in his frustration Glymour misrepresents it rather badly. And sometimes carelessly. My aim in this note is to address and correct some of his complaints. Several of them are welcome, in that they represent questions that any intelligent observer would pose. These will receive the bulk of my attention.

Glymour opens his critique by referring to my "recent connectionist crush." This phrase implies volumes, but unfairly. I have been arguing publicly for some kind of nonsentential, brain-based kinematics for cognition since 1971, and have been defending in particular a vector-coding/matrix-processing approach to animal cognition since 1982, before the term "connectionism" was ever coined. For me, the appeal of this vector/matrix approach derived from its virtues within empirical and theoretical *neuroscience*, not AI or philosophy. It embodied a powerful account of motor control, sensory processing, and sensorimotor coordination, an account that was both inspired by and explanatory of the brain's micro-organization in terrestrial animals generally (see, e.g., Churchland 1986, reprinted as chap. 5 of *A Neurocomputational Perspective* [hereafter *NCP*]).

By contrast, the true virtues of a vector/matrix approach to AI became publicly apparent only after the back-propagation algorithm found widespread use, after 1986, as an efficient means of *configuring* the weights (= the matrix coefficients) of artificial, multilayered networks. This allowed networks to be quickly trained to various cognitive capacities, which then allowed us to analyze in detail the computational basis of these acquired capacities. By then, some of us saw this (wholly welcome) development as a case of the AI community finally catching up, conceptually, with long-standing themes in the empirical and theoretical neurosciences.

In sum, my own approach to these issues is neither recent nor narrowly "connectionist" in character. Neither is it a "crush," in the sense that implies only a shallow and fleeting commitment. Having taxed the patience of my generous but skeptical colleagues for two decades now, I am sure that they will recognize a long and determined commitment to the kinematical themes that appear in *NCP*.

What about those themes? At the end of his third paragraph, Glymour expresses disappointment that they do not include any "new formal or computational results of any kind." Here I am inclined to some surprise, since four modest results occur immediately to me. The first concerns the unexpected size of the human *conceptual space*, as roughly calculated in *NCP*, and the explorational difficulties that this superastronomical volume entails (pp. 249–53). A second result concerns the character of *conceptual simplicity*, as neurocomputationally construed, and the reason why creatures blessed with it will generalize more successfully than others when confronted with novel sensory situations (pp. 179–81). A third result concerns the necessity for acknowledging at least two quite distinct processes of *learning* in highly developed creatures: a more basic process that is relatively slow, continuous, and inevitably destructive of prior concepts; and a secondary process possible only in recurrent networks, a process that is fast, discontinuous, and largely conservative of prior concepts (pp. 236–41). A fourth result concerns the presumptive *computational identity* of three phenomena counted as quite distinct from within commonsense, namely, perceptual recognition, explanatory understanding, and inference to the best explanation (chap. 10 of *NCP*). These results may yet prove to be flawed in some way, but it is wrong to imply their nonexistence.

Perhaps Glymour meant something more restrictively a priori by the phrase "formal or computational results," sufficiently restrictive to exclude the examples just cited. If so, then his original complaint is misconceived. For I take pains to emphasize at several places that my aims are those of an empirical scientist, not those of a logician or any other presumptively a priori discipline (see, for example, *NCP*, p. 198). In effect, Glymour is complaining that *NCP* does not approach the philosophy of science the way that he and others have traditionally approached it. But it is that very tradition whose methodological assumptions are now in question.

Let us turn to some of Glymour's specific criticisms. He asks after my account of a *specific* theory — electrodynamics, or thermodynamics, say — as opposed to a person's global theory of the world-in-general. My considered view of the latter is that it consists of the intricate set of partitions that learning has produced across a person's high-dimensional neuronal activation space. (This is outlined explicitly on pp. 232–34 of *NCP*, where the alternative explication in terms of a global set of synaptic weights is rejected on grounds that it puts impossible demands on the sameness of global theories across individuals.) From that more general perspective, a specific theory such as electrodynamics is naturally construed as a specific prototypical *subvolume* of that global activation space, a subvolume with its own partitional substructure, perhaps.

This account does put single concepts and entire theories on the same footing: a theory is just a highly structured prototype. However odd, this is indeed part of the story. (It should not seem *so* odd. The "semantic" view of theories [van Fraassen 1980] makes a similar move when it construes an entire theory as a single complex predicate.) In any case, this is the account that emerges on p. 236, where I refer to the phenomenon of multiple conceptual competence. The account is then put to systematic work in my subsequent discussion of mere conceptual redeployment versus outright conceptual change.

Glymour shows some comprehension of this shift in emphasis — from points in weight space to partitions across activation space — but he gets it garbled. He ascribes to me the view that a concept or theory is "a partition of *weight* space" (italics mine). This is nonsense. There are no relevant partitions in weight space, only positions and learning trajectories. It is the *activation* space that accumulates an intricate set of useful partitions in the course of learning.

Glymour complains that I leave entirely unaddressed the question of how two people share the same theory. He will find explicit discussions on pp. 171 and 234. My long discussion of conceptual redeployment (pp. 237–41) includes a further account of how two people can share a theory — the theory of waves, for example — while differing in the domains where they count it true. Huygens, for example, thought this theory true of light, while Newton did not. Huygens and Newton share the theory because they have a relevantly similar set of partitions across their activation spaces (i.e., they possess closely similar prototypes), which produce relevantly similar cognitive, verbal, and manipulative behavior when the shared prototype is activated. But they differ in their respective commitments to the theory because in Huygens the prototype is typically activated when he confronts water waves, sound waves, *and* light waves, while in Newton it is typically activated only in the former two cases. When Newton confronts optical phenomena, in the appropriate cognitive context, it is rather a *ballistic particle* prototype that typically gets activated.

As with duck/rabbit figures, one can choose, or learn, to apprehend some part of the world in quite different ways. One can systematically apprehend optical phenomena as wave phenomena, or as particle phenomena. And one can explore at length the subsequent rewards or disappointments that each mode of apprehension brings with it. In this way are theories "tested," whether one is a juvenile mouse or an aging human scientist.

This sketch leaves open all of the details of learning, to be sure. But that is deliberate. Just what theory of learning we should embrace as descriptive of normal, healthy brains is still a profoundly open question, as I emphasized on p. 187 in closing a survey discussion of the existing neurocomputational wisdom on the topic. Closing that question requires a great deal more empirical research into the functional plasticity of the nervous system. I cannot close that question a priori. And neither can Glymour. But one can draw a provisional lesson at this point about the kinematical framework within which any comprehensive theory of human and animal learning must function: it will be a kinematics of synaptic weight configurations and of prototype activations across vast populations of neurons. The familiar kinematics of sentences will come in only derivatively and peripherally, if it comes in at all.

I am puzzled to find Glymour claiming that none of the preceding questions is even addressed in chapters 9–11 of *NCP*. They are all addressed, repeatedly and in depth. And my answers are not hard to find. The index at the back of *NCP* will lead one to all of the page references cited above, and more.

On the issue of theory ladenness I shall here say little. My essay on the philosophy of Paul Feyerabend elsewhere in this volume contains a detailed discussion of the matter as reconstructed in neurocomputational terms. It will support the following brief response to Glymour's criticisms. The fact of many configurational alternatives is only one part of the story, which is why Glymour has trouble seeing the connection. The second part concerns the fact that no cognitive activity occurs save as information is processed through some one of those many possible configurations of synaptic weights. And the third part is that any weight configuration represents or realizes a speculative framework or theory as to the structure of some part of the world. It follows that any cognitive activity — perception included — presupposes a speculative theory.

This account does indeed portray unmodulated coding at the absolute sensory periphery as being free of theory, as Glymour and I both remark (*NCP*, p. 189; this volume, p. 355). But it also leaves it outside the domain of cognitive activity. Peripheral transduction is just that: transduction of one form of energy into another. Cognitive activity does not begin until the coded input vector hits the first bank of weighted connections. From that point onward, the input is (not just energetically but informationally) transformed into the speculative framework of the creature's acquired theories; it is represented as a point in the speculatively partitioned and antecedently structured space of possible neuronal activation patterns. In short, it is theory-laden.

Glymour concludes by remarking on the lack of any substantial methodological or normative contributions in *NCP*. There is some point to his complaint. My concerns are indeed primarily descriptive and explanatory. Yet I am again puzzled to hear him make the strong claim that normative questions are not even recognized, especially if he has in mind, as he says, the evaluation of *explanations*, and the matter of how good ones fit into a reliable empirical method. For I have discussed precisely this question at length, and proposed an explicit answer. It is cataloged in my index, under "Explanation: evaluation of, 220–223."

I there explore four dimensions in which explanatory understanding, construed as prototype activation, can be evaluated within a neuro-computational framework. The most important of these, in the present context, is the dimension of *simplicity*. *Ceteris paribus*, an activated prototype is better if it is a part of the most *unified* conceptual configuration. What simplicity and unity amount to is discussed in chapter 9 (pp. 179–81), where it is observed that networks that have formed the simplest or most unified partitions across their activation spaces are networks that do much better at generalizing their knowledge to novel cases. Very briefly, they do better at recognizing novel situations for what they are, because they have generated a relevantly unified similarity gradient that will catch novel cases in the same subvolume that catches the training cases.

The account thus proposes a novel account of what simplicity/unity is, and of why it is an epistemic virtue. This result seemed especially encouraging to me, since it constitutes an unexpected success on the *normative* front, and at a specific place, moreover, where the traditional approaches have proved a chronic failure. Be that as it may, the question was certainly addressed, and I would have been interested in Glymour's criticisms of the answer proposed.

As I see it, our normative and methodological insights are very likely to flourish as our explanatory insight into neural function and neural plasticity increases. But as I remark on p. 220, we are too early in the latter game to draw very much from it yet. Glymour complains that

I have drawn too little. My concern is that I may have tried to draw too much.

I conclude by pointing to an important background difference between Glymour and me, one that peeks out at two or three points in his essay, a difference that may do something to explain the divergent evaluations we display elsewhere. Glymour evidently regards as unproblematic the "elementary [distinction] between a theory, or a system of claims, and the physical/computational states someone is in when she or he has a system of beliefs." (I, in any case, am charged with missing it.) And to judge from his remarks in his fourth and seventh paragraphs, Glymour also regards the kinematics of beliefs and other propositional attitudes as unproblematic.

People familiar with my earlier work in the philosophy of mind will recall that I find both of these things to be problematic in the extreme, and for reasons quite independent of traditional issues in the philosophy of science. For reasons outlined in many places, including chapter 1 of *NCP*, I am strongly inclined toward a revisionary or eliminative materialism concerning the mind. From such a perspective one is more or less compelled to seek an epistemology and a philosophy of science of the character at issue. From Glymour's more conservative perspective, that approach is bound to seem mad. I quite understand this. But my answer is simple: Glymour is mistaken in both areas.

References

Churchland, Paul M. 1986. "Some Reductive Strategies in Cognitive Neurobiology." *Mind* 95:279–309.

———. 1989. *A Neurocomputational Perspective: The Nature of Mind and the Structure of Science.* Cambridge, Mass.: MIT Press.

van Fraassen, Bas. 1980. *The Scientific Image.* Oxford: Oxford University Press.

What the Cognitive Study
of Science Is Not

The overall aim of *Explaining Science* (1988) is to develop a "naturalistic" approach to understanding the nature of modern science that mediates between existing rival philosophical and sociological approaches. I reject philosophical attempts to explain science as the result of applying categorical principles of rationality, whether these be explicated in terms of formal logic, probability, or historical progress. For me, the only form of rationality that exists is the instrumental use of empirically sanctioned strategies to achieve recognized goals.[1] I also reject constructivist sociological approaches that portray scientific (theoretical) claims as arising out of social processes that allow these claims to be unconstrained by the world itself. The problem, then, is to explain how scientists sometimes produce tolerably good representations of the world without appealing to (for me nonexistent) categorical principles of rationality. In short, is it possible to understand science realistically without invoking special forms of rationality?

In conceiving *Explaining Science*, my major strategic problem was finding a framework in which to formulate my account of what science is and how it works. I had already rejected the frameworks of both logical empiricism and historical theorists. Nor could I adopt the framework of sociological theory. Cognitive science provided a framework that suited my purposes, so I used it. What I wanted to say about the nature of theories fit comfortably under the umbrella of "representation," while theory choice could easily be seen as the exercise of "human judgment" in a specialized context, particularly the context of experimentation.

Glymour focuses his attention on just two chapters of *Explaining Science*, complaining that they reveal no more than anyone could learn from their grandmother. These complaints may reflect his ignorance of the current general state of research, both philosophical and otherwise, into the nature of modern science. Or they may merely reflect his disdain for the philosophical, historical, and sociological criticisms of logical empiricism that now define the whole field of science and technology studies. Either way, his complaints are beside the point.

The first chapter to which Glymour refers is entitled "Realism in the Laboratory." It was inspired by the now classical laboratory studies of constructivists such as Latour and Woolgar (1979) and Knorr-Cetina (1981). They claimed that a close look at actual laboratory practice supports a constructivist approach. I set out to show that at least some laboratory practice supports a realist approach. I did present evidence that experimental nuclear physicists act and reason as if there really are such things as protons and neutrons. But that was just one *premise*, not the conclusion of my argument. My *conclusion*, based on the whole context of the laboratory, and not just on the acts and words of the participants, was that philosophers and sociologists, as students of the scientific enterprise, ought to reach the same conclusion regarding the reality of some "theoretical entities," protons in particular. We ourselves can safely conclude that these physicists are producing and using something like what they call protons in their ongoing research. And we can employ this conclusion in our own accounts of their research. Given the number of prominent philosophers of science, as well as sociologists of science, who reject these conclusions, they can hardly be regarded as trivial.

I fear I was not as clear as I should have been about the nature of the argument being presented. One philosophical reviewer, for example, characterized my argument as a simple application of the principle of inference to the best explanation, which principle he regarded as itself in need of justification. The result is a classical Humean regress. I explicitly denied appealing to any principle that would start the Humean game. If one nevertheless insists on a general principle of inference, it would have to be something like inference to the *only possible* explanation. But the notion of possibility employed here has to be something stronger than mere logical possibility. I would suggest something like "cultural possibility." That leaves one with the problem of specifying the relevant culture.

Within the culture of contemporary nuclear physics it seems pretty clear that there is no other possible explanation. But that is too narrow a culture. What I need is something like contemporary Western culture, with the proviso that we are talking about people with a reasonable degree of education and sophistication. That would have to include most professional philosophers and sociologists of science. I suspect Glymour agrees with my conclusion, though not with my form of argument. I know Latour, Woolgar, and Knorr-Cetina, as well as philosophers such as Laudan (1984) and van Fraassen (1980), reject both the conclusion and the argument. Both may be unsatisfactory, but neither is trivial.

The other chapter to which Glymour refers is titled "Models and Experiments" and has the modest goal of showing that scientific decisions by competent scientists do sometimes fit a "satisficing" model of scientific judgment developed in the previous chapter. It is part of this model that scientists' interests, such as those generated by individual "cognitive resources," often play a role in decisions regarding which models to pursue and which to accept as good representations of the world. Rival models of scientific judgment, such as those assuming that scientists intuitively calculate probabilities of hypotheses, or measure problem-solving effectiveness, typically deny any role for such interests in "rational" scientific decision making. My use of case-study material to support the claim that "interested" judgment is just fine in legitimate scientific practice is thus, when seen against the background of rival accounts, hardly equivalent to gathering evidence to show that two plus two equals four.

I suspect that, at bottom, Glymour's dissatisfaction with current cognitive approaches to the philosophy of science stems from his having a very different conception of what a cognitive philosophy of science should be like. This difference comes out when he claims that "there are philosophers whose work could mingle with the best of cognitive science and computer science." It has never been my goal that my work should "mingle with" cognitive science or artificial intelligence, good or bad. Rather, in Glymour's terms, my goal has been to produce work that could "mingle with" the best of science studies, that is, work in the history, philosophy, and sociology of science that seeks to develop a comprehensive understanding of the nature of science. What makes my approach "cognitive" is that it employs general frameworks found in the cognitive sciences, and, when applicable, specific results of research in the cognitive sciences. This does not require making original contributions to the development of the cognitive sciences, any more than it requires making original contributions to nuclear physics, geology, or any other science that one takes as an object of study.

One might fairly object that *Explaining Science* uses the cognitive sciences mainly just to provide an overall framework and does not make sufficient use of specific results of cognitive science research.[2] Ryan Tweney (1989) has suggested that the cognitive sciences may never be able to do much more than provide frameworks. I still think that we can do more, and so do others, including some represented in this volume.

Notes

1. The general thesis that scientific rationality is nothing other than instrumental rationality is further defended in a later article (Giere 1989).

2. It is thus fair enough of Glymour to point out that I did not make use of Paul Meehl's pioneering work on the reliability of clinical judgment. But his suggestion that this involved a failure to communicate with a colleague on my own campus is refuted by a few simple temporal facts: (1) I sent the completed manuscript of *Explaining Science* off to the University of Chicago Press in the fall of 1986. (2) I completed revisions on the manuscript in the spring of 1987. (3) I moved to Minnesota in the fall of 1987. (4) The book came out in the spring of 1988.

References

Giere, R. N. 1988. *Explaining Science: A Cognitive Approach.* Chicago: University of Chicago Press.

———. 1989. Scientific Rationality as Instrumental Rationality. *Studies in History and Philosophy of Science 20* 3:377–84.

Knorr-Cetina, K. D. 1981. *The Manufacture of Knowledge.* Oxford: Pergamon Press.

Latour, B., and S. Woolgar. 1979. *Laboratory Life.* Beverly Hills, Calif.: Sage.

Laudan, L. 1984. *Science and Values.* Berkeley: University of California Press.

Tweney, R. D. 1989. A Framework for the Cognitive Psychology of Science. In *Psychology of Science and Metascience*, ed. B. Gholson, et al., pp. 342–66. Cambridge: Cambridge University Press.

van Fraassen, B. C. 1980. *The Scientific Image.* Oxford: Oxford University Press.

Computing Coherence

The purpose of this note is to remedy some of Clark Glymour's (this volume) misconceptions about explanatory coherence and ECHO. After briefly responding to Glymour's challenges concerning the normativeness and necessity of ECHO and the nature of analogy and explanation, I shall show that his alternative approach to computing coherence is utterly inadequate.

First let us look at some of the issues Glymour raises about explanatory coherence. Is my theory descriptive or normative? This question is more complex than Glymour appreciates, since descriptions of scientific cases can have normative force. I have elsewhere discussed the general question of the relation of the descriptive and the normative (Thagard 1988, chap. 7) and the particular issue of the normativeness of explanatory coherence theory (Thagard 1989b, sec. 1). ECHO is intended to describe how people make inferences in accord with the best practices compatible with their cognitive capacities.

Glymour objects that my principles of explanatory coherence beg all the hard questions about explanation, analogy, and computation. Well, there are lots of hard questions here, but they cannot all be addressed at once. Keith Holyoak and I have developed a theory of analogy that complements the theory of explanatory coherence (Holyoak and Thagard 1989; Thagard et al. 1990). The theory of explanation is more problematic, but I see at least the possibility of a cognitive account of explanation being developed (Thagard, in press). Glymour asks for "empirical evidence that the account is correct." The numerous cases from the history of science to which ECHO has been applied only begin to suggest its adequacy as a theory of human cognition, but psychologists at several institutions are devising experiments to test it more thoroughly (Schank and Ranney 1991).

Glymour misrepresents ECHO, which does not assume that "all explanations of observations are equally important" (see Thagard 1989a, sec. 4.1). In addition, he misconstrues the discussion concerning alternatives to ECHO in section 6 of Nowak and Thagard (this volume). The point was to evaluate Hobbs's argument that since the naive method is as good as ECHO, then ECHO is unnecessary. Our argument demolished Hobbs's premise, but nowhere do we assert that "ECHO is necessary." It is of course possible that simpler algorithms than ECHO will suffice for judgments of explanatory coherence. The challenge for critics of ECHO is to produce one, and Glymour, like Hobbs, has failed to do so, as we shall shortly see.

Glymour claims that the historical simulations are "bogus," as if we made up the historical facts to suit ECHO. In fact, all the historical analyses have been closely based on scientific texts. In his months of research on the works of Copernicus and Ptolemy, Greg Nowak strove for both historical accuracy and a common level of detail that would not bias ECHO in any particular direction. Each simulation is intended to model a particular scientist such as Copernicus, not an entire debate.

Glymour thinks that ECHO is trivial and proposes a simple alternative that judges the acceptance of a proposition to be the sum of its pairwise coherences with other propositions in the system. I implemented his proposal in a few LISP functions and ran it on the Copernicus case as described in Nowak and Thagard (this volume). Call this program GECHO. To compute the acceptability of a hypothesis H, GECHO determines which other hypotheses participate in explanations with it. For example, P5 participates with: (PC3 PC1 P9 P17 P6 P12 P2 P3). Hence the acceptability of P5 is the sum of the coherence of P5 with PC3, of P5 with PC1, etc. GECHO determines pairwise coherence in accord with Glymour's equation. For example, the coherence of P5 with PC3 derives from their joint participation in two explanations, of E8 and E9. In the former case, there are a total of 5 explaining propositions, and in the latter there are 3, so the coherence between P5 and P13 is $1/5 + 1/3 = 8/15$. The coherences of P5, corresponding to the above list of propositions with which it participates, are: (8/15 47/60 7/12 1/4 47/60 1/5 89/60 59/60). This sums to the floating point number 5.6, representing GECHO's assessment of the acceptability of P5.

On Glymour's analysis, many Ptolemaic hypotheses come out as more acceptable than many Copernican hypotheses! There are 9 Ptolemaic units with acceptance greater than 5: (P2 P3 P5 P6 P12 P17 P22 P24 P34). And there are 21 Copernican units with acceptance less than 5: (C5 C7 C10 C11 C14 C15 C20 C21 C29 C31 C32 C32V C33 C34 C35 C23 C26 C27 C28 C40 C41). ECHO, in contrast, finishes with all Copernican units having activation well above 0 and all Ptolemaic units having activation well below 0. (ECHO's activations range between -1 and 1; Glymour's acceptances can be any number greater than 0.) Since we are modeling Copernicus, who accepted his own system while rejecting Ptolemy's, ECHO rather than GECHO produces the desired result.

The reason for the failure of Glymour's algorithm is that it neglects a fundamental factor in coherence calculations: the acceptability of a proposition depends not only on its coherences with other propositions, but on how acceptable those other propositions are. Moreover, GECHO does not take into account contradictions and competition between propositions. ECHO pulls minor Copernican hypotheses up because they cohere with major Copernican hypotheses, and drags major Ptolemaic ones down because they are incoherent (because of contradictions and competition) with major Copernican ones. To overcome this problem, Glymour would have to introduce some sort of iterative updating scheme that allows acceptance of a proposition to be adjusted by considering the provisional acceptances of the propositions with which it coheres. This is computationally nontrivial, since there is no guarantee that the updating will stabilize. Bayesian networks, for example, tend to be unstable in the general case, and various network manipulations are necessary to make the calculation of posterior probabilities tractable (Pearl 1988). ECHO networks generally stabilize in fewer than 100 cycles of updating because inhibitory links prevent activation from simply being passed round and round the network. The presence of inhibition, motivated by the explanatory coherence principles of contradiction and competition, turns out to have very desirable computational properties.

There are thus computational reasons to believe that Glymour's approach will not work even on straightforward cases. Moreover, he says nothing about how to implement consideration of analogy, and his suggestion about how to handle coherence in networks with hypotheses that explain hypotheses is ill specified, since two hypotheses may be connected through a hierarchy of explanations to more than one observation. Still, Glymour could probably implement such a search, since his pocket calculator evidently includes a list-processing programing language. Another flaw in GECHO is that it says nothing about the acceptability of observations, apparently assuming that they are unproblematic. ECHO is not so positivistic, and can in fact reject propositions describing observations if such propositions do not cohere with well-established hypotheses (see Thagard 1989a). ECHO also has dynamic properties that GECHO lacks, resisting new hypotheses that contradict ones that have already been established by a previous assessment of explanatory coherence.

Glymour's main argument against ECHO is merely that there *might* be a simpler algorithm for computing acceptability based on explanatory coherence. A *serious* challenge to ECHO would consist of full presentation of an algorithm that efficiently computes what ECHO does.

References

Holyoak, K., and P. Thagard. 1989. Analogical mapping by constraint satisfaction. *Cognitive Science* 13:295–355.

Pearl, J. 1988. *Probabilistic reasoning in intelligent systems.* San Mateo, Calif.: Morgan Kaufmann.

Schank, P., and M. Ranney. 1991. Modeling an experimental study of explanatory coherence. *Proceedings of the Thirteenth Annual Conference of the Cognitive Science Society.* Hillsdale, N.J.: Erlbaum, pp. 528–33.

Thagard, P. 1988. *Computational philosophy of science.* Cambridge, Mass.: MIT Press/ Bradford Books.

———. 1989a. Explanatory coherence. *Behavioral and Brain Sciences* 12:435–67.

———. 1989b. Extending explanatory coherence (reply to 27 commentators). *Behavioral and Brain Sciences* 12:490–502.

———. In press. Philosophical and computational models of explanation. *Philosophical Studies.*

Thagard, P., et al. 1990. Analog retrieval by constraint satisfaction. *Artificial Intelligence* 46:259–310.

Contributors

Gary Bradshaw is Assistant Professor at the University of Colorado. Along with Herbert Simon, Pat Langley, and Jan Zytkow, Bradshaw worked to develop computational models of scientific discovery. He is currently exploring the process of invention of complex devices.

Susan Carey received her Ph.D. from Harvard University in 1972. Her work concerns the nature and acquisition of mental representations in a variety of domains of knowledge. These include the mental representations of faces, of language, and of scientific language.

Michelene T. H. Chi is Professor of Psychology and Senior Scientist at the Learning Research and Development Center at the University of Pittsburgh. Her research interest in cognitive science centers on understanding a variety of changes in knowledge structures as one learns a new scientific domain, and how such changes affect the way the acquired knowledge is used. Her recent publications concern the way self-explanations mediate learning from examples. She has also edited *The Nature of Expertise* (1988).

Paul M. Churchland is Professor of Philosophy at the University of California, San Diego, as well as a member of the Cognitive Science faculty, the Science Studies faculty, and the Institute for Neural Computation. He is the author of *Scientific Realism and the Plasticity of Mind* (1979), *Matter and Consciousness* (1984), and *A Neurocomputational Perspective* (1989).

Lindley Darden is Associate Professor of Philosophy and History, a member of the Committee on the History and Philosophy of Science, and a member of the Committee on Cognitive Studies at the Univer-

489

sity of Maryland at College Park. She is a historian and philosopher of science interested in conceptual change in science and in the applications of techniques from artificial intelligence to scientific discovery. Her book, *Theory Change in Science: Strategies from Mendelian Genetics*, was published by Oxford University Press in 1991.

Eric Freedman has published several articles on the assumptions and methodologies appropriate to the study of the cognitive processes employed in science; his current research examines the relationship between scientists' mental representations and discourse. He is also engaged in a series of experimental studies designed to investigate how groups utilize competing hypotheses during induction and how discourse influences group decision making.

Steve Fuller teaches philosophy and the sociology of science in the Science Studies Center, Virginia Tech. He is the executive editor of *Social Epistemology: A Journal of Knowledge, Culture and Policy*, and the author of two books: *Social Epistemology* (1988) and *Philosophy of Science and Its Discontents* (1989). Fuller was the principal editor of the 1989 *Sociology of the Sciences Yearbook*, which was devoted to the recent "cognitive turn" in science studies.

Ronald Giere is Professor of Philosophy and Director of the Minnesota Center for the Philosophy of Science at the University of Minnesota. He holds degrees in physics (M.S.) and philosophy (Ph.D.) from Cornell University, and taught in the Department of History and Philosophy of Science at Indiana University before moving to Minnesota. In addition to many papers and several edited volumes, he is the author of *Explaining Science: A Cognitive Approach* (1988) and *Understanding Scientific Reasoning* (3d ed., 1991).

Clark Glymour is Alumni Professor of Philosophy at Carnegie Mellon University and Adjunct Professor of the History of Science at the University of Pittsburgh.

David Gooding lectures in history and philosophy of science at the University of Bath, England. He is the author of *Experiment and the Making of Meaning* (1990), co-author of *Faraday* (1991), and co-editor of *The Uses of Experiment* (1989) and of *Faraday Re-discovered* (1989).

Michael E. Gorman is Associate Professor of Humanities in the School of Engineering and Applied Science at the University of Virginia. He has published several experimental studies of scientific reasoning, including "Error, Falsification, and Scientific Inference: An Experimental Investigation," in the *Quarterly Journal of Experimental Psychology* (1989). His *Simulating Science* (forthcoming) fits this experimental work into a larger philosophical, historical, and sociological context. He is currently working with W. Bernard Carlson on a study of the cognitive processes of inventors.

Richard E. Grandy received his doctorate from the History and Philosophy of Science Program at Princeton in 1968. He has taught at Princeton and the University of North Carolina and is currently Professor of Philosophy and Cognitive Sciences at Rice University. He edited a collection of essays entitled *Theories and Observation in Science* and is the author of numerous articles in philosophy of science, language, and logic. His current research focuses on the application of cognitive sciences to philosophy of science and vice versa.

C. Keith Haddock is a Ph.D. candidate in clinical psychology and behavioral medicine at Memphis State University. His research focuses on quantitative methods of research synthesis, evaluation of treatments of childhood obesity, and behavioral psychologies of science.

Arthur C. Houts is Professor of Psychology at Memphis State University, where he works in the Center for Applied Psychological Research. He is a co-editor of *Psychology of Science and Metascience* and is currently conducting research on behavioral scientists who comprised the behavior modification movement in the United States from 1945–1975.

Nancy J. Nersessian is Assistant Professor in the Program in History of Science and the Department of History at Princeton University. She is also an associate member of the Department of Philosophy and a member of the Program in Cognitive Studies. Nersessian's publications include *Faraday to Einstein: Constructing Meaning in Scientific Theories* and the edited volumes *The Process of Science: Contemporary Philosophical Approaches to Understanding Scientific Practice* and *Selected Works of H. A. Lorentz.* She is the series editor of the Science and Philosophy series published by Kluwer Academic Publishers.

Greg Nowak is a graduate student in the Program in History of Science at Princeton University, currently completing a dissertation on the algebraic topology of Henri Poincare. His interests lie in the philosophy of mathematics and the history of modern mathematics.

C. Wade Savage is a member of the Department of Philosophy and member and former director of the Center for the Philosophy of Science at the University of Minnesota. He is the author of *The Measurement of Sensation* and the editor of several volumes in the Minnesota Studies in the Philosophy of Science series.

Paul Thagard is Senior Research Cognitive Scientist at Princeton University. He is the author of *Computational Philosophy of Science* (1988) and *Conceptual Revolutions* (1991), and co-author of *Induction: Processes of Inference, Learning and Discovery* (1986). His research primarily concerns computational models of scientific thinking.

Ryan D. Tweney is Professor of Psychology at Bowling Green State University. The author of many scientific and scholarly papers and the co-author of three books, he has focused much of his work on the application of methods and theories in cognitive science to problems in the understanding of scientific thinking. He has lectured widely on the life and works of Michael Faraday.

Index of Authors

Index of Subjects

501